Aerothermodynamics in Combustors

International Union of Theoretical and Applied Mechanics

R. S. L. Lee · J. H. Whitelaw · T. S. Wung (Eds.)

Aerothermodynamics in Combustors

IUTAM Symposium Taipei, Taiwan, 1991

Springer-Verlag
Berlin Heidelberg New York
London Paris Tokyo
Hong Kong Barcelona Budapest

Prof. Richard S. L. Lee

Dept. of Mechanical Engineering
State University of New York
Stony Brook, NY 11794-2300
USA

Prof. James H. Whitelaw

Mechanical Engineering Dept.
Imperial College
Exhibition Road
London SW 7 2BX
United Kingdom

Prof. T. S. Wung

Institute of Applied Mechanics
National Taiwan University
Taipei, Taiwan 106
Republic of China

ISBN-13: 978-3-642-84757-8 e-ISBN-13: 978-3-642-84755-4
DOI: 10.1007/978-3-642-84755-4

Library of Congress Cataloging-in-Pulication Data
Aerothermodynamics in combustors : IUTAM Symposium, Taipei, Taiwan, 1991 / R. S. L. Lee, J. H. Whitelaw, T. S. Wung, eds.
(IUTAM symposia)
ISBN-13: 978-3-642-84757-8
1. Combustion chambers--Congresses. 2. Aerothermodynamics--Congresses.
I. Lee, R. S. L. (Richard S. L.) II. Whithelaw, James H. III. Wung, T. S.
IV. International Union of Theoretical and Applied Mechanics. V. Series: IUTAM Symposien.
TJ254.7.A37 1992
621.402'3--dc20 92-13620

Softcover reprint of the hardcover 1st edition 1992

Typesetting: Camera ready by authors

61/3020-5 4 3 2 1 0 – Printed on acid-free paper.

Scientific Committee

Professor R S L Lee, *Chairman*, National Taiwan University, Republic of China

Professor M Barrere, ONERA, France
Professor D Bradley, University of Leeds, U K
Professor G Chernyi, Moscow University, U S S R
Professor F Durst, University of Erlangen-Nurnberg, Germany
Professor G J Hwang, National Tsing Hua University, Republic of China
Professor K K Kuo, the Pennsylvania State University, U S A
Professor A H Lefebvre, Purdue University, U S A
Professor Y Mizutani, Osaka University, Japan
Professor N Peters, RWTH Aachen, Germany

Professor J H Whitelaw, *Secretary*, Imperial College of Science, Technology and Medicine, U K

Sponsoring Organisations

International Union of Theoretical and Applied Mechanics
National Science Council, Republic of China
Ministry of Education, Republic of China
Society of Theoretical and Applied Mechanics, Republic of China
Chinese Society of Mechanical Engineers
Chinese Society of Engineers, Republic of China
Chinese Society of Chemical Engineers, Republic of China
Chinese Society of Aeronautics and Astronautics, Republic of China
National Taiwan University, Republic of China
U S Army European Research Office

Preface

A Symposium on ***Aerothermodynamics of Combustors*** was held at the Institute of Applied Mechanics of the National Taiwan University from 3 to 5 June 1991 and was attended by 130 delegates from eight countries. The topics of the forty formal presentations included measurements and calculations of isothermal simulations and of combusting flows with one and two phases, and with consideration of configurations ranging from simple diffusion to gas-turbine flows. The discussions inside and outside of the Symposium Hall were lively and an open forum session demonstrated the range of opinions currently and strongly held.

The International Union of Theoretical and Applied Mechanics initiated the Symposium under the chairmanship of Professor R S L Lee and with the Scientific Committee listed below. It benefited from sponsorship, again as listed below, and from contributors who presented interesting and up-to-date descriptions of their research. Invited lectures were delivered by Professors R Bilger and F Weinberg and set the scene in terms of quality of material and presentation.

The papers included in this Volume were chosen from those presented at the Symposium so as to provide a coherent and cohesive indication of the Symposium and of the Subject which it considered. They have been edited to ensure that they meet the standards of Journals and the editors are grateful to the authors for the patience with which they dealt with the various criticisms and requests and to Springer Verlag for their considerable assistance in publication matters. It is also a pleasure to aknowledge the many contributions of the Local Arrangements Committee for the Symposium and the many helpers who contributed so much to the success of the Symposium.

December 1991

R S L Lee
J H Whitelaw
T S Wung

Table of Contents

Part III Two Phase Flows

Part IV Unsteady Flows

Part V Gas Turbine Combustors

Part I

Invited Lectures

Advanced Laser Diagnostics: Implications of Recent Results for Advanced Combustor Models

R.W. Bilger

Department of Mechanical Engineering
The University of Sydney
NSW 2006, Australia

Summary
Measurements using laser Raman, laser Rayleigh and laser induced fluorescence in turbulent diffusion flames of hydrocarbon and non–hydrocarbon fuels indicate strong effects of stoichiometric mixture fraction and fuel kinetics on the flame structure and mechanism of extinction. Equations for the moments of species mass fractions conditional on mixture fraction are presented. They are shown to give insight into the various aspects of flame structure and the relative roles of kinetics and mixing in the extinction process and to be an effective basis for advanced combustor modelling.

Introduction

In association with colleagues at Sandia National Laboratories, Livermore, California, the combustion group at the University of Sydney has carried out a series of campaigns using the advanced laser diagnostics at Sandia to make measurements in turbulent flames designed to have chemical kinetic rates of the same order as the turbulent mixing rates. The piloted jet diffusion flame of Stårner and Bilger [1] and Masri and Bilger [2] has been the paradigm or model problem at the focus of these studies of strong turbulence/combustion interactions.

A better quantitative understanding of turbulence/combustion interactions is needed to improve the capability of combustor models for predicting extinction and pollutant emissions. Correa and Shyy [3] outline the features of combustor models currently in use and the problems in obtaining good modelling of kinetically dependent phenomena. Pope [4] reviews recent progress in model development including some attempts at modelling the Sydney/Sandia data. It is evident that there are several features of the piloted jet diffusion flame data, in particular features related to flame structure, that bring out the shortcomings of existing modelling approaches such as flamelet models and particle simulations of the joint pdf (probability density function) transport equation.

Recently Bilger [5] has formulated a new approach to modelling combustion in turbulent flow based on equations derived for the moments of reactive species conditional on

R. S. L.Lee · J. H. Whitelaw · T. S. Wung (Eds.)
Aerothermodynamics in Combustors
IUTAM Symposium Taipei, Taiwan, 1991

mixture fraction. The method gives good resuts for reaction in an isothermal scalar mixing layer [6]. Klimenko [7] has independently derived a similar model involving use of ideas from the theory of Brownian motion. This approach has several features which makes it potentially capable of addressing the flame structure questions that arise from the piloted jet diffusion flame data.

This paper reviews the laser diagnostic results obtained in piloted jet diffusion flames with particular attention being given to the significant questions of flame structure. Conditional moments are defined using the experimental data to illustrate their use. The transport equation for averages conditional on the mixture fraction are introduced and closure of the chemical source term in terms of the conditional averages discussed. For flames near extinction doubly conditional moments may be necessary to adequately define the reaction terms. The interrelationship of various aspects of flame structure are considered in terms of the fluctuations of the species from their conditional means and their correlation with fluctuations in the scalar dissipation and the kinetics rates. Finally the implications for the next generation of computer models are outlined.

Laser Diagnostics Results

Masri et al. [8–10] have used laser Rayleigh scattering (LRS) and spontaneous Raman scattering (SRS) to make point measurements of temperature and composition for the major species CH_4, O_2, CO_2, H_2O, CO and H_2 in piloted jet diffusion flames of methane near extinction. Conditional pdfs, conditional on mixture fraction, show strong bimodality for the reactedness of species and temperature for mixtures which are lean or near stoichiometric but they are monomodal for fuel–rich mixtures [10]. On the other hand piloted jet diffusion flames with $CO/H_2/N_2$ mixtures as the fuel shown this bimodality to a much lesser extent [11,12]. These findings have given rise to questions of whether the bimodality (which is commonly found in turbulent premixed flames) is associated with fluid dynamic aspects of the flame or with chemical kinetic effects. For undiluted methane in the fuel jet the stoichiometric mixture fraction, ξ_s, in air is small (0.055) so that the stoichiometric isopleth surface in the extinction zone some twenty diameters from the nozzle lies in the outer edge of the shear layer and even in the turbulent non–turbulent interface. The width of the reaction zone in mixture fraction space $\Delta\xi_R$ (≈ 0.05 for laminar counterflow methane flame in air and is expected to be much the same in turbulent flames) is only just less than the Kolmogoroff scalar scale

$$\Sigma_\xi \equiv (\bar{\chi}/\nu)^{\frac{1}{2}} \approx 1.4\ \xi' \mathrm{Re}_t^{-\frac{1}{4}} \tag{1}$$

which is about 0.06. Here $\bar{\chi} = 2\overline{\mathscr{D}\,\nabla\xi\cdot\nabla\xi}$ is the mean scalar dissipation, ν is the kinematic viscosity, ξ' the rms mixture fraction fluctuation and Re_t the turbulence Reynolds number based on the integral length scale and the rms viscosity fluctuations. With

$$\Omega \equiv \Delta\xi_R/\Sigma_\xi \tag{2}$$

much less than unity the turbulent flame structure could be expected to be in the flamelet regime [13] and this is marginally the case for undiluted methane. For the $CO/H_2/N_2$ mixtures $\xi_s \approx 0.3$, $\Delta\xi_R \approx 0.4$ and Σ_ξ remains at around 0.06 so that $\Omega \approx 6$, and a distributed reaction zone will be expected. Furthermore the stoichiometric isopleth will like much more in the centre of the shear layer. For jet diffusion flames using N_2– and air–diluted methane as fuel the bimodality is considerably less pronounced than for undiluted methane but is significantly more pronounced than for $CO/H_2/N_2$ mixtures [14]. The values of ξ_s, $\Delta\xi_R$ and Ω are much the same as for the $CO/H_2/N_2$ flame so that there is a strong suggestion of there being a kinetic effect on the bimodality. Methane is a very strong scavenger for radicals and extinction occurs in mixtures with much higher temperatures than when only H_2 and CO are present. The later measurements [12,14] include measurement of OH by laser induced fluorescence (LIF). Bimodality in the OH measurements is much more pronounced in the flames with diluted CH_4 as fuel. Kinetic effects are thus important in the bimodality aspects of flame structure.

Two–dimensional (2D) imaging of flame chemiluminescent emission and imaging of planar laser Rayleigh scattering (PLRS) and laser induced fluorescence of CH and OH have also been carried out [15,16] to give some elucidation of the spatial structure of these flames. For undiluted methane flames the reaction zones appear flamelet–like and local extinctions in the flame sheet occur. For fuels with higher ξ_s and Ω, reacting and extinguished zones appear to be associated with the large scale structures in the flow. Also zones of OH are broad compared with the fine scales of the turbulence but the CH appears in narrow, highly contorted sheet–like structures. These features of the OH and CH images are also apparent in the downstream regions of air–diluted methane flames. It appears that $\Delta\xi_R$ for CH, which corresponds to the consumption zone for CH_4, remains narrower than Σ_ξ while $\Delta\xi_R$ for OH corresponding to the slower three–body recombination reactions is quite broad.

A theoretical model for turbulent nonpremixed combustion must contain features which link the effects of stoichiometry and kinetics to the spatial structure and the pdf structure (bimodality, for example) of the flame. The conditional moment method appears to have the capacity to do this.

Conditional Moments

Figure 1 shows scatter plots of temperature and OH mass fraction in a piloted diffusion flame of methanol [17] at $x/D_j = 20$ and $u_j = 66\ ms^{-1}$, where D_j and u_j are the diameter and velocity of the methanol vapour jet. The data points are for data at all three radii, $r/D_j = 1.1$, 1.4 and 1.7. The full curves are for a laminar counterflow diffusion flame

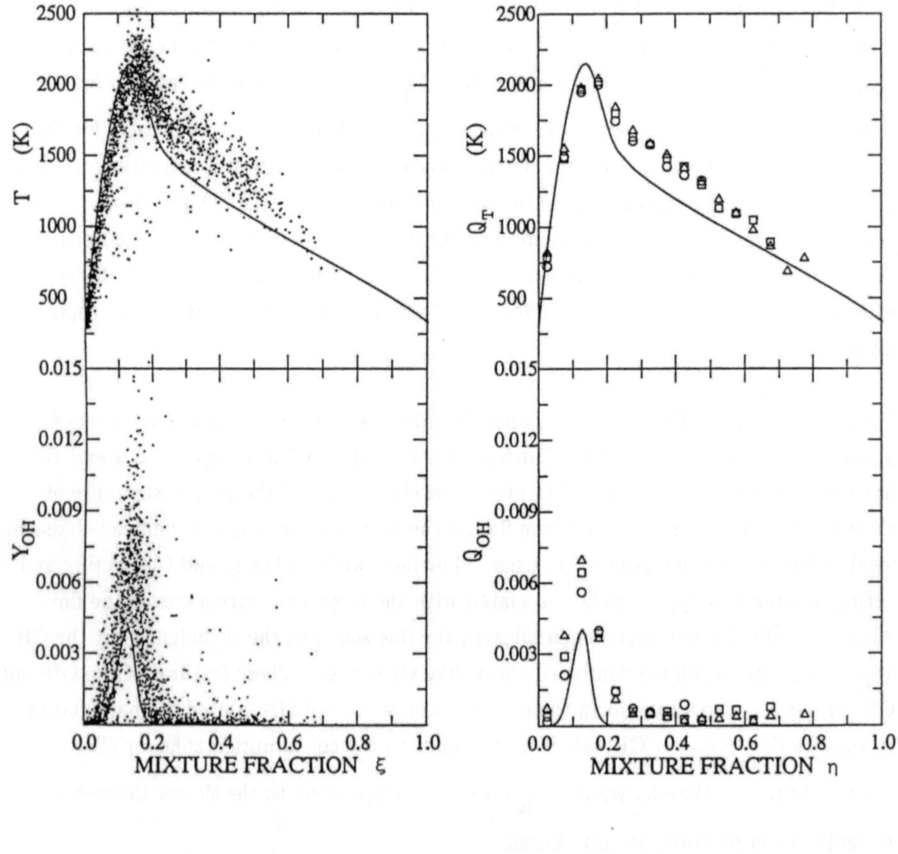

Fig. 1. Scatter plots of all data and conditional averages conditional on mixture fraction for temperature and OH mass fraction in a piloted diffusion flame of methanol [17] at $x/D_j = 20$, $u_j = 66\ ms^{-1}$ and r/D_j: ○ 1.7, □ 1.4, Δ 1.1. Full curves are for laminar counterflow diffusion flames as calculated with strain rate parameter $a = 5\ s^{-1}$.

calculation using Smooke's code [18] and for a strain rate parameter $a = 5\ s^{-1}$. At this jet velocity the flame is far from extinguishment and the data at any given mixture fraction are grouped over a narrow range about the conditional means Q_T, Q_{OH} where

$$Q_T(\eta,\underline{x}) = < T(\underline{x},t) \,|\, \xi(\underline{x},t) = \eta > , \tag{3}$$

$$Q_i(\eta,\underline{x}) = < Y_i(\underline{x},t) \,|\, \xi(\underline{x},t) = \eta >. \tag{4}$$

Here $\xi(\underline{x},t)$ is the mixture fraction at the spatial point $\underline{x}$, at the time t, Y_i is the mass fraction of species i, η is a sample space variable corresponding to ξ, and the angle brackets and the vertical bar denote averaging conditional on the conditions to the right of the bar, it here being assumed that the flow is statistically stationary. More general definitions of conditional moments are given in Bilger [5] including those suitable for density–weighted averaging. Conditional higher moments such as the variance may be defined in a similar manner and even conditional probability density functions [10].

Figure 1 also shows the conditional averages Q_T Q_{OH} for the data of Fig. 1 separated out for the three radii at which they were taken. It is seen that the data of Fig. 1 do, in general, lie close to these conditional means, and the conditional means are not a strong function of radial position in the flame. These two observations lie at the basis for the conditional moment closure method proposed for this type of flow. Of course, piloted turbulent diffusion flames which are close to extinction show much larger variance about the conditional mean. A doubly conditional moment closure is proposed for such flames.

Conditional Moment Closure

Bilger [5] derives equations for the variation of Q_i defined as in Eq. (3) and Q_H defined in Eq. (3) but for the standardized enthalpy H. The derived equations involve several terms which are difficult to evaluate at this stage, but which are hypothesized to be negligible. Furthermore, the assumption of negligible variation of Q_i with radius yields

$$<\rho u \,|\, \eta> \frac{\partial Q_i}{\partial x} = <w_i \,|\, \eta> + \frac{\mathscr{D}_i}{2\,\mathscr{D}} <\rho\chi \,|\, \eta> \frac{\partial^2 Q_i}{\partial \eta^2} \tag{5}$$

where ρ is the mass density, u the velocity component in the x (streamwise) Cartesian coordinate direction, w_i the net mass rate of formation of species i per unit volume, χ the instantaneous scalar dissipation

$$\chi \equiv 2\mathscr{D}\nabla\xi\cdot\nabla\xi, \tag{6}$$

$\mathscr{D}$ the species diffusivity associated with ξ, and $\mathscr{D}_i$ the diffusivity of species i. The equation for the standardized enthalpy conditional moment Q_H has no chemical source term but, instead, a radiation loss term. Methods for evaluating the conditional mass flux $<\rho u|\eta>$ and scalar dissipation $<\rho\chi|\eta>$ are given [5]. Klimenko [7] derives a result very similar to Eq. (5).

Closure of the set of Eq. (5) for all the species i requires evaluation of the conditional average reaction rate $<w_i|\eta>$. For flames far from extinguishment it is proposed that this be done using the conditional averages Q_i in the usual rate expressions together with a representative conditional temperature T_η deduced from Q_i and Q_H. The errors involved will be much less than using the conventional mean values $\bar{Y}_i$ and $\bar{T}$ in determining the conventional mean reaction rate $\bar{w}$. From Fig. 1 it can be seen that most of the variation of T and Y_{OH} can be associated with variation in the mixture fraction.

Equation (5) together with appropriate boundary conditions [5] represent a set of parabolic partial differential equations in x, η space and as such are relatively straight forward to solve. A similar transformation may be made for species mass fractions (unaveraged) in laminar flames [19] and an analogous form to Eq. (5) has been obtained for transient flamelets [20]. Solutions of Eq. (5) for turbulent flows will be different from those of the steady laminar counterflow diffusion flame due to the term on the LHS of Eq. (5), particularly in the near nozzle region of the flame where strong axial variations are obtained. Transient laminar flamelet models of turbulent combustion [20] incorporate such a term, but there is difficulty in correctly parameterizing this effect. At this stage codes are being written for solution of Eq. (5) for turbulent jet diffusion flames and solutions have been obtained for hydrogen fuels in the first instance [21]. The results show good agreement with measurement. Solutions have also been obtained for the case of an isothermal reacting scalar mixing layer [5], and agree quite well with measurements [6].

The conditional moment method as described so far for the first moment, Eq. (5), does not distinguish between flamelet or distributed reaction zones. The fluctuations about the conditional mean are taken as small, so that the conditional mean reaction rate is given with sufficient accuracy by the use of the conditional mean composition and temperature. Only the conditional average $<\chi|\eta>$ of the scalar dissipation is needed in Eq. (5) so that the conditional means Q_i and Q_T do not depend on either the fluctuations in χ or on their being variation or lack of variation of χ through the reaction zone; that is, they do not depend on whether the reaction zones are flamelet like or distributed.

The effects of turbulence on the reaction zone structure will, however, be important in the balance equations for the conditional second moments. It is possible to derive equations for the conditional second moments, that is the conditional variances for fluctuations of the species and temperature about their conditional means, and the conditional covariances of these fluctuations, all conditional on mixture fraction. These equations contain interesting terms such as

$$\tfrac{1}{2}(\mathscr{D}_i/\mathscr{D}) < \chi' y_i | \eta > \frac{\partial^2 Q_i}{\partial \eta^2}$$

where χ' is the fluctuation of χ from its conditional mean $<\chi|\eta>$ and y_i is the fluctuation of Y_i from its conditional mean Q_i. The conditional correlation $<x' y_i|\eta>$ will be very strong in the flamelet regime when $\Omega << 1$, for then χ will not vary through the reaction zone. For $\Omega >> 1$ there will be considerable variation of χ through the reaction zone and the conditional correlation will be weaker than in the flamelet case.

There will also be effects on the conditional moment equations of the conditional correlation between species fluctuations and reaction rates, that is terms of the form

$$< w'_i \, y_j | \eta >$$

Closure at the conditional second moment level involves evaluation of these terms. This can be done by presuming the form of the conditional joint pdf which is then scaled to match the computed conditional first and second moments. This conditional joint pdf is then used to evaluate the conditional species reaction–rate correlations. This approach becomes cumbersome when a large number of species is involved, and may be quite sensitive to the presumed form of the pdf when the flame is near to extinction and the fluctuations are large. The bimodality that becomes prevalent under such conditions makes the second order closure approach inherently inaccurate. The use of a doubly conditional moment approach [5] then becomes preferable. Before passing on to consideration of this, however, it is worthwhile to note that the method of Eq. (5) using first conditional moment closure could be very useful in predicting pollutants such as NO, CO and even soot in turbulent flames and combustors.

Doubly Conditional Moments

Figure 2 shows conditional scatter plots of species mass fractions and temperature in a piloted turbulent diffusion flame of air–diluted methane. The data shown are for that

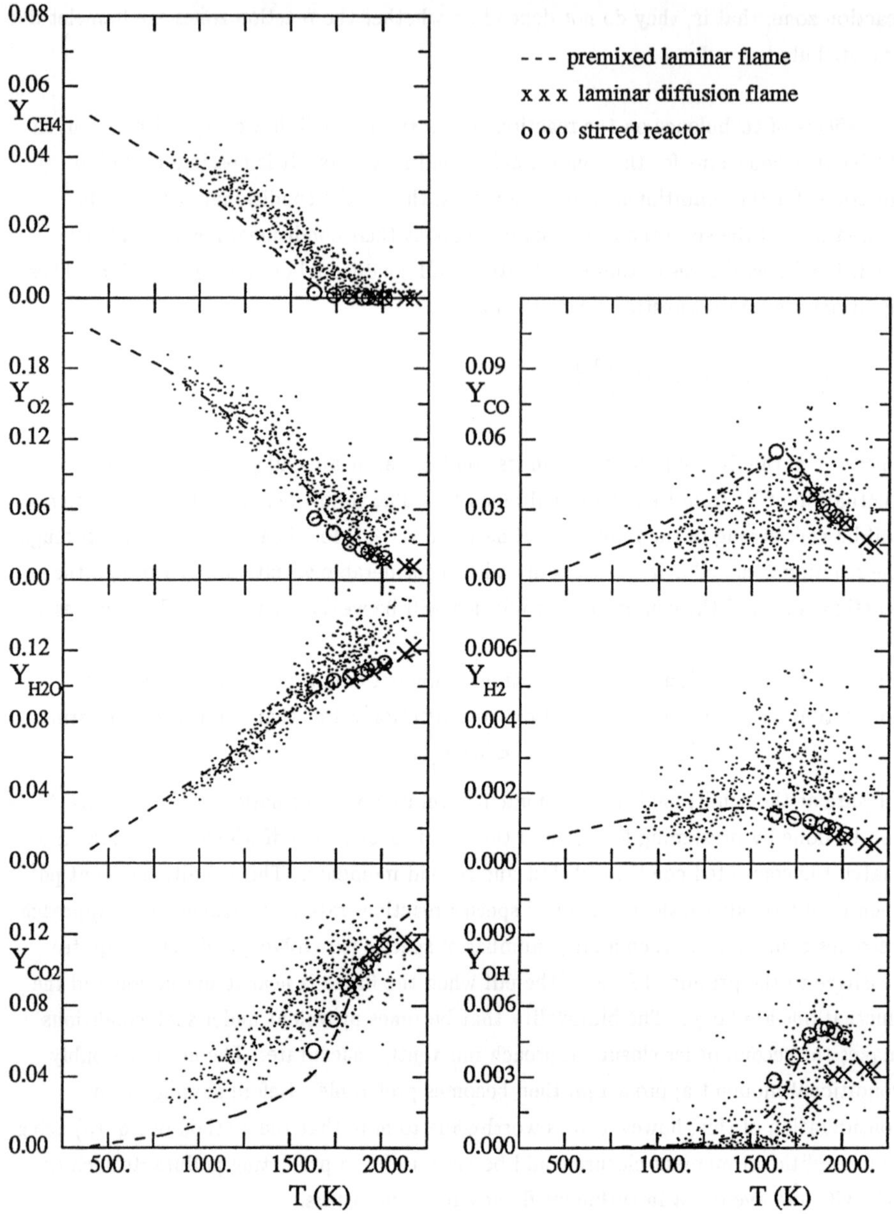

Fig. 2. Correlation plots of species mass fraction with temperature conditional on mixture fraction being close to stoichiometric for a piloted diffusion flame of methane diluted with air [14]. Correlations for laminar premixed and diffusion flames and for stirred reactors are also shown.

fraction of the total data which lies close to stoichiometric. It is apparent that there would be a large scatter of the data about the singly conditional mean values and the closure of Eq. (5) will not be satisfactory. It is seen, however, that the data for the major species, CH_4, O_2, H_2O and CO_2 lie close to a mean line that could be drawn as a function of temperature. This line would be that through the doubly conditional averages

$$Q_i(\eta,\underline{x}) \equiv \langle Y_i(\underline{x},t) \mid \xi(\underline{x},t) = \eta, c(\underline{x},t) = \zeta \rangle \tag{7}$$

where c is a reaction progress variable, such as temperature is in the present case, and ζ is its associated sample–space variable. The correlation for CO, H_2 and OH is not quite so good but this may be due to experimental error [5].

An equation is derived [5] for the doubly conditional averages, $Q_i(\eta,\zeta,x)$, which for statistically stationary flow, becomes, after neglecting terms hypothesized to be small,

$$\langle \rho u \mid \eta,\zeta \rangle \frac{\partial Q_i}{\partial x} = \langle w_i \mid \eta,\zeta \rangle + \langle w_c \mid \eta,\zeta \rangle \frac{\partial Q_i}{\partial \zeta} + \langle \rho \mathscr{D}_i \nabla\xi \cdot \nabla\xi \mid \eta,\zeta \rangle \frac{\partial^2 Q_i}{\partial \eta^2}$$

$$+ \langle \rho \mathscr{D}_i \nabla c \cdot \nabla c \mid \eta,\zeta \rangle \frac{\partial^2 Q_i}{\partial \zeta^2} + \langle 2\rho \mathscr{D}_i \nabla\xi \cdot \nabla c \mid \eta,\zeta \rangle \frac{\partial^2 Q_i}{\partial \eta \partial \zeta}. \tag{8}$$

Here it is assumed, again, that only variations in the direction of the streamwise coordinate, x, are significant. Closure for the conditional average reaction rates $\langle w_i \mid \eta,\zeta \rangle$ and $\langle w_c \mid \eta,\zeta \rangle$ is made at the order of $Q_i(\eta,\zeta,x)$ neglecting higher order terms. The modelling of the conditional dissipation–like terms $\langle \rho \mathscr{D}_i \nabla c \cdot \nabla c \mid \eta,\zeta \rangle$ and $\langle 2\rho \mathscr{D}_i \nabla c \cdot \nabla \xi \mid \eta,\zeta \rangle$ presents a challenge. Using appropriate boundary conditions [5] the modelled form of Eq. (8) may be solved to provide predictions of $Q_i(\eta,\zeta,x)$. It can be expected that these predictions will depend on the nature of the flow and other factors such as the amount of premixing. Thus $Q_i(\eta,\zeta,x)$ could be quite different in the stabilization region a lifted diffusion flame than in the extinction region a piloted diffusion flame of the same fuel at jet velocities producing local extinction. The $Q_i(\eta,\zeta,x)$, while interesting in themselves and important for determining $\langle w_i \mid \eta,\zeta \rangle$ and $\langle w_c \mid \eta,\zeta \rangle$, will tell us nothing about the joint pdf of ξ and c which is needed to complete the modelling of the combustion process. The essential features of this joint pdf are embodied, however, in the conditional mean of c, $Q_c = \langle c \mid \eta \rangle$, and its conditional variance, $V_c \equiv \langle c'^2 \mid \eta \rangle$, where the prime here denotes fluctuation from the conditional mean. Equation (5) written for Q_c can be solved to provide predictions of Q_c, except that now $\langle w_c \mid \eta \rangle$ is obtained by presumed form pdf closure

$$<w_c|\eta> = \int_c <w_c|\eta,\zeta> \, p_{c|\eta}(\zeta)d\zeta \tag{9}$$

where $p_{c|\eta}(\zeta)$ is the conditional pdf of c conditional on $\xi = \eta$. This conditional pdf will be evaluated using a presumed form and its predicted moments Q_c and V_c. In Eq. (9) $<w_c|\eta,\zeta>$ is evaluated using the doubly conditional first moment closure as in Eq. (8).

A balance equation for $V_c = V_c(\eta,x)$ has yet to be rigorously derived. It is apparent, however, that the major terms will include

$$<\rho u|\eta> \frac{\partial V_c}{\partial x} + \frac{\partial}{\partial x}\{<\rho u' c'^2|\eta>\} + 2<\rho u' c'|\eta> \frac{\partial Q_c}{\partial x} = 2 <w'_c c'|\eta>$$

$$- 2\rho \mathscr{D}_i \nabla c' \nabla c' |\eta> + <\rho(\mathscr{D}_c/\mathscr{D})\chi' c'|\eta> \frac{\partial^2 Q_c}{\partial \eta^2}. \tag{10}$$

In this equation primes on w_c and χ denote fluctuation from their conditional mean value. Gradient modelling of the second and third terms on the LHS of Eq. (10) presents problems due to the well–known existence of countergradient fluxes in premixed flames [22]. The reaction rate correlation term $2<w'c'|\eta>$ can be evaluated by presumed form pdf closure as in Eq. (9). The second term on the RHS is the now familiar conditional scalar dissipation of the reactive scalar which poses some challenge to model correctly so that it is valid, both at the flamelet and at distributed reaction zone limits.

The last term on the RHS embodies some interesting physics of the reaction zone structure. With c defined to be zero for unreacted mixing and positive when the mixture is reacted, $\partial^2 Q_c/\partial\eta^2$ is negative. Also the correlation between χ and c will be negative, since high values of χ will lead to mixing being faster than reaction locally and c being less than its conditional mean. This term is thus positive and a source contributing to the increase of V_c. The correlation between χ and c can be expected to be higher at the flamelet limit. When there is turbulence within the reaction zone local diffusion will smooth out c′ to lower absolute values than if it followed the peaks and valleys of χ throughout the reaction zone.

Flame Structure

The conditional moment theory gives us a basis for understanding the interrelationship between different aspects of flame structure in nonpremixed flames. The parameter Ω defined in Eq. (2) is probably sufficient to determine whether the reaction zone is flamelet–like or whether it is distributed by the fine scale turbulence. Bimodality of the conditional pdfs, however, is not simply a question of whether conditions are in the flamelet or distributed regimes. Rather, it is dependent on the history of the flow and the interplay of the various terms in Eq. (10). When $\Omega << 1$ the reaction zones are flamelet–like and the last term on the RHS of Eq. (10) becomes large and the tendency for bimodal structure is strong. When $\Omega >> 1$ this term is much smaller and the tendency to bimodality is reduced. For hydrocarbon fuels the tendency to bimodality is also increased by the reaction rate correlation term, which is the first term on the RHS of Eq. (10). For this to be positive Q_c must be less than the value $c_p(\eta)$ where $<w_c|\eta,\zeta>$ peaks along the ζ coordinate. For hydrocarbon fuels this lies close to the maximum value of c possible at that η, whereas for $CO/H_2/N_2$ mixtures $c_p(\eta)$ is much more toward the middle of the range. Bimodal flame structure is thus a result of several factors operating through Eq. (10) rather than having a simple parametric dependence.

The thin reaction–sheet structures for CH found in hydrocarbon flames can be explained in terms of the value of Ω for the inner reaction zone where CH is produced and consumed. This inner zone is much narrower in mixture fraction space, $\Delta\xi$, than for the OH reaction zone. Furthermore, laminar flame calculations indicate that $\Delta\xi_R$ decreases with χ while $\Delta\xi_R$ for OH is much less sensitive to χ.

Advanced Combustor Modelling

For prediction of pollutant emissions such as CO, NO_x, and soot, it is evident that the conditional moment method with closure at the first conditional moment level may suffice. Such emissions are usually of most interest under conditions far from extinction. The method has been demonstrated [21] for simple geometry and gives good predictions for NO in the turbulent hydrogen diffusion flames studied. There is no inherent difficulty in including reactions for prompt NO or reactions from fuel–bound nitrogen. For more complex flow geometries such as gas turbine combustors methods for imperfectly stirred reactors and for flows with curved streamlines have been outlined [5].

For prediction of extinction it is apparent that it will be important to predict the spread of reactedness in the mixture and in particular the amount of bimodality in the conditional pdfs of reactive species and temperature. The experiments [2,8] indicate that combustion can continue under conditions for which the mean temperatures are quite low: it is the isolated patches of well–burnt mixture that keep the flame alight. For modelling such problems it would be desirable to solve the modelled form of Eq. (8) to obtain the doubly conditional means of the species composition and temperature with closure for the reaction rate term in terms of these doubly conditional means. Solution of Eq. (3) and (10) for Q_c and V_c will then provide the diagnostic for extinction. In some cases it may be possible to use empirically determined values for the doubly conditional averages (e.g. as in Fig. 2) and so avoid the solution of Eq. (8). The solution of Eq. (3) and (10) would then use the empirical values of the doubly conditional averages for evaluating the reaction rate terms.

Conclusions

It can be concluded that the conditional moment approach shows considerable promise as a theoretical basis for understanding questions of flame structure, particularly the relationships between reaction zone structure, chemical kinetics and bimodality of conditional pdfs. It is also a promising basis for advanced computer models which can adequately handle the combustion/turbulence interactions involved in systems using complex chemical kinetic schemes and yet be quite tractable on currently available computers.

Acknowledgements

This work is supported by the Australian Research Council. The contributions of S.B. Pope, S.H. Stårner, A.R. Masri, R.W. Dibble, R.S. Barlow and J.–Y. Chen to this work are gratefully acknowledged.

References

1. Stårner, S.H.; Bilger, R.W.: Characteristics of a piloted diffusion flame designed for study of combustion turbulence interactions. Combust. Flame 61 (1985) 29–38.

2. Masri, A.R.; Bilger, R.W.: Turbulent non–premixed flames of hydrocarbon fuel near extinction: mean structure from probe measurements. Twenty–first Symposium (International) on Combustion, The Combustion Institute, Pittsburgh PA (1988) 1511–1520.

3. Correa, S.M.; Shyy, W.: Computational models and methods for continuous gaseous turbulent combustion. Progr. Energy Combust. Sci. 13 (1987) 249–92.

4. Pope, S.B.: Computations of turbulent combustion: progress and challenges. Twenty–third Symposium (International) on Combustion, The Combustion Institute, Pittsburgh PA (1991) 591–612.

5. Bilger, R.W.: Conditional moment methods for turbulent reacting flow using Crocco variable conditions. Charles Kolling Research Laboratory TN F–99, Department of Mechanical Engineering, The University of Sydney 1991.

6. Bilger, R.W.; Saetran, L.R.; Krishnamoorthy, L.V.: Reaction in a scalar mixing layer. J. Fluid Mech. (in press).

7. Klimenko, A. Yu.: Multicomponent diffusion of various admixtures in turbulent flow. Fluid Dynamics 25 (1990) 327–334, translated from Isv. Akad. Nauk SSSR, Mekhanika Zhidkoshi i Gaza, No 3 (1990) 3–10.

8. Masri, A.R.; Dibble, R.W.; Bilger, R.W.: Turbulent nonpremixed flames of methane near extinction: mean structure from Raman measurements. Combust. Flame 71 (1988) 245–266.

9. Masri, A.R.; Dibble, R.W.; Bilger, R.W.: Turbulent nonpremixed flames of methane near extinction: probability density functions. Combust. Flame 73 (1988) 261–285.

10. Masri, A.R.; Dibble, R.W.; Bilger, R.W.: Turbulent nonpremixed flames of methane near extinction: conditional probability density functions. Combust. Flame 74 (1988) 267–284.

11. Masri, A.R.; Dibble, R.W.: Spontaneous Raman measurements in turbulent $CO/H_2/N_2$ flames near extinction. Twenty–second Symposium (International) on Combustion, The Combustion Institute (1989) 607–618.

12. Stårner, S.H.; Bilger, R.W.; Dibble, R.W.; Barlow, R.W.: Piloted diffusion flames of CO CH_4N_2 and CO H_2N_2 near extinction. Combust. Flame 83 (1991) 63–74.

13. Bilger, R.W.: Structure of turbulent nonpremixed flames. Twenty–second Symposium (International) on Combustion, The Combustion Institute, Pittsburgh, PA (1989) 475–488.

14. Stårner, S.H.; Bilger, R.W.; Dibble, R.W.; Barlow, R.S.: Piloted diffusion flames of diluted methane near extinction: detailed structure from laser measurements. Combust. Sci. Technol 72 (1990) 255–269.

15. Barlow, R.S.; Dibble, R.W.; Stårner, S.H.; Bilger, R.W.; Fourguette, D.C.; Long, M.B.: Reaction zone structure in dilute methane jet flames near extinction. AIAA Paper 90–0732 (1990).

16. Stårner, S.H.; Bilger, R.W.; Dibble, R.W.; Barlow, R.S.; Fourguette, D.C.; Long, M.B.: Joint planar Rayleigh/LIF imaging in piloted turbulent jet diffusion flames near extinction. Combust. Sci. Technol. (in preparation).

17. Masri, A.R.; Dibble, R.W.; Barlow, R.S.: The structure of turbulent nonpremixed flames of methanol over a range of mixing rates. Combust. Flame (submitted).

18. Smooke, M.D.; Puri, I.K.; Seshadri, K.: A comparison between numerical calculations and experimental measurements of the structure of a counterflow diffusion flame burning diluted methane in diluted air. Twenty–first Symposium (International) on Combustion, The Combustion Institute, Pittsburgh, PA (1987) 1783–1792.

19. Bilger, R.W.; Stårner, S.H.: A simple model for carbon monoxide in laminar and turbulent hydrocarbon diffusion flames. Combust. Flame 51 (1983) 155–176.

20. Haworth, D.C.; Drake, M.C.; Pope, S.B.; Blint, R.J.: The importance of time–dependent flame structures in stretched laminar flamelet models for turbulent jet diffusion flames. Twenty–second Symposium (International) on Combustion, The Combustion Institute, Pittsburgh, PA (1989) 589–597.

21. Smith, N.S.; Chen, J.–Y., Bilger, R.W.: Modelling of nonpremixed hydrogen jet flames using a conditional moment method. Twenty–fourth Symposium (International) on Combustion, The Combustion Institute, Pittsburgh, PA (submitted).

22. Moss, J.B.; Simultaneous measurements of concentration and velocity in an open premixed turbulent flame. Combust. Sci. Technol. 22 (1980) 119–129.

Combustion Research for the 21th Century - Some Speculative Extrapolations

FELIX WEINBERG

Department of Chemical Engineering and Chemical Technology,
Imperial College,
London SW7 2BY, England

Introduction

Seventeen years ago, at the Tokyo International Combustion Symposium, I was asked to give a personal overview of the past and future of combustion research in a plenary lecture (1). Since then many profound and important changes which affect our subject have taken place. It was undoubtedly very rash of me but, having again been offered the opportunity to air my prejudices, I could not resist the occasion not only to hold up my former forecast to the cold light of hindsight but, more particularly, to assess the effects of developments and constraints which have arisen since then on projections for the future of combustion research. From the many new developments, I have tried to focus on those relevant to the topics of this meeting.

(This is a concise summary of a one hour Plenary Lecture which was illustrated by many slides and diagrams of practical examples that could not be reproduced in the present format.)

Developments, Constraints, Trends and Opportunities

As regards external constraints, we are, at the time of writing, in the throes of the fifth oil crisis in the last half-century (the Abstract was written at the height of the Gulf crisis and the emphasis on "the time of writing" was prompted by the surprising intermittency of this phenomenon, which is such that the prevailing state of euphoria may well have reasserted itself by the time of the Symposium), we have ushered in public environmental consciousness, we have discovered a hole in our ozone layer, and we are in the process of adding a relatively inert product of all fossil fuel combustion - carbon dioxide - to our list of potentially hazardous pollutants. All these are our concerns. NO generated during combustion in aircraft gas turbines (especially SSTs) is a major threat to ozone. Insecurity in oil supplies as well as recognition of carbon dioxide as a pollutant emphasises the need for research on combustor efficiency. Indeed energy conservation has become an anti - pollution measure. Europe has set itself the task of reducing carbon dioxide down to 1988 levels by the year 2005; this should be perceived in terms of the 5 fold increase in the world's

R. S. L. Lee · J. H. Whitelaw · T. S. Wung (Eds.)
Aerothermodynamics in Combustors
IUTAM Symposium Taipei, Taiwan, 1991

energy consumption since the 2nd World War - almost entirely due to the developed countries. In addition to research on highly efficient, fuel saving, combustion methods generally, the combustion of hydrogen, and hydrogenation of fossil fuels (subject to hydrogen production methods which do not make large energy demands) are seen as important topics for research. There are substantial moves towards natural gas as replacement for petroleum products, wherever possible. Fuel technology is not on the agenda of this meeting but the above has important implications for combustion research. Also, there is renewed interest in burning hydrocarbons, such as methane at very lean equivalence ratios because leakage into the atmosphere of such species increases the greenhouse effect disproportionately more than the equivalent amount of carbon dioxide.

As regards research developments, while some avenues of research have been attenuating, other new opportunities are emerging; the time available for combustion experiments in micro-gravity, for example, has progressed from the seconds of drop towers, over the tens of seconds of parabolic trajectory aircraft flights, into the realm of measurements of extended duration with the advent of orbiting laboratories. Considering major trends, however, there has been quite a remarkable and continuing escalation in available computing power, as well as great strides forward in optical diagnostics - both of which have had the effect of prompting a substantial growth in our knowledge of fundamental kinetics. On the other hand, some of the tools of this trade are very expensive, so much so as to place them almost beyond the means of researchers financed entirely by universities in the absence of research contracts - a situation which has given rise to new kinds of research organisations.

The concentration of expensive equipment in corporate research establishments has not led, and shows no sign that it is going to lead, to the greater general and inexpensive availability of unaffordable research tools and computing facilities. The requirement that such organisations maintain themselves financially, together with the research aspirations of their guardians, ensure that they are fully employed. Affluent industrial research laboratories are well equiped to deal with these challenges but their studies tend to be largely "misson - oriented". It is not unusual to see costly optical installations in industry used for very qualitative visualisation - for example in engines to pinpoint pre-ignition sites or optimise the location of ignition plugs. Yet, innovative and fundamental research is both necessary and often relatively inexpensive. Modern tools sometimes offer a choice between solving problems and trampling them to death. For example, verifying particular steps in proposed reaction mechanisms by fitting computer predictions to complex combustion behaviour, is likely to be much less efficient and reliable than studying them individually.

Our subject has also been invigorated by periodic transfusions of new blood from other research disciplines. We have always been a pretty mixed crowd, which has more recently been enriched by an influx of laser spectroscopists, modellers and even environmentalists and climatologists. On the political side also, the research pattern has been affected in disparate ways in different countries. In many of the Western democracies, governments have resolved that state - supported research should be motivated by profit rather than by curiosity and the rest of the world may be heading towards the same ideology. Whilst I shall attempt to explore research achievements and strategies in response to such constraints, in the absence of a reliable crystal ball, the aim is to provoke discussion rather than to offer any infallible insights - to which I can make no claim whatsoever.

Computing - Kinetics - Diagnostics

It may seem a little odd, at first sight, to group three such apparently disparate topics together under one heading. However it is quite remarkable how large strides these fields have made, each partly in response to the advance in the others. The structure and properties of laminar premixed flames of simple fuels are now routinely modelled with the detailed kinetics of a full complement of reaction steps (e.g. ref 2). These have indeed been occasionally used to confirm (or otherwise) the kinetics of critical steps in particular reaction schemes. In the equivalent turbulent case (e.g. ref 3), as yet, only much reduced mechanisms can be handled but a number of entirely new approaches have emerged. These include the concept of dealing with convoluted flames in complex flow fields by the use of "libraries" of strained flames (e.g. 4,5), at least until the Kolmogorov scale of turbulence becomes small enough to result in a distributed flame structure.

The striking and continuing rapid progress in computer modelling raises the question of whether we may eventually reach the state of being able to predict combustor performance without recourse to experimentation (and whether it might be less costly to do so). It is remarkable that no one in the field I have spoken to thought that was as yet on the horizon, in spite of the circumstance that growth in computer memory and speed of has been almost exponential hitherto. Part of the reason lies in the computational time being proportional to the cube of the Reynolds Number of the flow system (6), even without taking chemical kinetics into account. However there are major gaps in our knowledge and ability to model as soon as we move away from the simplest of academic sytems. A recent discussion meeting on this topic identified the major hurdles holding up progress as the need for improvements in modelling swirl, entrainment and particle diffusion. More information is required on transport properties, not least with respect to the effects of higher hydrocarbons. Other desiderata lie in grid generation and adaptive gridding as well as availability to modellers of strained flame (diffusion and premixed) libraries in a suitable form. The kinetics interface with modellers (including validity ranges and sensitivity analysis) needs to be developed and kinetic data for higher hydrocarbons, including petroleum and coal fractions, requires much further study.

Massive progress has also been made in optical diagnostic methods - particularly those based on laser stimulated spectroscopic techniques (e.g. 7). In the last few years this has culminated in a group of two dimensional imaging methods (e.g. 8). Whilst time - resolved point measurements are of great value in the study of turbulence, the ability to obtain an instantaneous snapshot is particularly important in investigating flame structure. We have already heard at this meeting of the excellent use to which OH and CH distribution records have been put in the study of turbulent flames (9, 10). In the laminar case there is a particular problem because of the very large gradients that occur, particularly within premixed flames. Quite mild - e.g. stoichiometric hydrocarbon/air - flames harbour temperature gradients of the order 10^5 ^{o}/cm , so that an error in position of 10μm can correspond to one of 100 ^{o} in temperature, which militates against the use of point-by-point traverses. I am not convinced that any flame can be maintained stationary to such minute tolerances (nor are diameters of focused laser beams infinitesimal on that scale). Moreover, one of the parameters exhibiting such large gradients is refractive index and this can lead to large beam deflections both in stationary flames (11) and within fluctuating phase boundaries (13). All this makes methods such as planar laser induced fluorescence (8,13), which produces instantaneous records of two dimensional distributions of species excited to fluoresce, so attractive. They may be expected to be increasingly combined with simultaneous other coincident imaging methods using light sheets for scattering, excitation, incandescence, together with self luminosity, schlieren and interferometry to combine maximum information content in a single (or superimposable) record, even of irreproducible events. Interesting recent newcomers to this field are degenerate four wave mixing (14) and laser induced incandescence of particles (15). The former in particular promises the facility of subtracting out abberations and beam deflection distortions because it acts as a phase conjugate mirror.

A curious thought concerning these methods is that some of them are not "non-invasive" in the former meaning of that term. The incident beam may well be intense enough to excite reacting species, however, the image has been acquired before the system can respond. If and when we proceed to high speed cine recording, we may have to reappraise the "non-invasiveness" of some of the techniques in this light.

Advanced Combustion Methods

Many of the former (1) predictions have come to pass, occasionally somewhat obliquely. Above all, combustion techniques have become much more sophisticated. Diffusion flames are being phased out wherever the type of fuel permits. The merits of lean premixed burning, often employing some means of increasing reaction rate, are being investigated in a wide variety of combustion systems. Staged combustion, in which the radical pool of an initially fuel - rich burning zone is utilised by following with the admixture of excess air in a second region, is increasingly being adopted as a pollution abatement method in more difficult systems. Developments and opportunities for future research in advanced combustion methods are cursorily reviewed below. By way of illustrating developments, this section is organised under the chapter headings of a recent book of the same title (16): combustion in fluidised beds, pulsating combustion, combustion in heat-recirculating burners, catalytic combustion, electrical discharge - augmented flames and plasma jets in combustion, and the control of combustion processes by electric fields. Several of these techniques have passed from the research to the development stage; others offer major new research opportunities.

Even 17 years ago, the major attractions of **fluidised bed combustion** were evident (e.g. 1). In addition to allowing the less ideal fuels - and coal in particular - to be burned homogenously and efficiently, and greatly improving heat transfer to steam pipes buried within the bed especially when pressurised, with alkaline oxides such as limestone added to the bed they are able to remove gaseous acidic oxide pollutants such as sulphur dioxide, which is retained as calcium sulphate. Today, pressurised fluidised bed combustors are proposed to power the flagships of the future cleaner coal burning power generating industry, at least in the UK. They are at the heart of both, the proposed "combined" and "topping" cycles of future electricity generation schemes (e.g. 17). These add gas turbines fed by gaseous hot combustion products from a pressurised fluidised bed to the conventional steam turbines and, in the latter scheme, an additional coal gasification stage, also accomplished in a pressurised fluidised bed. The practical problems remaining to be solved probably concern erosion of steam pipes and turbine blades more than combustion fundamentals, though there are still some very active areas of combustion research, particularly as regards rates of burning (e.g. of CO and coke particles - e.g. 18, 19) kinetics of sulphur removal by calcium oxide (20, 21) and the study of staged combustion in fluidised beds as a means of further reducing NO_x emissions.

The principles and background of **pulsating combustion** have been covered in the paper by Professor Zinn (22) much more thoroughly than I could aspire to, in the given time. Considering that the first - and for a very long time the sole - practical application was the V-l flying bomb of World War II - it is most encouraging to note that the subject has now reached the stage at which furnaces and space heaters based on this principle and burning natural gas can be bought over the counter for domestic use. They are of simple construction and regulate their own supply of air and fuel, as well as venting their combustion products without the need of an extractor or flue. This saves enegy (even a flue requires a substantially elevated product temperature for venting). They manifest very high heat transfer rates and, since they operate slightly lean, produce relatively low levels of NO_x. Silencing these devices has been successful to the point where noise pollution is reduced to a hum which is certainly acceptable - in a garage if not in one's living room.

It is thought likely that the next major step forward in this field, now under active study (23), is towards tunable pulsating combustors. The object is to synchronise the burner with the resonant frequency of the system to be heated. For example, heat transfer to the contents of a dryer can greatly be enhanced by the large amplitude oscillations induced by such resonance.

Great strides have been made in **heat - recirculating burner systems** (16) in which the reactants (or the combustion air alone) are preheated using heat "borrowed" from beyond the flame zone. In the absence of heat abstraction or losses the final exit temperature would, of course, remain unchanged, in the steady state, but in between a greatly elevated flame temperature and hence reaction rate (and thermodynamic efficiency, in the case of a converter) can be realised - subject only to the limitations of the heat exchanger. There is very great variety of possible systems, all based on a combustor in between the two limbs of a heat exchanger (indeed today very few new practical combustion systems are designed which do not involve some element of heat recirculation). At one extreme we have extremely efficient heat exchangers such as the double spiral "Swiss roll" burner which may recycle 5 times the (low) combustion enthalpy and can burn very poor fuels and lean mixtures such as ventilation air from mines, down to 1/5 of the limit of flammability. Burners of this kind are suitable for waste gas incineration. Some other burners are simply designed to save fuel. Thus, the British Gas "self-recuperative" ("self" because the heat exchanger is built into it) burner is designed to replace a conventional burner in order to heat up a furnace to a given temperature. Instead of raising the flame to a higher temperature, it saves the corresponding amount of fuel and devices of this kind have already notched up massive natural gas savings. To achieve higher temperatures than counterflow recuperators allow, ceramic "regenerators" have been developed. The hot products heat up a packed bed, after which the flow is reversed so that the bed preheats the incoming air. Packings of refractory spheres have yielded thermal efficiencies in excess of 85%; spouted or fluidised refractory beds may also be used.

Recirculating particulate beds are also being developed to extend the principle of heat recirculating burners to solid or difficult liquid fuels. (Such fuels and their pyrolysis products are apt to clog up the fine passages of counterflow heat exchangers, whereas refractory particles heated to high temperatures by the flame products in, for example, a spouted bed and travelling downwards to preheat the incoming reactants, are ideally suited to this task.) Their use for incineration has been widely studied. It is to be anticipated that, in the future, much more advantage will be taken of the potential for increased thermodynamic efficiency of these devices.

Such heat recirculating burners have the following important major combustion features:

they enable lean mixtures to be burned at higher rates and can extend the definition of what is a fuel;

they are able to pre-gasify (or pre-vaporise), premix and preheat solid or liquid fuels prior to burning so that a blue "Bunsen burner" type of flame can result from burning very unpromising feedstock;

they also tend to separate the product exit path from that by which the energy is extracted and this may become particularly important in the future, if it becomes necessary to extract pollutants from the products.

A promising pointer to the future is that German regulations for emission standards take into account fuel efficiency of equipment and therefore have more lenient limits for recuperative burners. (They range from 300ppm NO_x at 350°C preheat to around double that value at 600°C.) For very high temperature flames, staged combustion can again be employed. It has been shown (24) that NO_x from British Gas burners fall substantially below the levels demanded by these regulations. When heat recirculating burners are used to burn ultra lean, NO_x levels tend to be inappreciable.

Promising areas for future research on heat recirculating burners include a renewed interest in MHD "topping" of power generation within the EEC; ceramic regenerators promise to produce the increased flame temperatures and hence ionisation required. We have already had one paper in this meeting (25) dealing with the hazards of burning hydrogen, otherwise regarded as an ideal fuel. In addition to its predilection for detonating, in stoichiometric mixtures with air it also produces oxides of nitrogen. Both these problems are removed where it is burned in very lean mixtures with air. We are currently working on the use of recuperative burners to burn hydrogen/air mixtures below the normal limit of flammability with air, so that ignition away from the burner is impossible. One final example of encouraging new developments is provided by the work at Princeton on the "TONE" engine (26) which features a heat exchanger between hot products and incoming air. It is capable of burning extremely lean mixtures - well below the conventional misfire limit - and has yielded very favourable efficiency and emission characteristics.

As regards **catalytic burners**, those operating on diffusion flames have been available for fuels such as liquified petroleum gas for some time; recently new long-life catalysts have been developed in Japan which are suitable for natural gas. In the field of cleaning up combustion products, catalysis has received a major commercial boost from the general trend towards exhaust catalysts in internal combustion engines. The consequences for CO_2 emissions of this approach, in preference to lean burn engines, are further discussed below. (Pollutant removal subsequent to completion of the combustion process, whether by ammonia injection, with or without catalyst, scrubbing, absorption, etc. is not dealt with here.) The great potential of catalytic burners themselves for lean combustion by increasing burning rates on and near their surfaces, however, is increasingly being realised in practice. We have heard one interesting paper on modelling catalytic ignition (28). Flame stabilisation in fast premixed reactant flows by catalysts has been a major subject for study in recent years (see e.g. Kesselring, Chapter 4 in ref. 16). This subject merges into surface combustion and, for instance, the Alzeta "Pyrocore" burner composed of a porous matrix of ceramic fibres on whose surface combustion occurs, operates at a temperature some 400°C below that of conventional gas burners (29). I shall presently mention the use of electric discharges, in the form of plasma jets, for the destruction of some pollutants. An entirely new application of catalysts has recently emerged in this field when NO reduction by corona discharges was investigated and found to be strongly dependent on the type of catalytic coating covering the plane electrode (30).

Of all the methods available for injecting specific radicals into combustion processes, the most direct is based on the use of **plasma jets** (16). Major progress has been made in studies on the use of both, pulsed plasma jets for enhanced ignition and continuous flow plasma jets for flame stabilisation. The drive towards using the greatly enhanced ignition and flame propagation induced by pulsed plasma jets for lean-burn internal combustion engines has been attenuated (along with combustion - driven jets and other means of producing radical - rich initiation) in favour of stoichiometric combustion with added catalytic clean up combustion - generated pollutants. However, these decisions pre-date concern about carbon dioxide emission and ought speedily to be reconsidered in this light; the additional costs of changes at the refinery involved in the production of unleaded fuel, together with decreasing compression ratio, alone add up to an additional 15% of fuel used up and greenhouse gases produced. In the meantime we have learned a great deal, using some of the new diagnostic methods mentioned above, about the manner in which free radicals from plasma jets promote ignition and how centrifugal fields concentrate them in vortices which can be projected by the igniter (16, 30, 31).

Considerable improvements have also been achieved in converting the much more energetic surface discharge igniters used for jet engine combustors into plasma jets, particularly for dealing with the critical problem of relight following extinction at high altitudes (32, 33). It emerges that this kind of radical initiation works well even in a spray of aviation fuel. In the meantime, continuous plasma jets have been studied for their potential as flame holders in scram jets both in the USA and in Japan (34 - 36) A few years ago it seemed highly unlikely that the same principle might be applied in conventional jet propulsion. However, the pressure for NO_x reduction in aircraft emissions with the aim of protecting the ozone layer , and particularly so for SSTs, are changing the industrial outlook. Jet engine combustor design for the next century is being based on leaner combustion, often employing multiple injectors with the aim of varying thrust by shutting down some of them; designs with plasma jet pilots are being put forward (37) because of the hazard posed by flame - out in such systems.

Plasma jets offer at least three further interesting new study areas. The radicals they generate have been shown capable of destroying certain pollutants - notably soot and NO (the latter by the addition of nitrogen atoms - e.g. ref. 16). More recently sulphur dioxide has been efficiently converted to SO_3 by this means (38). Current work is concentrating on treating a cocktail of pollutants and higher electrical potentials. Another interesting new scheme is based on using plasma jets as generators of nucleii to increase radiation in natural gas burner flames (39). In addition to previously observed increases in flame stability, the soot so formed increases heat transfer and reduces flame temperature - and hence NO_x formation. Finally, plasma jets offer a potential and much neglected tool for the study of fundamental kinetics. Instead of the enormous number of different radicals produced in flames, they are capable of providing a steady flow of just one or two labile species with appreciable life-times. Some mechanistic studies have been carried out by Dr D. Lucas at the Lawrence Berkely Laboratory but this field of research offers unfathomed opportunities.

The **use of electric fields in combustion** is still very much at the research stage, with several novel developments promising well for the future. It has been shown that the force exercised on neutral gas by flame ions in electric fields can be used to increase flame stability against blow-off substantially on industrial burners (40). This allows not only larger throughputs for a given equivalence ratio but the use of leaner mixtures with concomitant reductions in NO_x emissions. The same principle has been used for controlling flames in the absence of earth gravity (41). Under normal gravity, we rely on natural or forced convection to replenish reactants and direct hot products to the surface to be heated. Forced convection requires a compressor for keeping fuel under pressure in cylinders, and thus entails a considerable weight penalty. Natural convection is not available under micro-gravity conditions, making heat transfer from flames difficult to achieve.

Diffusion flames, indeed, will tend to become spherical and will eventually extinguish as a result of blanketing by their own products. It has been shown that controlling flames in zero gravity by the application of electric fields which can achieve the equivalent of 800 g using compact, light weight, equipment can provide very efficient and economical heating of specified areas by small flames, making optimum use of the oxygen available in the working environment and using liquid or solid fuel requiring no pressurised containment.

As regards transposing charge carriers, numerous attempts have been made to use ion currents in the exhaust products of IC engines to monitor the combustion process with the ultimate aim of feedback management. Another area which has recently resurfaced (42, but see also 43) is the manipulation by electric fields of charged refractory particles in order to control their deposition and size in manufacturing processes. This may well prove important in future in conjunction with the recent development of a whole family of new combustion systems (eg 44-46) which lead to the production of valuable refractories and metals in the form of such particles. Examples are the formation of titanium nitride from titanium and liquid nitrogen (and, analogously, silicon and boron nitrides and carbides), the production of Si (Ti, etc.) from silicon tetrachloride by reaction with metallic Na (K, Mg, etc.). (Such flame reactions in themselves are likely to open an important new chapter in combustion research, albeit one which is too "chemical" to fit easily into the frame of reference of this meeting.)

In Closing

I should like to begin this section with an apology; all this is, of course, an entirely personal view of various promising lines of research. I cannot pretend to have covered all the important new developments in combustion and I hope that I might be forgiven for concentrating on aspects which are of particular interest to me. I suppose we choose our fields of study largely by what seems most important to each one of us.

Notwithstanding the great strides which have been made in computing, diagnostics, chemical kinetics and advanced combustion methods, it is clear that much more remains to be done. Members of the Combustion Institute may recollect having heard Walter Berl recount how his old teacher advised him to chose for his field of research a subject that would not be solved in his life time. Combustion research has been engaging mankind for longer than any other, yet there always seems much more to be done.

It is, no doubt inevitable that in future we shall have to rely even more heavily on industrial sponsorship and mission oriented research support. However, as research workers we are aware of some important facts wich may not be generally recognised by the powers that be and which we should lose no opportunity to propagate. We are, after all concerned citizens of our various countries, quite apart from being scientists and engineers - and some of us sit on a variety of committees as well. Let me make a few suggestions:

we may be hard to convince, in view of past experience, that some new line of fundamental research may not lead to an important unforeseen practical application;

we may well think it unreasonable of goverments to expect, for example, fuel manufacturers to fund research on improved fuel economy;

we know that there is a steady supply of brilliant young researchers who are, at least initially, attracted by curiosity-driven rather than commercially driven research and whom the subject can ill afford to lose;

we observe that the development and implementation of innovations generally arises not from their invention but out of urgent practical needs, or regulatory pressures. Even though our needs appear to be escalating by the day, basic science can only advise technology as to what is feasible; we cannot force development of our ideas, no matter how good we think they are;

I am sure you can think of further such points. It is up to us to make our voices heard.

Perhaps the major change in our subject is the importance it has acquired. When I started, my research seemed very academic; I do not recollect even being aware that I was dealing with the process by which the overwhelming majority of the world's energy is released and I cannot honestly say that I was very worried about our environment. With the world's concern focused on fuel resources and combustion - generated pollution, it may have become the most important field of endeavour and what we are about may have far-reaching consequences. We may anticipate that that burning fuels inefficiently will be regarded as a barbaric habit in the next century and I would like to think that our grandchildren will be proud of the part we played in putting this to rights.

References

1. Weinberg, F.J.: "The First Half-Million Years of Combustion Research and Today's Burning Problems", 15th Symposium (Int.) on Combustion, The Combustion Institute, Pittsburgh 1975, Plenary Lecture.

2. Dixon-Lewis, G.: 23rd Symposium Symposium (International) on Combustion, The Combustion Institute, Pittsburgh 1991, p.305

3. Jones, W.P; Whitelaw,J.H.: 20th Symposium (International) on Combustion ,The Combustion Institute, Pittsburgh 1985, p.233

4 . Peters, N.: 21st Symposium (International) on Combustion,The Combustion Institute, Pittsburgh 1987, p1231.

5. Law, C.K.: 22nd Symposium (International) on Combustion, The Combustion Institute, Pittsburgh 1989, p.1381

6. Jones, W.P.: Imperial College, Private communication, (1990).

7. Eckbreth, A. C.: 18th Symposium (International) on Combustion ,The Combustion Institute, Pittsburgh 1982, p.1471.

8. Hanson R.K. : 18th Symposium (International) on Combustion ,The Combustion Institute, Pittsburgh 1982, p. 1677

9. Chen,J.-Y.; Dibble, R.W.: This Symposium .

10. Bilger, R.: This Symposium .

11. Weinberg, F.J.: Optics of Flames, Butterworths, London 1963.

12. Hong, N.S.; Jones, A.R.; Weinberg, F.J.: Proc. Roy. Soc.A353. (1977) 77.

13. Cattolica, R.J.; Vosen S.R.: Twentieth Symposium (International) on Combustion, The Combustion Institute, Pittsburgh 1984, p. 1273.

14. Ewart, P.; Snowdon P.; Magnusson, I.: Opt. Lett.,14 (1989) 563.

15. Dec,J.; Siebers,D.; Loye, A.: Sandia National Laboratories, Combustion Research Facility Newsletter, 13 (1991) 3.

16. Weinberg, F.J. (Ed) "Advanced Combustion Methods", Academic Press, London and New York 1986.

17. Dawes, S.G. :Energy World, November 1990, p.9.

18. Hayhurst, A. N.: Combustion and Flame, 85 (1991) 155.

19. Hustad, J. E.; Vareide, D; Sonju, O. K.: Combustion and Flame, 85 (1991) 232.

20. Allen, D; Hayhurst, A. N.: 23rd Symposium (International) on Combustion, Pittsburgh:,The Combustion Institute 1991 p. 935.

21. Iisa, K.; Hupa, M.: 23rd Symposium (International) on Combustion, Pittsburgh:,The Combustion Institute 1991 p 943.

22. Bai, T.;Shani, S.; Daniel B.R.; Zinn, B. T.: This Symposium.

23. Zinn, B. T.: Georgia Tech., Private communication, 1991.

24. Davies, R.M.; Masters, J.: "Gas Engineering and the Environment", The Institution of Gas Enginers, March 1990.

25. Haibel, M.; Mayinger, F.; Strube, G.: This Symposium.

26. Ronney, P.D.: Princeton University, Private Communication (1991).

27. Tien, T. C.; T'ien, J.S., This Symposium.

28. Gas Research Institute Digest,. 12 (1989) 22, 38

29. Suhr,H.; Weddigen, G.: Combust. Sci. and Tech., 72 (1990) 101.

30. Vosen, S.R.; Cattolica R.J.; Weinberg, F.J., Proc Roy Soc A418 (1988) 313.

31. Weinberg F.: Combustion and Flame 78, (1989) 121.

32. Low, H.C.; Wilson, C.W.; Abdel-Gayed, R.G.; Bradley, D. 25th Joint Propulsion Conference, A.I.A.A. - 89 -2944 (1989).

33.Cheriyan, G.K.; Krallis,K.; Weinberg, F.J.: Combustion Science and Technology, 70 (1990) 171.

34. Wagner, P.C.; O'Brien W.F.; Northam, G.B.; Eggars, J.N.: 23rd JANAF Combustion Meeting (October 1986).

35. Sato, Y.; Kimura I.; et al.; 25th Joint propulsion Conference, paper no AIAA 89 -2564 (1989)

36. Masuya, G; Kimura, I.; et al.: 26th Joint Propulsion Conference, AIAA paper No. 90 -2098 (1990)

37. Wilson, C.W. ; Rolls Royce, Bristol, Private Communication (1991).

38. Carabine, M.D.; Clay, P.G.; Itauma, C.: SO2 Control Symposium (1990) EPA/EPR New Orleans.

39. Hilliard, J.C.; Strassburger, R.; Velosa J.; Ruiz, R.: Fossils Fuels Combustion Symposium PD-Vol. 25, ASME, 1989, p.95.

40.Calcote H.F.; Berman, C.H.: Ibid, p. 25.

41. Carleton, F.B.; Weinberg, F.J.; Nature, 33O (1987) 635.

42. Katz J.F.; Hung, C..H. : 23rd Symposium (International) on Combustion, The Combustion Institute 1991, p.1733.

43. Hardesty, D.R.; Weinberg, F.J.: 14th Symposium (International) on Combustion, The Combustion Institute, Pittsburgh 1973, p. 907.

44. Merzhanov, A. G.: 23rd Symposium (International) on Combustion, The Combustion Institute, 1991 (Paper No. 226).

45. Calcote, H.F.; Felder, W.; Keil, D.G.; Olson, D.B. : Ibid. p. 1739.

46. Davis, K.A.; Brezinsky, K.; Glassman, I: Combustion Science and Technology, 77 (1991), 171.

Part II

Experiments and Calculations of Combusting Flows

Transition of Turbulent Premixed Flame Structure with Turbulence Reynolds Number and Damkohler Number

A. Yoshida, Y. Ando, M. Narisawa and H. Tsuji

Department of Mechanical Engineering
Tokyo Denki University, Tokyo, Japan

Abstract

Various regimes of turbulent premixed combustion are identified by the turbulence Reynolds number and the Damkohler number. In the present study, attention is focused on the transition of the flame structure from a wrinkled laminar flame to a distributed reaction zone. Schlieren photographs were taken for the qualitative measurements of the transition of the flame structure and the time-resolved temperature measurements were made by using a fine-wire thermocouple. Laser tomography was applied to the fractal analysis of the wrinkled laminar flame front. It was confirmed experimentally that the transition from a wrinkled laminar flame to a distributed reaction zone is caused by an increase in Reynolds number or a decrease in Damkohler number. The wrinkled laminar flame can be represented by fractal surfaces with a fractal dimension and inner and outer cutoffs for the fractal behavior. In the distributed reaction zone, vortex stretching eradicates the non-uniformity of mean flow and creates a uniform turbulent flow field of isotropy through the cascade of turbulent energy transfer. In the transition regime, the turbulent premixed flame has characteristic features of both wrinkled laminar flame and distributed reaction zone.

1. Introduction

The structure of the turbulent premixed flame depends on the turbulence Reynolds number defined by $R_\lambda = (u'\lambda)/\nu$ or $R_L = (u'L)/\nu$ and the Damkohler number, $D_a = (L/u')/(\delta_L/S_L)$, where u' is the turbulence intensity, ν the kinematic viscosity, λ the Taylor microscale, L the integral scale of turbulence, δ_L is the laminar flame thickness and S_L the laminar burning velocity. Here, we can assume approximately that R_L is proportional to $R_\lambda{}^2$. Various regimes of turbulent premixed combustion is shown in Fig. 1 [1]. When the Damkohler number is large, combustion reactions are rapid compared with the turbulent processes, so that the combustion tends to occur in a thin laminar flame. For small Damkohler numbers, the turbulent mixing is augmented by the small vortices and is rapid compared with the chemistry, so that the combustion reactions are distributed throughout the volume occupied by the turbulent flame zone. Between these two extremes, there exists a broad transition regime. In the

R. S. L. Lee · J. H. Whitelaw · T. S. Wung (Eds.)
Aerothermodynamics in Combustors
IUTAM Symposium Taipei, Taiwan, 1991

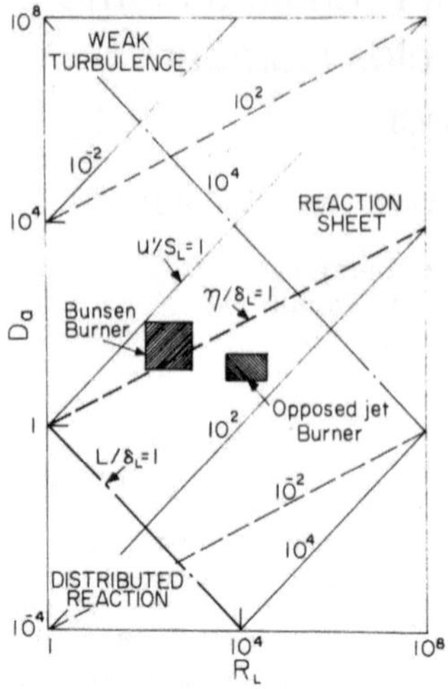

Fig.1. Various regimes of turbulent premixed combustion

present study, the transition from the former to the latter was studied by the Bunsen burner flames and the opposed jet burner flames.

2. Experimental Apparatus

A conical turbulent premixed flame with wrinkled laminar flame structure was produced by a conventional cylindrical burner of 30 mm in I.D. Turbulence producing grid was installed at 55 mm upstream from the exit. The propane-air premixed flame was stabilized at the exit by a small annular diffusion flame. The equivalence ratios of the mixture are 0.7 and 0.8. The average velocity of mixture at the burner exit ranged from 2.68 m/s to 5.37 m/s. The values of turbulence characteristics at

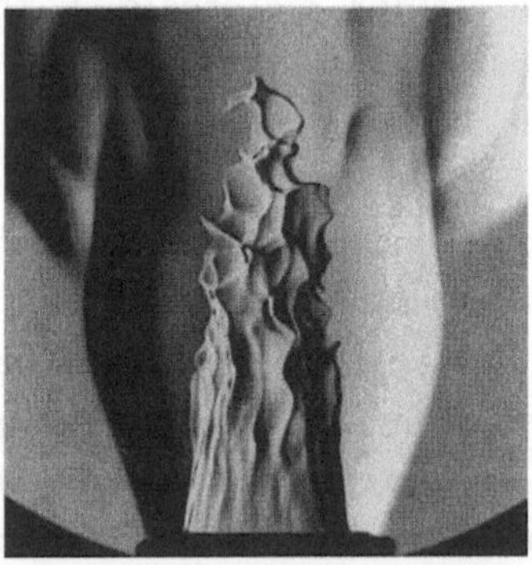

Fig.2. Schlieren photograph of wrinkled laminar flame

the burner exit were used in the estimation of turbulence Reynolds number and Damkohler number. The turbulence intensity was changed from 0.04 m/s to 0.57 m/s. The turbulence Reynolds number based on the Taylor microscale ranged from 4.9 to 74.4. The turbulence scale could not be changed significantly and a typical Damkohler number based on the integral scale is of the order of 20.

Figure 2 shows a typical schlieren photograph of the wrinkled laminar flame [2]. The mean velocity of the unburned mixture is 2.68 m/s, the turbulence intensity 0.15 m/s and the equivalence ratio 0.7. The turbulence Reynolds number based on the Taylor microscale is 17.4.

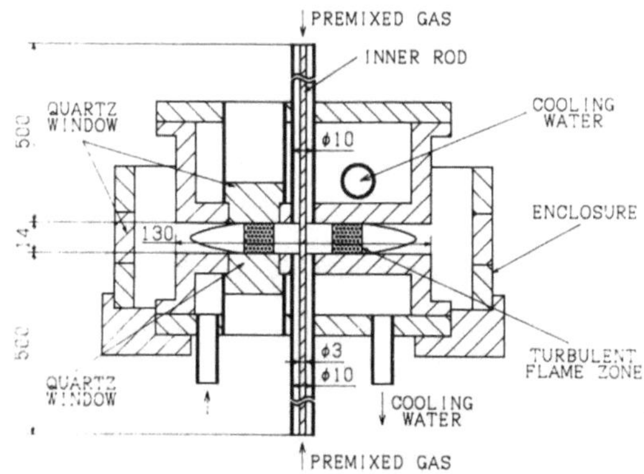

Fig.3. Opposed jet burner

Extremely high intensity turbulence with small scale was found to be achieved by an opposed jet burner [3]. The burner used in the present study is shown in Fig. 3. This burner consists of two mixture supply tubes of 10 mm in I.D., which are placed oppositely. The distance between two mixture supply tubes is variable from 5 to 30 mm. Dilution by the surrounding atmosphere is eliminated by two water-cooled square plates of 130 mm X 130 mm. Equal flow of a premixed propane-air mixture was supplied to each tube. A stable annular turbulent flame was produced in the range of the mixture supply velocity of 9 m/s and 41 m/s and in the range of the equivalence ratio, ϕ, greater than 0.90. The stability was found to be almost independent of the flow velocity.

In the estimation of the turbulence Reynolds numbers and the Damkohler numbers, there is one problem, because the turbulence characteristics vary significantly in space. We used in the estimation the values of u', L and λ at the center plane and just in front of the turbulent flame zone. Therefore, all of the information pertain to order of magnitude estimates, not to uniquely defined, precise quantities. A typical value of R_λ ranges from 300 to 1300 and that of D_a from 5 to 25.

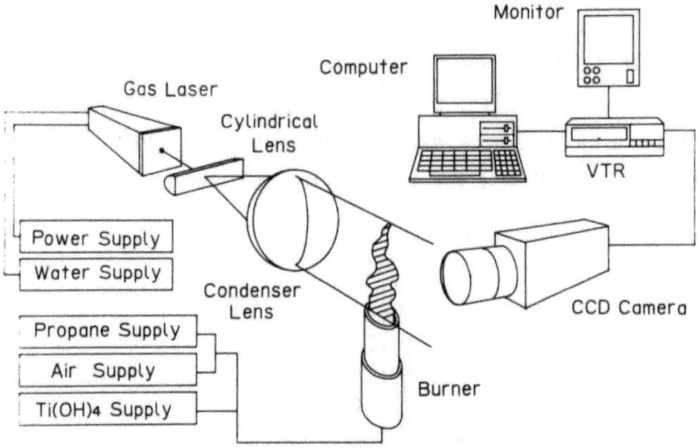

Fig.4. Laser tomography and data processing system

Figure 4 shows the laser tomography and the data processing system. An Ar ion laser of 0.8 W operating at 514.5 nm was expanded to a two dimensional laser sheet by a cylindrical lens and then focused by a condenser lens. The focal lengths of these lenses were 15 mm and 1000 mm, respectively. The nominal thickness of the resulting sheet was about 0.2 mm across the measuring area. The unburned mixture was seeded by fine TiO_2 particles. Since the effective scattering cross section of particles is greatly reduced at high temperature [4], only the unburned mixture region is visualized by Mie scattering of these particles. The effect of seeding on the flame process was found negligible. The scattered light was collected at the right angle to the sheet and was recorded on the video tape through a CCD camera with an electronic shutter of 1/2000 s duration. The image recorded on the video tape was sent through an interface to the frame memory (512 X 512 pixels) of a computer. The overall spatial resolution was around 1.5 mm.

The instantaneous temperature was measured by a compensated fine-wire thermocouple of which diameter was 50 μm. The thermal inertia of the thermocouple was compensated electrically as described in Ref. 3. The turbulence characteristics were measured by a hot wire anemometer.

3. Wrinkled Laminar Flame

3.1 Fractal Analysis of Wrinkled Laminar Flame

The flame image recorded on the video tape was processed to afford the instantaneous shape of wrinkled laminar flame. The algorithm used in the fractal analysis is same as that used in Ref. 5, in which the length of flame boundary is obtained by covering it with circles of different diameters. The processed image is shown in Fig. 5 with the procedure for obtaining the length. Successive 10-100 images were analyzed to obtain the fractal dimension and the inner and outer cutoffs.

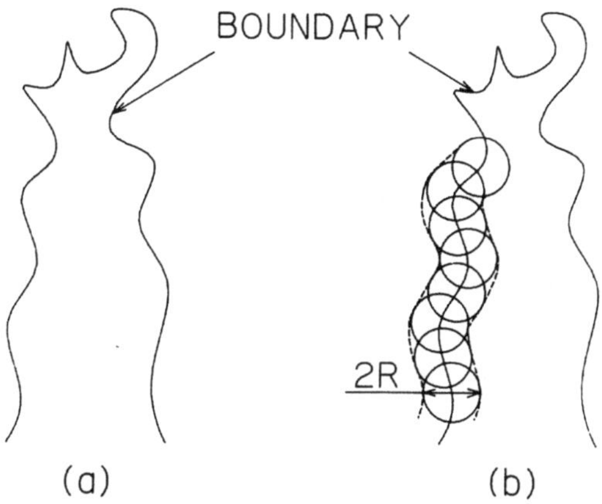

Fig.5. Image processed instantaneous flame shape

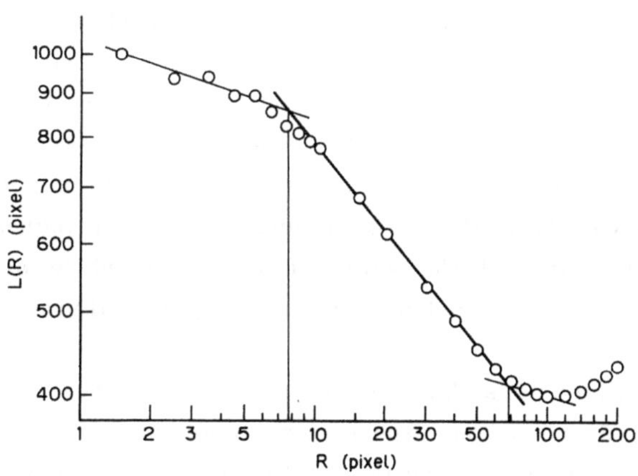

Fig.6. Typical length plot for fractal analysis

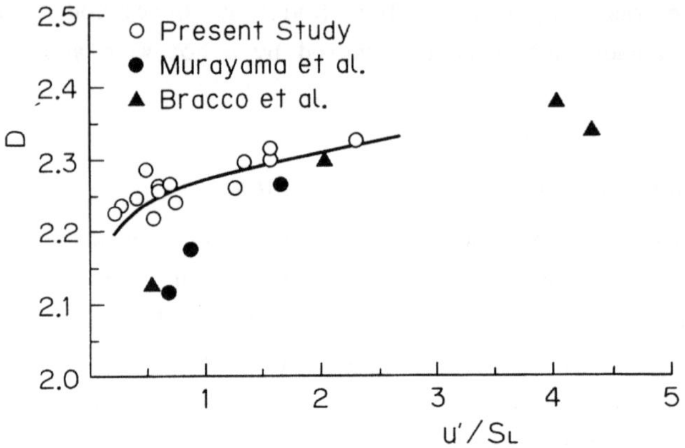

Fig.7. Fractal dimension versus u'/S_L

The logarithm of the length of flame boundary is plotted against the logarithm of the radius of circle. An example of such plots is shown in Fig. 6. It was found that there exist three regions, in each of which data are expressed by a straight line segment. The first and third line segments are nearly horizontal and the second is diagonal in this log-log plot. The intersection point of the first and second line segments represents the inner cutoff and that of second and third line segments represents the outer cutoff. In the vicinity of each cutoffs, a smooth transition between the straight line segments is observed.

From the slope of the second line segment, the fractal dimension D was determined. Since the wrinkled laminar flame geometry changed every instance, we used arithmetic means of the values for 10-100 instances. Figure 7 shows that the fractal dimension measured in the present study ranges from 2.21 to 2.32 depending on the characteristics of the unburned mixture turbulence. It was found that the fractal dimension depends solely on the turbulence intensity normalized by the laminar burning velocity and that the dimension is almost independent of the turbulence Reynolds number. The fractal dimension increases with the turbulence intensity and the value of 2.36 obtained in a internal combustion engine is approached at high intensity turbulence [6].

However, the dimensions for low intensity turbulence are larger than those of Refs. 5 and 6. The experimental conditions of Ref. 5 were almost same as the present study except for the mixture velocity. In Ref. 5, the mixture was supplied at a relatively low velocity and the turbulent flame oscillates periodically. This flame oscillation was synchronized with the vortex shedding in the mixing region of the burned gas and the ambient air. These large scale vortices stretch the wrinkles of the flame front and the roughness of the wrinkled laminar flame surface is reduced significantly. On the other hand, in Ref. 6, the fractal dimension was measured in an internal combustion engine and the characteristics of the mixture turbulence and of the laminar flame were roughly estimated by empirical equations. Therefore, at present, the exact comparison with the present data is impossible.

Recent measurements in turbulent shear flows indicated that constant property surfaces in these flows are fractal with a D between 2.35 and about 2.60 [7] which is larger than the fractal dimension of the wrinkled laminar flame. Therefore, the wrinkled laminar flame does not act as a constant-property scalar surface, especially for low intensity turbulence.

3.2 Inner and Outer Cutoffs of Fractal Behavior

For turbulent shear flows, the inner and outer cutoffs appear at scales approximately equal to the orders of the Kolmogorov microscale and the integral scale, respectively [7]. In general, the laminar flame front is expected to be smoother than that implied by the unperturbed turbulent flow field. The smoothing of the flame surface is predominant in the small scales whereas the larger scales tend to remain unaffected. This suggests that the inner cutoff tends to shift to higher values than

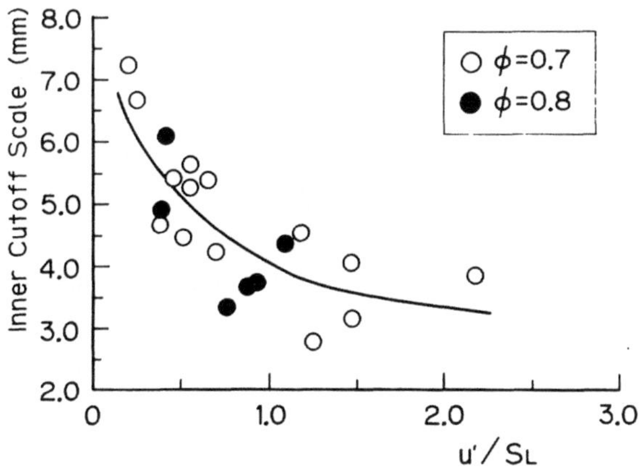

Fig.8. Inner cutoff versus u'/S_L

the Kolmogorov microscale. Figure 8 shows the effect of u'/S_L on the inner cutoff. The inner cutoff obtained in the present study is significantly larger than the Taylor microscale which is much larger than the Kolmogorov microscale and decreases with the turbulence intensity. This fact suggests that the inner cutoff decreases with the roughness of wrinkled laminar flame surface.

The outer cutoff is much larger than the integral scale of the unburned mixture turbulence and is about 2/3 of the burner diameter independently of the characteristics of the unburned mixture turbulence. The burner diameter represents the possible maximum scale in the system, which determines the outer cutoff of the fractal behavior.

4. Distributed Reaction Zone

4.1 Opposed Jet Burner Flames

With increasing the turbulence Reynolds number and decreasing the Damkohler number, the local structure of the wrinkled laminar flame is destroyed by small dissipative eddies and the distributed reaction zone is produced. Figure 9 shows a direct photograph of the opposed jet burner flame. The propane-air mixture of the equivalence ratio of 0.95 was supplied at the mean velocity of 18.8 m/s. The distance between two plates is 14 mm, and the inside and outside diameters of the turbulent flame zone are 20 and 60 mm, respectively. A stable annular turbulent premixed flame was produced around the impinging jets.

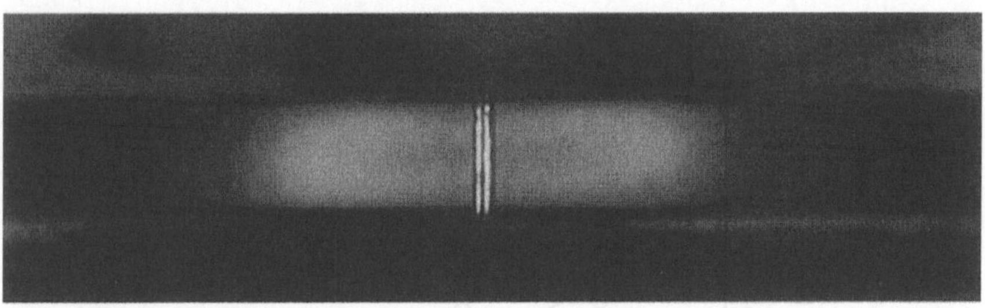

Fig.9. Direct photograph of opposed jet flame (V = 18.8 m/s, ϕ = 0.95)

Figure 9 shows the schematic diagram demonstrating the typical features of the flow field. Mixture jets with equal flows impinge with each other on the center plane between two plates and a stagnation plane is formed. The inner rod makes each jet annular. The annular jets are easier to be stabilized than the pipe flow jets, when they impinge with each other. The impinging jets change their directions and the shear stress caused at the interface between two flows generates a strong turbulence. Recirculation zones are produced around both jets. Main turbulent flame is stabilized by these recirculation zones.

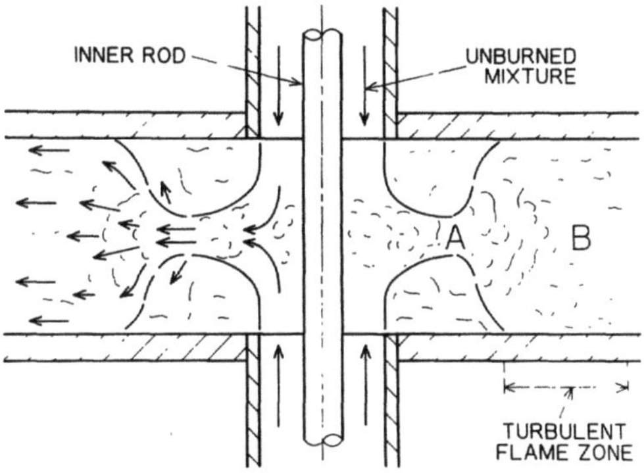

Fig.10. Schematic diagram of impinging jet flow field

4.2 Schlieren Photograph

Figure 11 shows the schlieren photographs of the opposed jet premixed flames in the transition regime from a wrinkled laminar flame to a distributed reaction zone. Boundaries between unburned mixture jets and recirculation zones are clearly shown in each schlieren photograph. It is well known that, for Bunsen burner flames, schlieren photographs show sharp boundaries which separate the unburned mixture from the completely burned products as shown in Fig. 2. For V = 9.1 m/s, Fig. 11(a), shows the same features as the Bunsen burner flames and the flame zone includes many large scale wrinkles. For this case, the turbulence Reynolds number based on the Taylor microscale, R_λ, is 280 and the Damkohler number based on the integral scale, D_a, is 26. With increasing the mixture velocity to 18.8 m/s, Fig. 11(b), R_λ and D_a become 520 and 12, respectively. With an increase in R_λ and a decrease in D_a, the scale of wrinkles becomes smaller on the unburned gas side and the flame zone consists of a lot of knob-like images. Concurrently, thin horizontal stringy images appear on the burned gas side. Further increase of the mixture velocity to 30.4 m/s, Fig. 11(c), leads to an increase in R_λ to 1300 and a decrease in D_a to 5. The wrinkles on the unburned gas side disappear and whole combustion flow field is filled with the bundles of the stringy microstructure. It should be noticed that this is the direct evidence of the transition of the turbulent premixed flame structure from a wrinkled laminar flame to a distributed reaction zone.

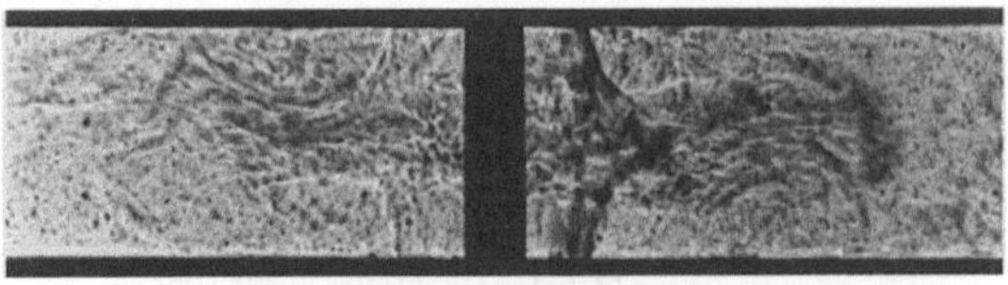

(a) V = 9.1 m/s

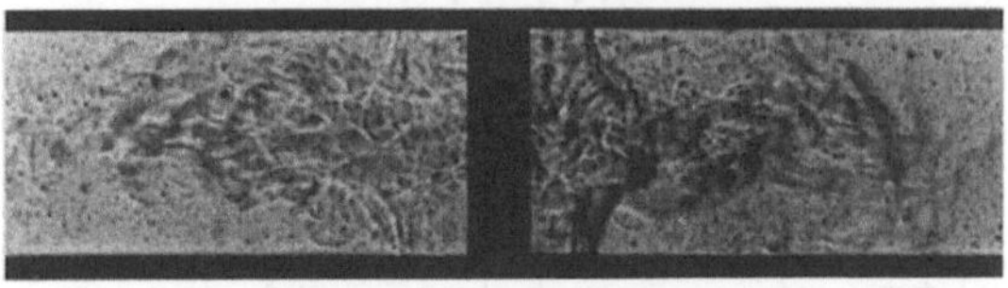

(b) V = 18.8 m/s

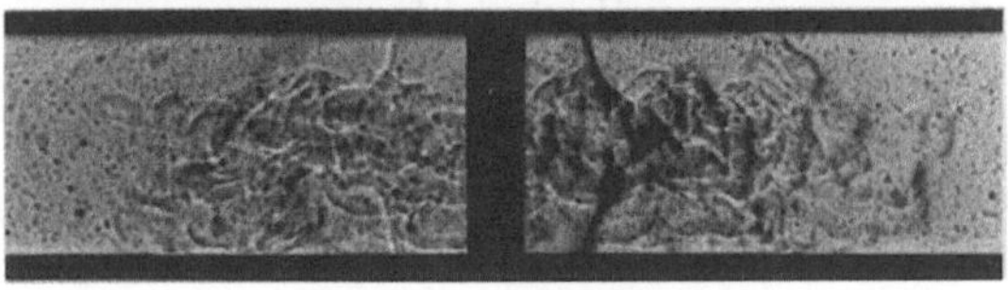

(c) V = 30.4 m/s

Fig.11. Schlieren photographs of opposed jet flames (ϕ = 0.95)

4.3 Laser Tomography

Schlieren pictures display an integration of the density gradient along the optical path. Therefore, the observation is a little bit ambiguous. On the other hand, the tomographic method with a thin laser sheet reveals the instantaneous shape of the cross section of the unburned mixture region. Figure 12 shows three examples of the image processed tomographic pictures taken at different instances. The experimental conditions are almost same as those of Fig. 11(b).

As in the schlieren photographs, the boundaries between the unburned mixture jets and the recirculation zones are recognizable. It is seen that a tongue of the unburned mixture spreads into the burned gas region. The light scattering region suddenly disappears on the unburned gas side of the turbulent flame zone. This fact suggests that the temperature increases significantly on this boundary. Therefore, the unburned mixture changes to partially reacted gas or completely burned gas across the boundary. Comparing these three pictures, it was found that this boundary fluctuates with relatively large scales.

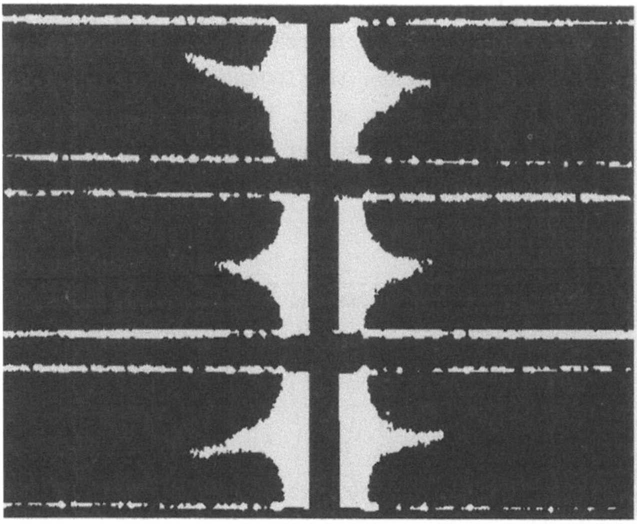

Fig.12. Laser tomographic pictures of opposed jet flame(V = 20.5 m/s, ϕ = 1.0)

4.4 Mean and Fluctuating Temperature

Figure 13 shows the mean temperature contours for the mean velocity of 18.8 m/s. For this case, the turbulent flame zone extends from R = 10 mm to 30 mm. On the unburned gas side of the flame zone, relatively large recirculation zones are produced around the exits of the mixture jets. Therefore, the turbulent flame zone cannot be assumed to be cylindrical on the unburned gas side. However, on the burned gas side, the isotherms of the mean temperature are nearly parallel and the mean struc-

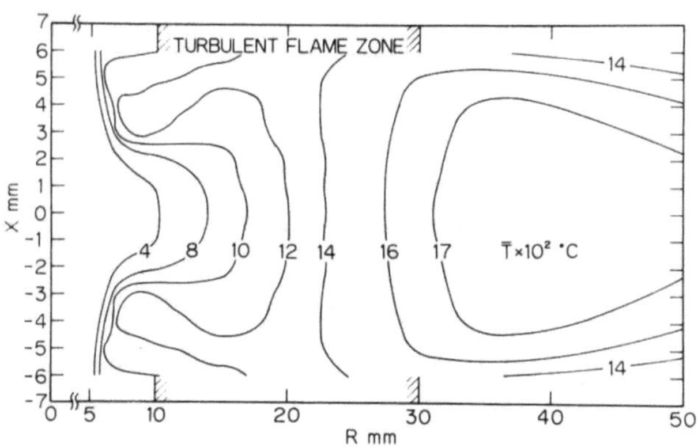

Fig.13. Mean temperature contours (V = 18.8 m/s, Φ = 0.95)

ture of the turbulent flame can be assumed to be cylindrical. It is clear that the mean temperature increases from 400 to 1200 C on the unburned gas side. In this region, the tongue of the unburned gas moves up and down. On the burned gas side, the mean temperature increases gradually from 1400 to 1700 C. This fact suggests that the mean reaction rate is faster on the unburned gas side than that on the burned gas side.

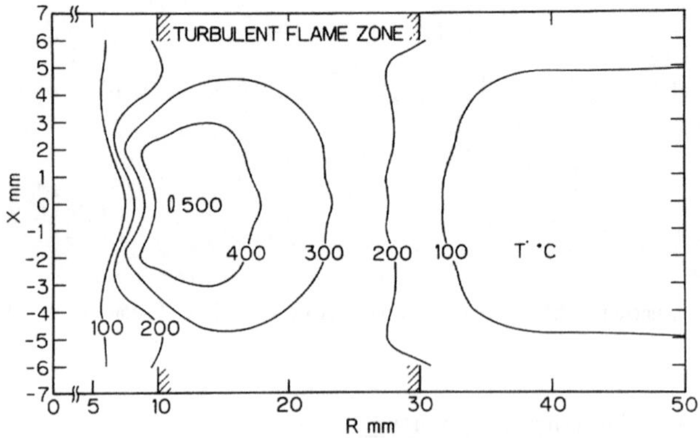

Fig.14. Fluctuating temperature contours (V = 18.8 m/s, ϕ = 0.95)

Figure 14 shows the fluctuating temperature contours measured under the same conditions as Fig. 13. The fluctuating temperature is high on the unburned gas side, suggesting that the instantaneous temperature changes from nearly unburned gas temperature to that of fully burned products. On the other hand, it should be noticed that the fluctuating temperature decreases on the burned gas side. Such a small temperature fluctuation is a typical feature of the distributed reaction zone [8]. As shown by the schlieren photographs, the flame zone in this region is filled with the bundles of stringy microstructure. These stretched vortices in which chemical reaction occurs eradicate the non-uniformity of the mean flow. In the recirculation zones, the fluctuating temperature is relatively low, suggesting that the uniform high temperature gas is recirculating.

4.5 Probability Density Functions of Temperature Signal

Figure 15 shows the radial variations of the probability density functions of the temperature signal measured along the center plane between two plates. The mean velocity of unburned mixture was 18.8 m/. At the inner surface of the flame zone, each PDF has a peak at the unburned

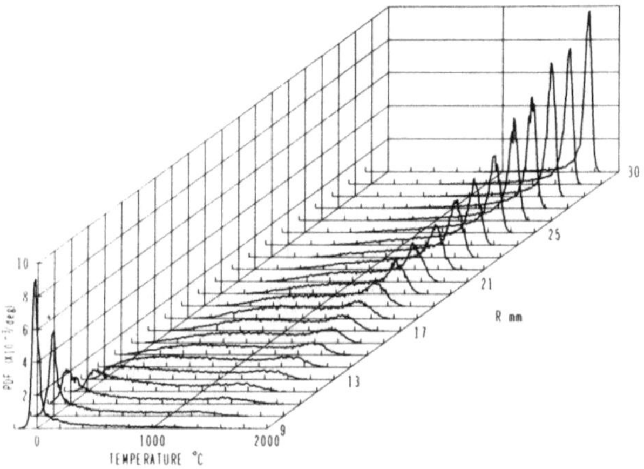

Fig.15. Probability density functions of temperature signal (V = 18.8 m/s, ϕ = 0.95)

gas temperature with a long tail to the intermediate and burned gas temperature. Therefore, in this region, the thermocouple is exposed to the unburned mixture and partially reacted gas. The probability of the intermediate and burned gas temperature increases with the radial distance and essentially bimodal profiles appear on the unburned gas side of the flame zone. As compared with the PDF's of the true wrinkled laminar flame, the probability of the intermediate temperature is significantly high. This fact suggests that the turbulent flame zone consists of unburned mixture, completely burned gas and partially reacted gas. Even if the wrinkled flame feature remains in the flame zone, this wrinkled flame should be much thicker than the wrinkled laminar flame.

In the middle and on the burned gas side of the flame zone, the peak at the unburned gas temperature disappears and each PDF has a peak at the burned gas temperature with a long tail to the intermediate temperature. Therefore, the flame zone consists of the partially reacted gas and the completely burned gas. The unburned mixture does not exist in this region.

5. Transition of Turbulent Flame Structure

From the experimental results obtained with the conical turbulent flame, it can be said that the so-called wrinkled laminar flame is produced within the range of R_λ below 100 and D_a above 20.

Even if R_λ increases to about 300, the wrinkled laminar flame is also achieved in the opposed jet burner, when D_a remains of the order of 20. When R_λ increases to 520 and D_a decreases to 10, a large temperature fluctuation was observed on the unburned gas side of the flame zone. The large temperature fluctuation is a typical feature of the wrinkled flame, because the discontinuity surface of temperature fluctuates in the flame zone. Therefore, the wrinkled flame feature remains on the unburned gas side. The role of the energy containing large eddies is to corrugate a reaction zone. Such a reaction zone would be locally extinguished by strong stretching and the wrinkled flamelets are produced. The Damkohler number based on the integral scale is of the order of 10 and greater than unity. Therefore, there should remain the wrinkled flame feature on the whole. Here, we must notice that the turbulence spectra are very broad and the eddies smaller than the integral scale are also included in the mixture turbulence. If the characteristic time scale is estimated based on the Kolmogorov microscale which typifies the smallest dissipative eddies, the Damkohler number is reduced to 0.1. Therefore, the local structure of the wrinkled laminar flame is destroyed by these dissipative eddies. As a result, a thick wrinkled flame, as described in 4.4, should be produced in which small reacting eddies are embedded. On the burned gas side, the flame zone consists of the stringy microstructure and the discontinuity surface does not exist. In this region, the fluctuating temperature decreases and also the mean temperature becomes uniform. These are the characteristic features of the distributed reaction zone as described below.

With further increase in R_λ to about 1300 and decrease in D_a to about 5, the whole combustion flowfield consists of the folded and tangled vortex tubes. Through the cascade of turbulent energy transfer, the vortex stretching enhances the small scale turbulent mixing and eradicates the non-uniformity of the temperature field. On the other hand, The strong stretching of vortices reduces the chemical reaction within the vortices and the mean reaction rate decreases even though the mass and heat transfer is enhanced.

Conclusions

1. The wrinkled laminar flame is produced, when the Damkohler number is of the order of 20, and the turbulence Reynolds number is smaller than 300.
2. The wrinkled laminar flames stabilized on the cylindrical burner can be represented by fractal surfaces with a fractal dimension and inner and outer cutoffs for the fractal behavior.
3. The transition from a wrinkled laminar flames to a distributed reaction zone is caused by an increase in R_λ or a decrease in D_a.
4. The distributed reaction zone is produced at the turbulence Reynolds number of the order of 1000 and the Damkohler number of 5. In the distributed reaction zone, the vortex stretching eradicates the non-uniformity of the mean flow and creates a uniform turbulent flow field of isotropy through the cascade of turbulent energy transfer.
5. At the turbulence Reynolds number of 500 and the Damkohler number of 10, the turbulent premixed flame in the transition regime was achieved. In the transition regime, the turbulent flame shows the characteristic features of both wrinkled laminar flame and distributed reaction zone.

References

1. Yoshida, A.; Kubozuka, S.: Structure of highly turbulent opposed jet premixed flame. Joint International Conference, Australia/New Zealand and Japanese Sections, 163-169, The Combustion Institute 1989.
2. Yoshida, a.: Characterization of turbulent premixed flame structure for mathematical modeling of combustion, Mathematical modeling in combustion science (Buckmaster, J. D. and Takeno, T., eds.). Springer-Verlag 1988.
3. Yoshida, A.: Structure of opposed jet premixed flame and transition of turbulent premixed flame structure. Twenty-Second Symposium (Int.) on Combustion pp. 1471-1478 The Combustion Institute 1988.
4. Kennedy, I. M.: Some aspects of seeding flames with refractory oxide particles. Comb. Sci. Tech. 27 (1982) 247-252.
5. Murayama, M.; Takeno. T.: Fractal-like character of flamelets in turbulent premixed combustion. Twenty-Second Symposium (Int.) on Combustion pp. 551-559 The Combustion Institute 1988.
6. Mantzaras, J.; Felton, P. G.; Bracco, F. V.: Three-dimensional visualization of premixed-charge engine flames: islands of reactants and products; fractal dimensions; and homogeneity. SAE paper No 881635 (1988).
7. Sreenivasan, K. R.; Meneveau, C.: The fractal facets of turbulence. J. Fluid Mech. 173 (1986) 357-386.
8. Summerfield, M.; Reiter, S. H.; Kebely, V.; Mascola, R. W.: The structure and propagation mechanism of turbulent flames in high speed flow. Jet Propul. 25 (1955) 377-384.

Laser Velocity and Density Measurement of a Flat Counter Flow Diffusion Flame

F. Durst[1], E. Dornberger[1], G. Jakob[1], J.C. Rolon[2], D. Veynante[2] and J.P. Martin[2]

[1] Lehrstuhl für Strömungsmechanik, Universität Erlangen-Nürnberg, Cauerstraße 4, 8520 Erlangen, Germany

[2] Laboratoire E.M2.C., C.N.R.S. et Ecole Centrale des Arts et Manufactures, 92290 Chatenay-Malabry, France

ABSTRACT

This paper presents non-intrusive measuring techniques for velocity and density measurements of a flat counter flow diffusion flame. Classical LDV-measurements become difficult near such flames because of strong refractive index gradients existing on both sides of the flame location. A method is preposed in which the resultant distortions of the interference fringes can be eliminated. A new technique for simultaneour measurements of the velocity and the density fields in axially symmetric flows is also suggested and is applied to the present flame study. For the local density measurements, a laser beam deviation (LBD) is detected by a position sensor diode and data are processed by the Abel integral transform using the Gladstone-Dale approximation.

INTRODUCTION

The accuracy of laser techniques in gas flows with combustion yielding heterogeneous temperature and species concentration gradients can be limited by beam deviation when the light passes through a flowing fluid with non-homogeneous distribution of the refractive index. The measuring volume is desplaced by refractive index gradients and deforms, and may even not exist in the case of laser-Doppler anemometry, the displacement of one of the beams is such that the beams do not intersect (asymmetric displacements). Precise velocity measurements become very difficult in this situation. This work proposes a way to overcome this problem in the case of a flat axisymmetric counter flow diffusion flame, (Figure 1). In a preliminary work the cold counter flow situation has been studied. (Rolon, Martin and Durst (1986)). To allow measurements in the combusting flow case, the present study was performed.

R. S. L.Lee · J. H. Whitelaw · T. S. Wung (Eds.)
Aerothermodynamics in Combustors
IUTAM Symposium Taipei, Taiwan, 1991

Plane, counter flow diffusion flames are of great interest to combustion research yielding a closer understanding of combustion mechanisms. They also constitute suitable test cases for the investigations of the interactions between fluid mechanics transport properties of combusting gases and the resulting chemical kinetics. Indeed, the overall shape of the resultant flame is simple, the flow is laminar and the resultant flame is fairly easy to stabilize in the stagnation region of the counter flowing jets. Figure 2 shows an example of such a flame.

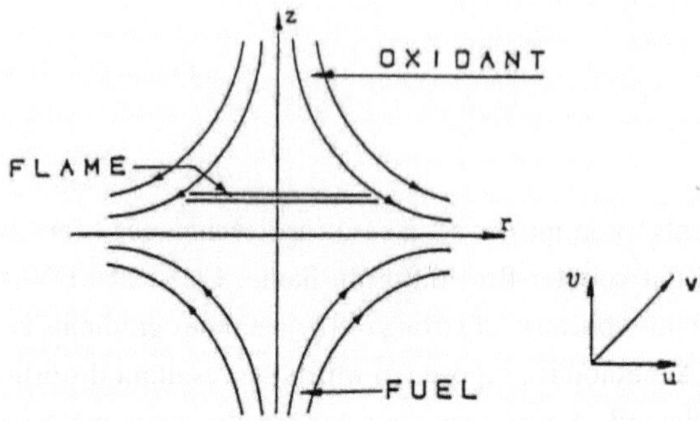

Fig. 1: Geometry of the flow. The flame is formed in the vicinity of a stagnation plane established by the counter flows of oxydant and fuel.

Fig. 2: Propane-air counter flow diffusion flame

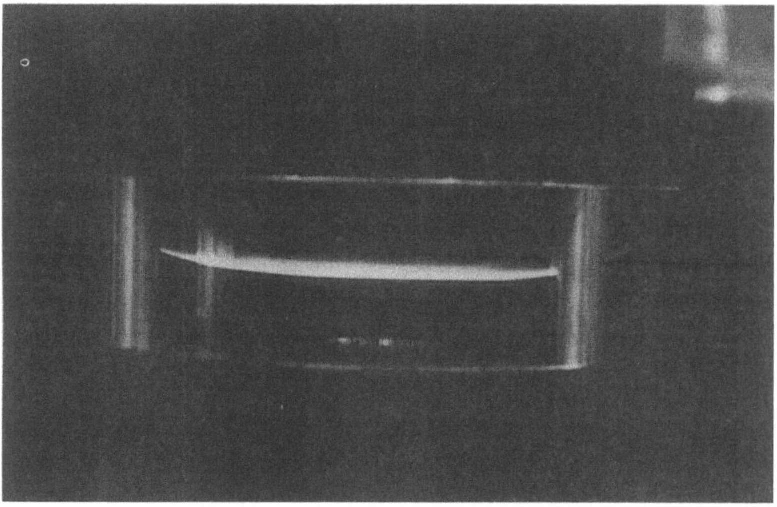

Fig. 3: Flat propane-air counter flow diffusion flame

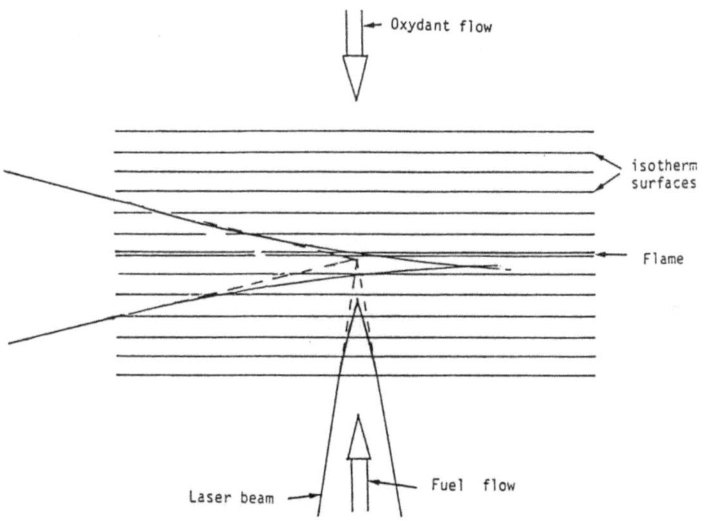

Fig. 4: Beam orientation of LDA measurements in a flat diffusion flame and measuring volume displacement under refractive index gradients: --- in isothermal homogenous flow; ----- in refractive flow with flame

Plane, counter flow diffusion flames have been extensively employed to study the effects of straining on combustion and have provided useful information for the so called "flamelet model" (see Peters (1984)), and particularly for the "Coherent Flame Model" (Marble and Broadwell (1977)), now used to predict turbulent combustion.

Experimental and theoretical aspects of flat counter flow diffusion flames have been studied by many authors (Potter and Butler (1959), Pandya and Weinberg (1964), Hanh and Wendt (1981), Spalding (1961), Tsuji (1982), Linan (1974), Giovangigli and Smooke (1987), to only mention some of them. Theoretical studies are generally based on a potential flow theorey description of the flow field, as proposed by Leclerc (1950), Spalding (1961), or Rosenhead (1966). The potential flow theory yields the following relationships for the velocity distribution.

$$u = a \cdot r \tag{1}$$

$$v = -2 \cdot a \cdot z'^{*} \tag{2}$$

where u and v are the radial and axial velocity components, respectively, r is the radial and z' the axial coordinate (with the origin at the stagnation point). The strain rate a is given as:

$$a = \partial u / \partial r \tag{3}$$

and turns out to be constant. Spalding (1961) proposed for a the value

$$a = V_o / D \tag{4}$$

or

$$a = 2V_o / H \tag{5}$$

respectively, where V_o is the jet exit velocity, D the jet diameter and H the distance between the two injectors.

With the above idealized flow field, the combustion is treated separately, neglecting heat release effects and hence is treated in contradiction to the physics of the true flow and combustion processes. As a matter of fact no precise experimental velocity data are yet available in the flow when combustion takes place. Measurements with a Pitot tube are too intrusive and destroy the flow structure, and a hot wire is not employable when combustion is present. A speciall difficulty is also encountered with laser-Doppler-velocimetry: near flames with flame thicknesses of about 0.1 mm, (see photographic representation in Figure 3), the resultant thermal gradients are very strong (about 1000 K per mm). They induce refractive index gradients that deviate laser beams from their straight-lined trajectories (Fig. 4).

---

*) z' differs from z used later on to present the experimental results

EXPERIMENTAL ARRANGEMENT

Test Facility

The experiments described here were conducted with two different sets of jet producing nozzles. They correspond to Type I burners as defined and introduced by Tsuji (1982) to produce counters flow diffusion flames; Type I flames are produced between two counter flowing jets that are formed by oppositely located axisymmetric nozzles (Figure 5). The central nozzles that produce the counter flows of air and gaseous fuel, are surrounded by N_2-annular flows protecting the actual flame form disturbances by chemical reactions due to the interference with the surrounding air. The protecting N_2-gas streams also help to reduce externally caused disturbances of the resultant flame sheet and are essential to establish flat diffusion flames as those shown in Figure 3.

The complete test facility employed in this study is shown in Figure 5 in form of a block diagramme. The figure indicates the two separate flow control lines set up to supply the oxygen and the fuel to the oppositely located nozzles. Osygen and fuel were supplied from oxygen and fuel bottles. H_2, CH_4, and C_3H_8 were employed as fuels. Each line also comprised mixing devices to yield well mixed fuel-N_2 or air-N_2 mixtures prior to feeding the flow, through the nozzles. To control the mass flow rates of the gases, pressure regulators and minute, chocked, sonic nozzles were employed.

The output voltage of piezo resistive transducers, measuring the pressures before and after the flow controlling sonic nozzles, were digitized and recorded by a microcomputer. Flow rates and chemical equivalence ratio were computed and displayed. The computer provided helpful assistance to adjust the flows and stabilize the flame with a given strain rate and also to control the relevant experimental conditions to yield stable flow and combustion conditions.

In order to obtain about 1000 validated laser-Doppler signals per second, two home-made particle seeders were employed. Silicon oil was employed as spraying liquid, for all measurements in the cold flows. To study the flow with combustion two fluidized beds with cyclone generators were used to introduce titanium dioxide or magnesium oxide scattering particles to both flows. The particles used in all flows were found to be sufficiently small to follow both flows, e.g. the mean diameter was approximately 1 μm, so that thes followed the small velocities which caracterizes the flow field between the burners.

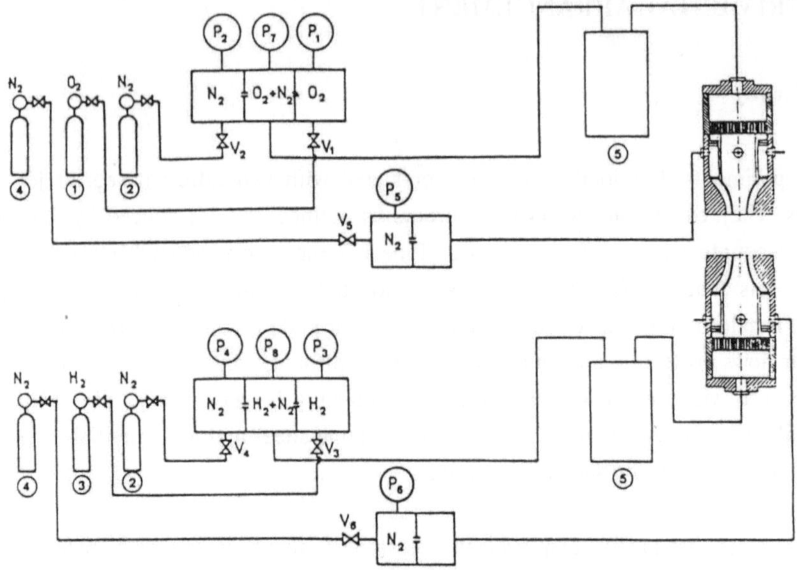

Fig. 5: Flow devices and control: 1 oxygen supply, 2 nitrogen supply, 3 fuel supply; 4 outer ring flow nitrogen supply 5 particle seeders for LDV measurements; P_1 to P_8 pressure captors for flow regulation and controls.

To perform complete velocity profile measurements, the burner test rig was traversed. The displacement of the burner assembly relative to the spatially fixed optical system was provided by two stepping motors monitored by the computer through a serial line and a special motor driver.

Laser-Doppler System

The employed laser-Doppler anemometer (LDA) was set up with two different configurations and lasers. In the cold flow, foreward scattering with a 15 mW He-Ne laser was employed (Fig. 6) and single channel OEI-LDA optics. This set-up was used to measure successively the radial (u) and axial (v) velocity components by simply rotating the transmission optics of the LDA-system. In the flame, the forward scatter LDA-system of Fig. 6 was not used due to the deviation problems for the laser beams described in the introduction section.

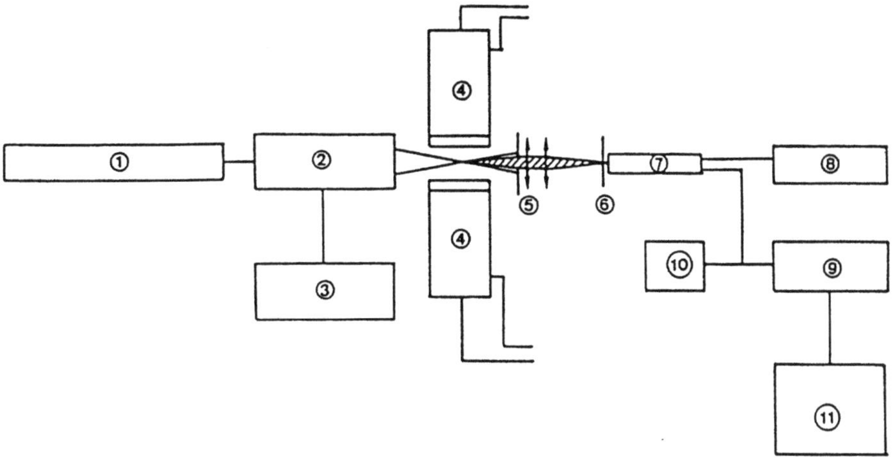

Fig. 6: Sketch of experimental set-up including LDV-system, burners and electronic equipment for the cold flow: 1 He-Ne laser; 2 LDV transmission optic with double Bragg cell; 3 Bragg cell power supply; 4 burners; 5 light collecting lens with light stop for the incident laser beams; 6 pinhole; 7 photodetector; 8 photodetector and preamplifier power supply; 9 LDA counter; 10 oscilloscope; 11 computer.

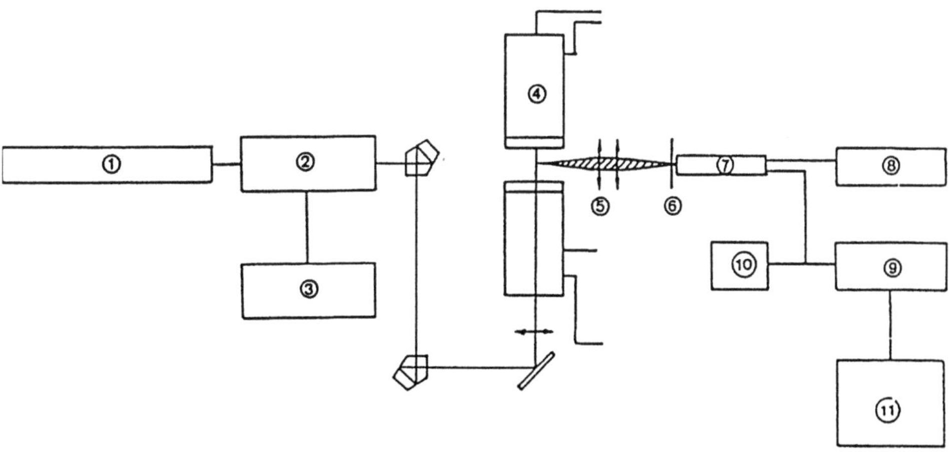

Fig. 7: Sketch of the optical arrangements used to measure the radial (u) velocity component in the case of flat diffusion flame: 1 He-Ne laser; 2 LDV transmission optic with double Bragg cell; 3 Bragg cell power supply; 4 burners; 5 light collecting lens with light stop for the incident laser beams; 6 pinhole; 7 photodetector; 8 photodetector and preamplifier power supply; 9 LDA counter; 10 oscilloscope; 11 computer; 12 glass inlet wall.

A second optical system was emloyed using a periscope (see Fig. 7) which allowed the optical axis of the LDA-optics to be perpendicular to the flame surface. In order to assure sufficient scattered light for the 90° scattering arrangement for the receiving optical systems a 4 Watt Argon-Ion laser was used. The incident beam separation of 50 mm was reduced to 20 mm in order to cross the lower burner through a glass wall after going through a 300 mm focal length lens; the length of the measuring volume was about 6 mm. Calculations showed that the displacement, due to the presence of refractive index gradients, of that part of the measuring volume (diameter about 300 μm), that was focused onto the photomultiplier, was less than 70 μm in the vertical direction and could be neglected compared to the length of the control volume. This can be envisaged from Fig. 8 which shows the laser beams arrangement to measure the radial velocity component in the range of $r = \pm 11$ mm and $-H<z<H$.

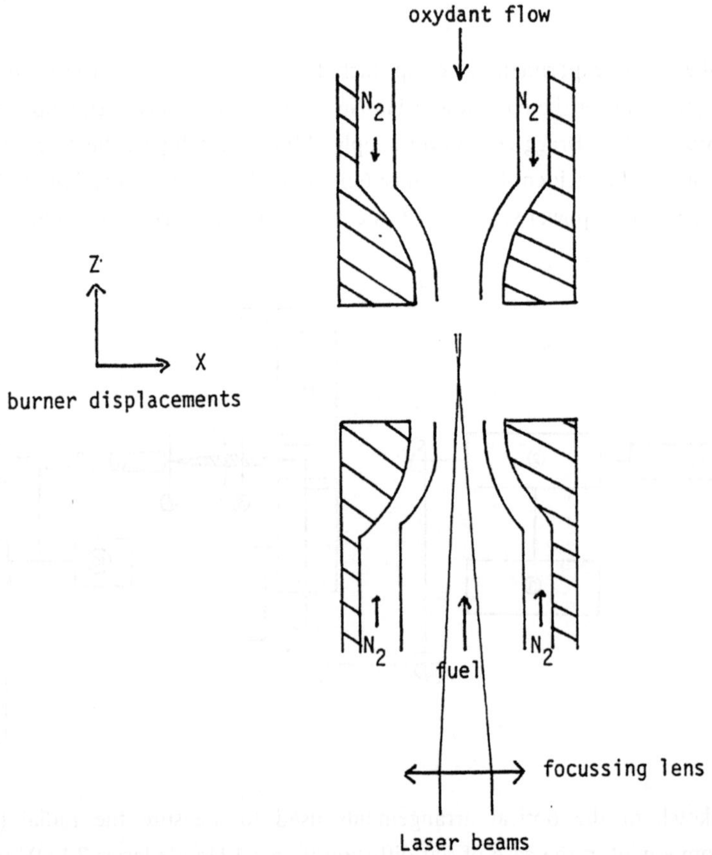

Fig. 8: Detailed burner configuration with the paths of the laser beams crossing the lower nozzle.

In order to measure the axial component of the velocity in the flow with combustion the optical arrangement shown in Fig. 6 was employed. Close to the flame front (+ 1.0 mm), no Doppler bursts were detected by the photomultiplier. Thermophoresis effects may be the cause of the apparently particle free zone close to the flame.

Electronic Signal Processing and Data Acquisition

To carry out the velocity measurements, the scattered light from the measuring volume was collected by a double lens system and focussed onto a pinhole in front of the photomultiplier. Its output signal was transfered to a DISA-counter (model No. 55L90), which was operated with a band pass filter setting of 64 kHz and 2 MHz. The Doppler signal frequency was of the order of 200 kHz and the frequency shift applied through the Bragg cells was 200 kHz. These settings were found to be sufficient to avoid any loss of information by the filter settings but to assure a high percentage of signal acceptance. About 200 to 1000 validated signals per second were obtained at each measuring point depending on the instruments settings and on the location in the flow.

To evaluate the final velocity, the digital signals from the counter were recorded by a PDP 11/23 through a 16 bits Input/Output interface triffered by the data ready signal of the counter. For this purpose, the Disa counter was directly connected to the microcomputer without buffer interface. Mean velocity, "turbulence intensity" and mean particle rate were computed. Because of laminar flows, no bias corrections needed to be applied to the particle average that was computed at each measuring point.

RESULTS AND DISCUSSION

The flow between the nozzles (Fig. 8) was studied extensively by the present authors. In this paper a collection of experimental results are presented that correspond to different burner separations H, both in the cold field and in the flow with the flame. For the presentation of the results, the point $r = 0$, $z = 0$ is chosen to be located on the flow axis and at equal distances $H/2$ from the burners (see Fig. 8). For the results prescribed here, the flow rate was controlled in order to reproduce the following outlet conditions:

- strain rate $a = 24\ s^{-1}$ as defined by formula (4)
- equivalence ratio $\phi = 1$
- equal flow momentum for both jets

Flow field for the cold flow case

Velocity profiles were taken in a radial range of $r = \pm 35$ mm. This region was considered as being of major interest: In it the flat flame stabilized and took the shape shown in Fig. 2 by laser flow visualization. Starting at the center of the flow, measurements were taken at various distances from the centre, mapping out between the two burners' top surfaces. For each measuring plane, the *z*-position and radial location were adjusted by an automatic traverse. Measurements were taken every millimeter in the radial direction. Following the axis, variable position steps were used and near the stagnation plane, 0.5 mm position stepts were selected to resolve the strong velocity gradients occurring in this part of the flow. Particular attention was given to this region in order to see how experimentally obtained velocity distributions agreed with the theoretical flow fields used in the modelling of flat diffusion flames. In all known numerical modelling attempts, the flow field is taken to have a constant strain rate that is hot affected by the heat release in the flame.

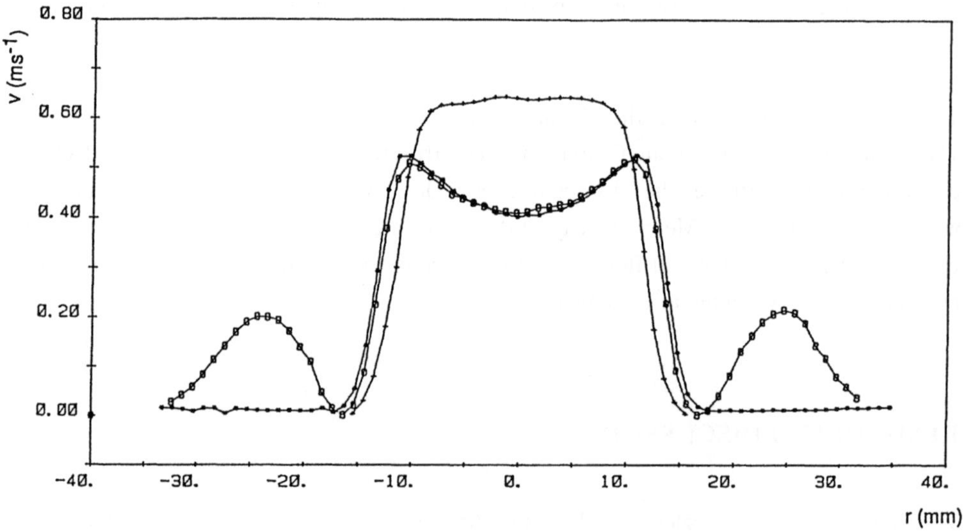

Fig. 9: Axial velocity component for single nozzle at 7.7 mm out of the exit of nozzle for three cases:

(+) Free jet produced by one nozzle without opposed jet or surface and without the surrounding ring flow

(*) Counter jet flows produced by two identical opposed jets without surrounding ring flow

(o) Counter jet flows with surrounding ring flow

The fluid was, in all cold flow cases, air under laboratory conditions. Experiments were performed for which the burner separation was successively H = 15 mm, 25 mm, 28.6 mm and 35 mm. Under these conditions, the radial (u) and axial (v) velocity components were measured. Figs. 9 to 16 show the major features of the flow and its dependence on the distance between the front surfaces of the jet producing nozzles.

In Fig. 9 the symbols (+) show the axial v velocity component, measured at 7.7 mm away from the exit of a single nozzle. It shows a very constant central part of the velocity profile as expected.

When the two nozzles are opposed to each other, the velocity profiles, still recorded at 7.7 mm away from the nozzle orifice, change drastically: the v component is reduced and the u component increased according to mass conservation; at the same time the constant v velocity component region around the flow axis disappears and a M or W shaped profile is observed (see Fig. 9 symbols (*)). One can note (symbols (o)) that the nitrogen ring does not modify the central velocity profile.

Fig. 10 shows a vector representation of the entire flow field obtained for H = 25 mm. It is found that two different asymmetric stable flows exist even if the same equal momentum condition are set up for each of the opposite outlet flows. A detailed analysis of the flow field show that, if the stagnation plane of one flow is located at z_o, the other one is located at $-z_o$. Hence it is sufficient to discuss only one flow configuration. The experimentally observed asymmetry can be caused by disturbances and results in asymmetric recirculating flow regions that establish themselves near the horizontal burner surfaces. The larger of these regions is located further away from the stagnation plane of the flow. The experiments confirm the recirculating flow regions always form if both nozzle exits are surrounded by plane surfaces that are positioned at a certain distance from each other. No recirculating flow region exists when the surfaces are removed so that the nozzle exits finish with slim wall rings.

Details of the radial profiles of the v (axial) component of the velocity field at different z locations are provided in Figure 11. For this flow z_o is equal to 1.5 mm. In the central flow region (-10 mm $< r <$ 10 mm), v decreases from the outlets of the nozzles to the stagnation plane. When approaching z_o, the M shape of the velocity profile is smoothed and v does not depend too much on r when -1.5 mm $< z - z_o <$ 1.5 mm.

Fig. 12 provides the axial profiles of the v component of the velocity for various r locations. Only profiles in the central part of the flow are provided, and v increases linearly with z. A careful analysis of the experimental data, shows that for $r <$ 3 mm the slope is constant, noted k_2 in the following sections. Out of the axis, the slope increases slightly with r.

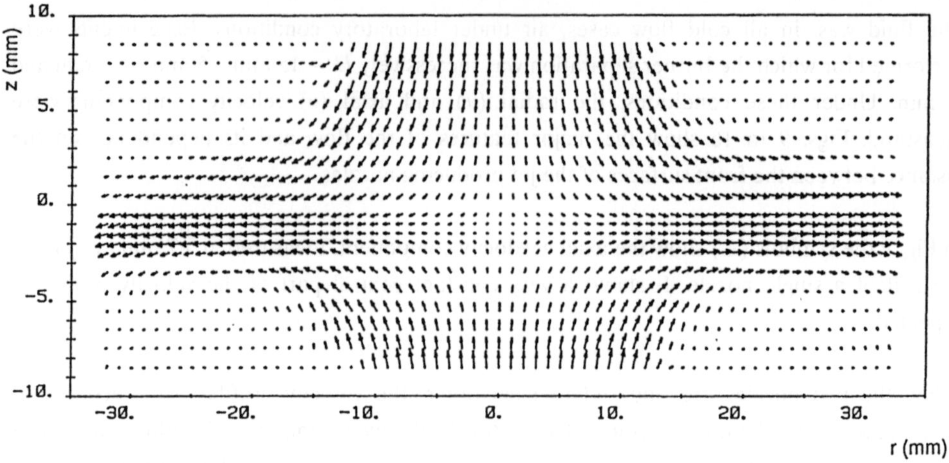

Fig. 10: Vector representation of the measured flow field for flow case II, with lower position of the stagnation point z_o = -1.5 mm

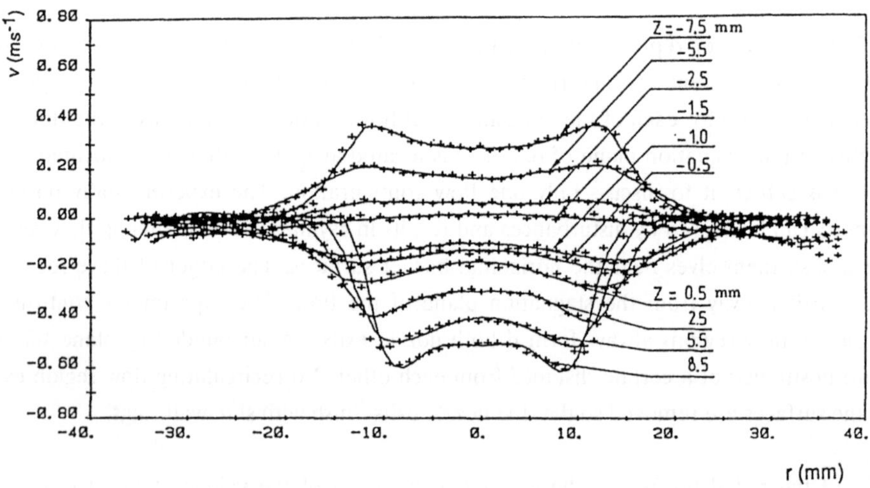

Fig. 11: Radial profiles of the v (axial) component of the velocity field at different z locations

Fig. 13 provides the radial distributions of the u (radial) component of the velocity as a function of r for various z locations. These curves can be devided into three different parts. Around the axis (-7 mm < r 7 mm) u increases linearly with r with a characteristic slope, this slope will be noted k_1 for $z = z_o$; for larger r we observe a second linear increase with a larger slope. At a given point (depending on z) u reaches a maximum after which it decreases. For

z_o -1 mm < z < z_o+1 mm, the decrease is observed only after a constant u region: this is caused by the stretching in the stagnation plane. Out of this region the stretching is reduced and the interaction between the flow and the recirculating zones is more pronounced. In the z_o-1 mm < z < z_o+1 mm region, the slope is equal to k_1 and independent of z. Further out of the stagnation plane this slope decreases with increasing z.

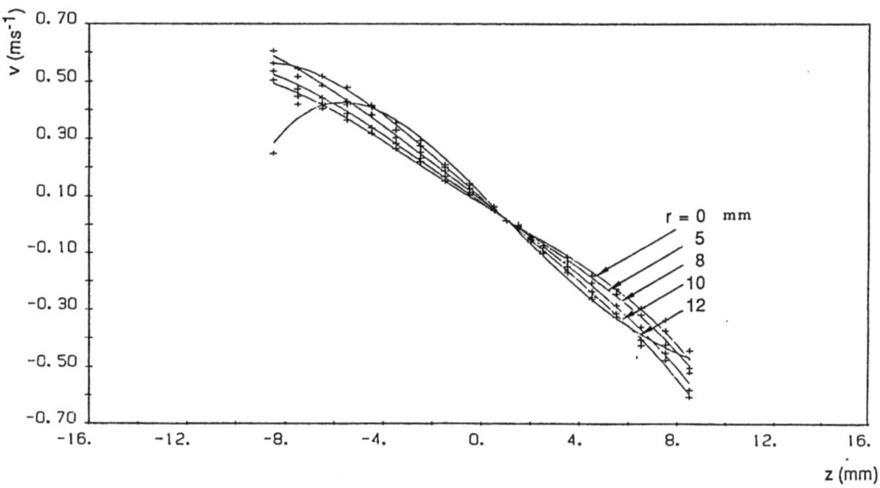

Fig. 12: Axial profiles of the v (axial) component of the velocity field at different r locations

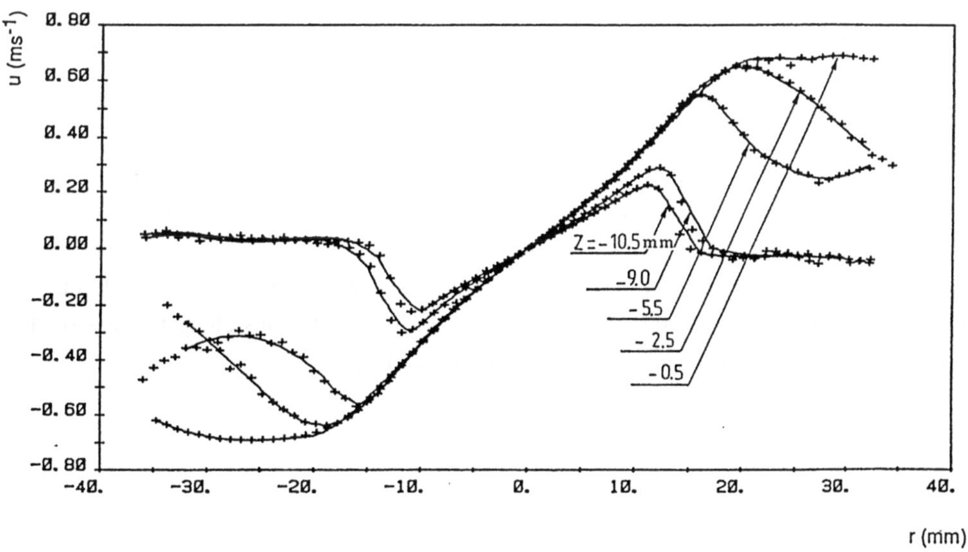

Fig. 13: Radial profiles of the u (radial) component of the velocity field at different z locations

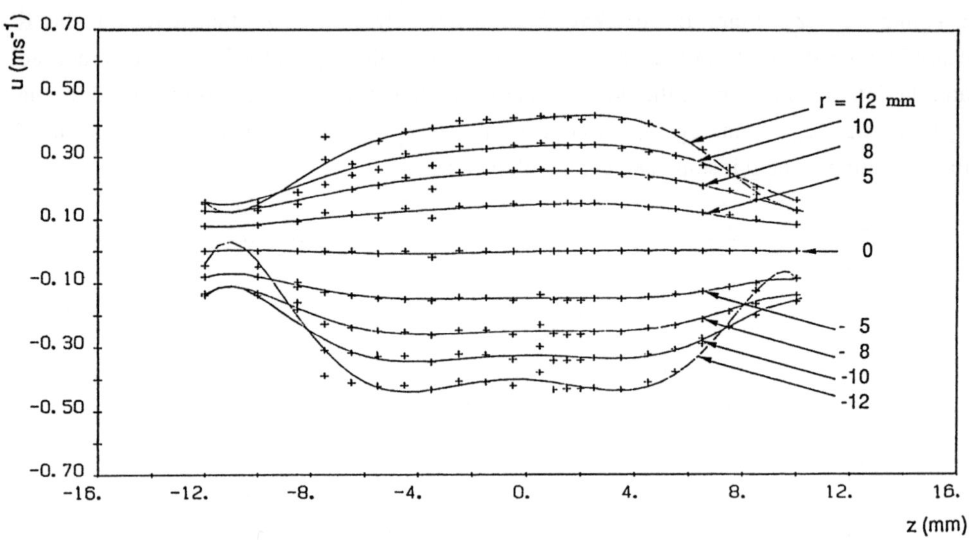

Fig. 14: Axial profiles of the u (radial) component of the velocity field at different r locations

Fig. 14 shows the u-distribution as a function of z for various r locations. Around z_o and for -12 mm $< r <$ 12 mm, a constant $\partial u/\partial z = 0$ region is observed.

The above described general features of the flow remain valid for all the flow cases. The z_o position of the stagnation plane is different in each of the cases as seen from Table 1. The major and fundamental difference between the flow characteristics is discussed in the following paragraph.

Figs. 15 and 16 show the radial distribution of u and the axial evolution of v respectively for various burner separations H. One can see that the central part slope of each curve depends on H. Thus this shows that k_1 and k_2 are functions of the burner separation. In the centre part of the flow the slope increases with decreasing distance between the burner nozzle ouztlet planes. On the other hand, the slopes of the v-velocity distributions along the axis of the burner arrangement show relatively constant values, see Fig. 16.

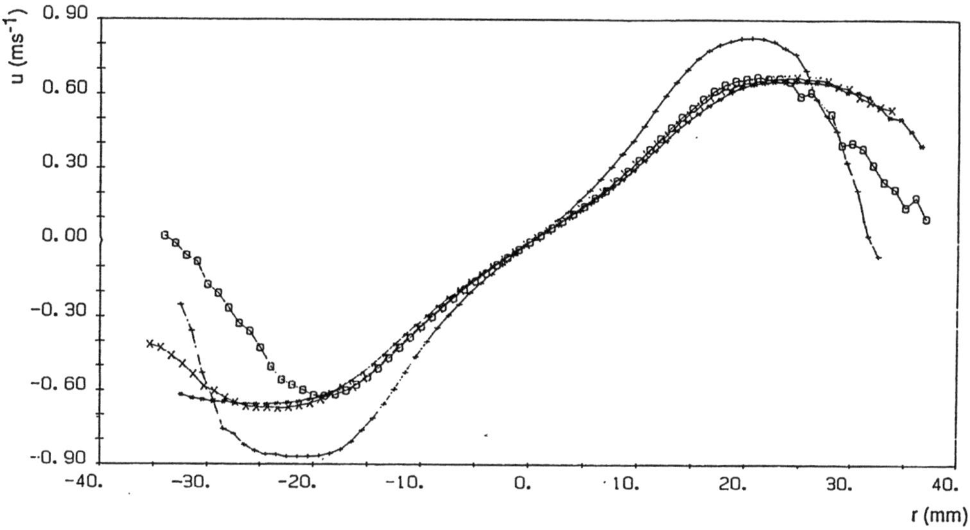

Fig. 15: Radial evolution of the *u* (radial) component of the velocity for various burners separation *H*

(+) -*H* = 15.0 mm; (x) -*H* = 28.6 mm; (o) -*H* = 25.0 mm; (*) -*H* = 35.0 mm

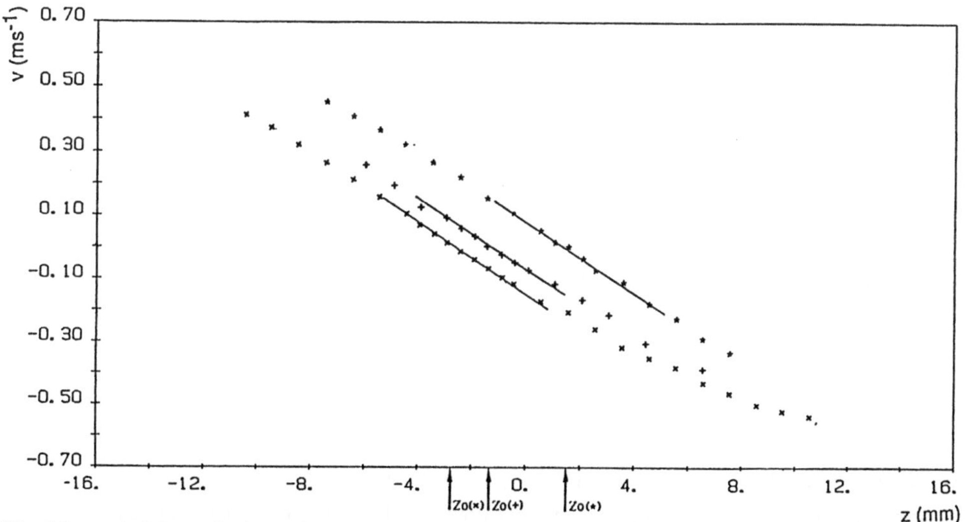

Fig. 16: Axial evolution of the *v* (axial) component of the velocity for various burners separation *H*

(*) -*H* = 28.6 mm; (+) -*H* = 25.0 mm; (x) -*H* = 35.0 mm

Comparison between experimental data and theoretical prediction

The flow gradient (dv/dz) can be computed by numerically differentiating the measured v-velocity distribution with respect to z and by utilizing the information to check the continuity equation in the following form:

$$r \cdot u = -\int_0^r r \frac{\partial v}{\partial z} dr$$

Taking into account, that in the flow region close to the stagnation plane, the two velocity components u and v exhibit linear variations with r and z, repsectively, the above equation can easily be integrated to yield:

$$r \cdot u = k_2 \cdot r^2 \qquad \text{with} \qquad \frac{\partial v}{\partial z} = -2 \cdot k_2$$

Comparing the experimental data via this equation shows that the continuity equation is satisfied in the centre part of the flow to better than $\pm 1\%$. Similar agreement was observed if the continuity equation was verified in the following form:

$$\int_0^{\xi} \frac{\partial u}{\partial r} dz + \int_0^{\xi} \frac{u}{r} dz = v$$

where x was the coordinate in the z-direction with the origin in the stagnation plane. Introducing again

$$k_1 = \frac{\partial u}{\partial r}$$

permits the first and second integration to be performed to yield for the centre part of the flow

$$v = 2 \cdot k_1 \cdot z$$

In Table 1 the data for k_1 and k_2 are presented together with theoretical data proposed by Spalding (1961) (eq. 4) and Williams (1981) (eq. 5) for the constant strain rate deduced from a potential solution for the counter flowing jet flow. Small differences exist between k_1 and k_2 reflecting small errors of $\pm 1\%$ in the measurements. The data show, however, that the theoretical value of the strain rate proposed by Williams is notably higher than the experimental results. The value imposed by Spalding (1961) is independent of the distance H and underestimates the experimental value by about 20%. Nevertheless, it seems that the experimentally found k_1 and k_2 values approach the theoretical value of Spalding when H increases, thus the flow field approaches the potential flow solution if the two opposing nozzles are far away from each other.

Considering flow solutions for the counter jet stagnation flow that takes the viscosity of the flow into account, Rosenhead (966), suggests that the flow field can be written as:

$$u = k \cdot r \cdot f'(\zeta) \qquad \text{and} \qquad v = -2 \cdot (kv)^{\frac{1}{2}} \cdot f(\zeta)$$

where k is a constant, ν is the kinematic viscosity and

$$\zeta = z \cdot (k/v)^{\frac{1}{2}}$$

This solution yields, however, in the centre part of the flow the same gradients as those calculated by the potential flow theory.

	H (mm)	z_o (mm)	k_1 (s)	k_2 (s)	V_o/D (s)	$2Vo/H$ (s)
0	15.0	0.4	37.0	...	24.0	80.0
I	25.0	-1.4	29.2	27.6	24.0	28.0
II	28.6	-1.5	29.0	28.4	24.0	45.7
III	35.0	-2.8	28.6	28.2	24.0	37.4

Table 1: k_1 and k_2 are the mean values for the experimentally determined strain rate; z_0 is the stagnation point position; V_0 is the air mean outlet velocity of the vurners; D is the nozzles diameter and H the distance between them.

The findings summarized in Table 1 readily suggest that the discrepancy between the experimentally found slope of the velocity profile in the centre of the stagnation plane and the theoretical results is entirely due to the choice of small values of H/D in the experiments. To demonstrate this, a test case was run for H = 15 mm and the results in Figure 15 and Table 1 show that the slope increases even faster. This readily suggests that the theoretically found central slope of the velocity profiles is only reached by the experiment for large values of H/D.

Flow field with combustion

The results now presented are obtained with a hydrogen-air diffusion flame. The theoretical strain rate chosen is 24 s^{-1} and the chemical equivalence ratio is unity.

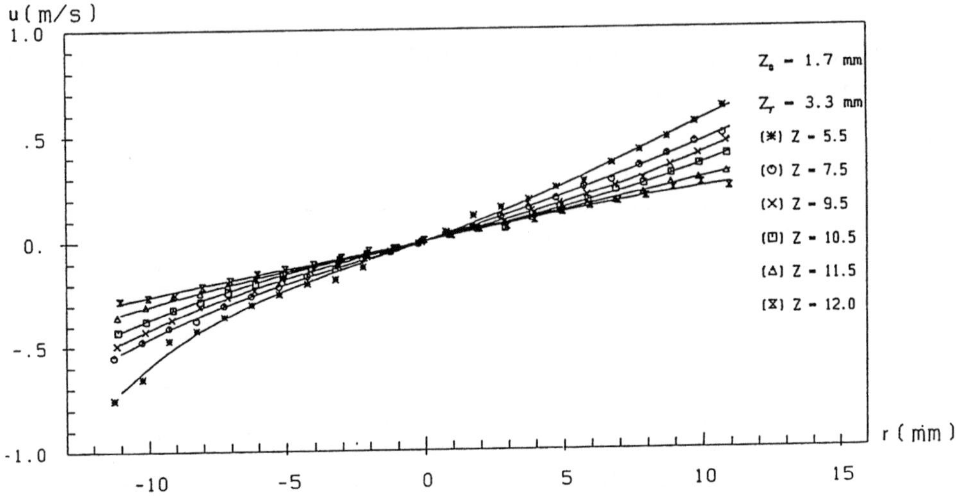

Fig. 17: Radial profiles of the v component of the velocity, z_o and z_f are the stagnation point and the flame position respectively, in a coordinate system with the origin $z = 0$ at $H/2$.

Fig. 17 shows radial velocity profiles as a function of radius r for several values of the axial coordinate z. Experimental data and fitted curves are displayed. The strain rate (i.e. the radial velocity gradient), is found to be a linear function of the radial position, for z constant, as predicted by the theory. The detailed measurements show that the local strain rate is also a function of z.

Fig. 18 provides the distribution of the v (axial) component of the velocity field, in radial direction at different z locations. The profiles were obtained with the optical arrangement of Fig. 6. At the nozzle outlets they present the same behaviour as the cold flow measurements. In approaching the reaction region they become more uniform due to the increasing strain rate.

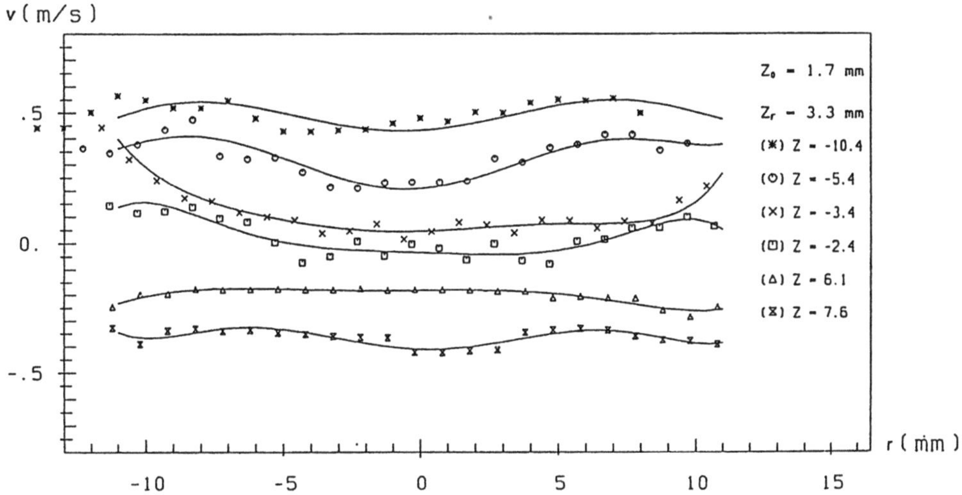

Fig. 18: Radial profiles of the v component of the velocity at different z locations

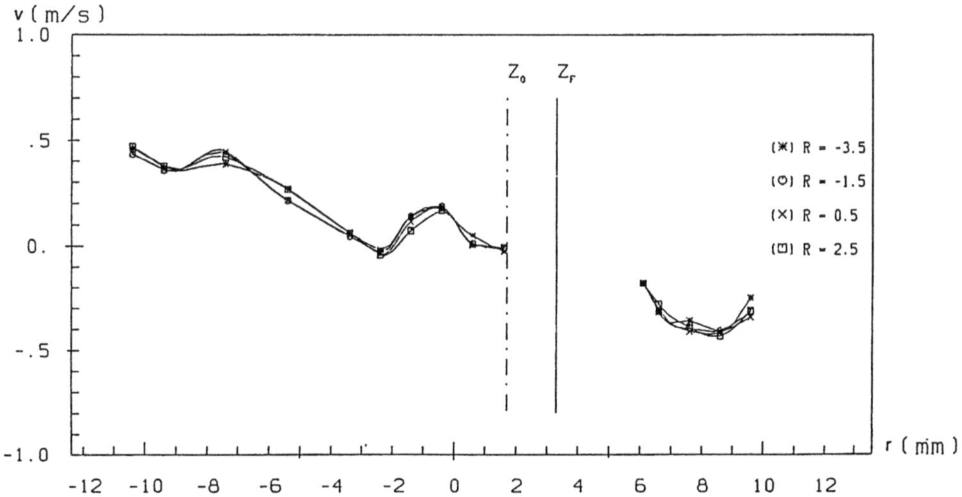

Fig. 19: Axial profiles of the v component of the velocity at different r locations

Fig. 19 displays the axial (v) component of the velocity in z direction at different r locations. The velocity increase due to thermal expansion is recognized at both sides of z_o and z_f, which are the axial coordinates of the stagnation plane and the flame position respectively. The interrupted range of these curves shows the area where Doppler bursts do not exist due to the thermophoresic effects.

LDV-LBD COUPLED SYSTEM

With the current optical configuration, only the velocity components parallel to the flame front could be measured directly in the reaction region. To determine the axial velocity component in this zone, the mass conservation equation could be applied in conjunction with direct measurements of the radial u-component of the velocity and density measurements. For a simultaneous measurement of the velocity and density fields, a two colour laser was applied (argon-ion laser). The green beam of the laser was used for velocimetry as described above and the density was determined from the blue beam deviation (Fig. 20), detected by a position sensor. The first experiments with the complete system are now under way (Fig. 20). Local density can be calculated from the laser beam deviation with an Abel integral transform in an axisymmetric flow field.

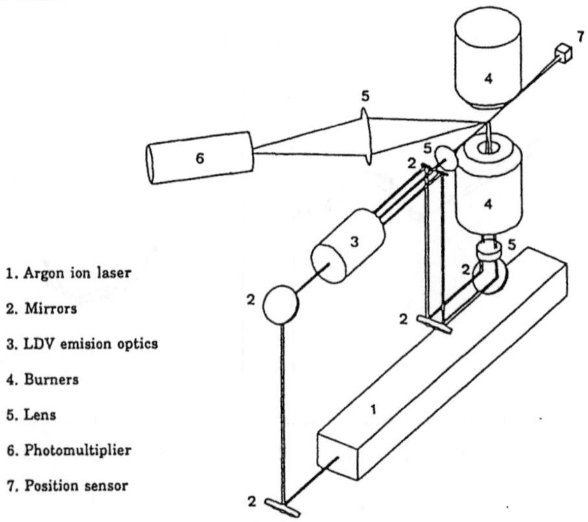

Fig. 20: LDA optical arrangement (green Ar^+ laser line) with laser beam position sensor (blue Ar^+ laser line) for simultaneous velocity and density measurements.

Laser Beam Deviation as a Local Density Probe: Application to a Counterflow Diffusion Flow

The laser deviation technique was used to measure density in one-dimensional combustion flows and for spectral analysis of turbulence. In axisymmetric flows, the laser beam deviation was utilized together with an Abel inversion calculation to give the local density and its gradients. The theoretical and experimental methods are discussed below together with the density and density gradients profiles measured in the structure of a flat counter flow diffusion flame.

Laser Deviation Theory and Abel Inversion

Assume a light (laser) beam to cross a gradient of refractive index $n(x,z)$, then according to Weinberg (1963), a deflection $d\Theta$ occurs

$$\frac{d\Theta}{ds} = \frac{1}{n}\frac{\partial n}{\partial \xi}$$

where $\partial n/\partial \xi$ is the component of the gradient normal to the laser beam (Fig. 21). A good approximation is that refractive index for gases is a linear function of the local density (see Weinberg (1963))

$$n - 1 = \beta \cdot \left(\frac{\rho}{\rho_0}\right)$$

where ρ_o is the density under standard conditions and β is a constant depending on the gas nature and light wavelength. The deflection gradient $d\Theta/ds$ can be written

$$\frac{d\Theta}{ds} = \frac{1}{n}\frac{\beta \partial \rho}{\rho_0 \partial \xi}$$

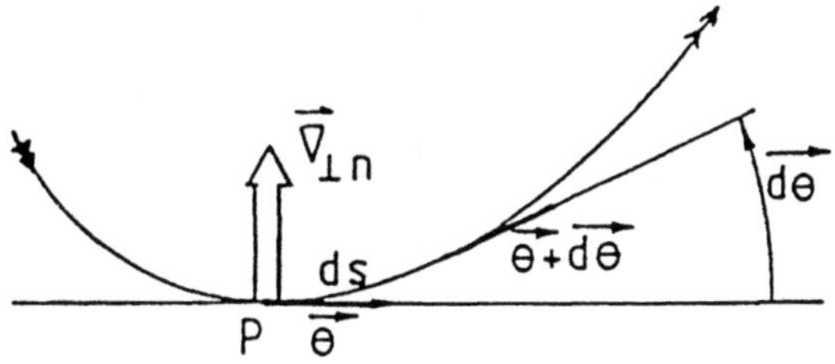

Fig. 21: Principle of laser deviation

When a laser beam crosses an axisymmetric flow field with variable density $\rho(x,r)$, it deviates from the normal straight-line trajectory. Assume a new coordinate system, in which O is the symmetry axis of the flow field, O_y is parallel to the laser beam and O_z is the normal axis to them. As the deviation of the laser beam is small in gases (see Weinberg), the beam entering the density field at $(x,-y,z)$ leaves it at point (x,y,z) with a final deviation which projections in the O_{yz} and O_{xy} planes are respectively equal to

$$\Theta_z(x,z) = \frac{\beta}{n\rho_0}\int_{-Y}^{Y}\frac{\partial\rho}{\partial z}dy$$

and

$$\Theta_x(x,z) = \frac{\beta}{n\rho_0}\int_{-Y}^{Y}\frac{\partial\rho}{\partial x}dy$$

If the flow field is limited in the region $r < R$ (see Fig. 22), and because of the cylindrical symmetry it can be written

$$\Theta_z(x,z) = \frac{2\beta}{n\rho_0}z\int_{z}^{R}\frac{1}{r}\frac{\partial\rho}{\partial r}\frac{r\,dr}{\sqrt{r^2-z^2}}$$

$$\Theta_x(x,z) = \frac{2\beta}{n\rho_0}\int_{z}^{R}\frac{\partial\rho}{\partial r}\frac{r\,dr}{\sqrt{r^2-z^2}}$$

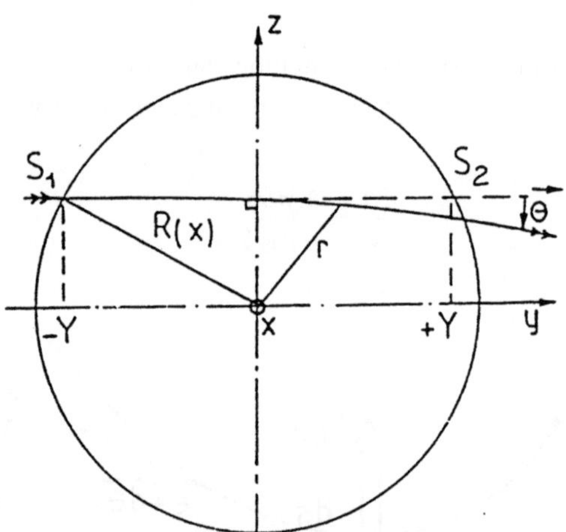

Fig. 22: Laser beam deviation in axisymmetric flow field

Application of the Abel inversion shows that if $\phi\,(x,z)$ and $G\,(x,r)$ are related by

$$\phi(x,z) = 2\int_z^R G(x,r)\,\frac{r\,dr}{\sqrt{r^2 - z^2}}$$

then

$$G(x,r) = \frac{1}{\pi}\int_r^R \frac{\partial(x,z)}{\partial z}\,\frac{dz}{\sqrt{z^2 - r^2}}$$

These relations show that the Abel inversion of

$$n\rho_0\beta^{-1}\frac{d\Theta_x(x,z)}{dz} \qquad \text{and} \qquad \frac{n\rho_0}{\beta z}\Theta_z(x,z)$$

leads to ($-\partial\rho/\partial x$) and to ($-1/r\;\partial\rho/\partial r$), which are the density gradients along and across the axis of the flow field respectively (AZIN-HEIRA (1986)). If the surrounding density $\rho_1 = \rho(r > R)$ is known a further integration

$$\rho(x,r) = \int_r^\infty -\frac{\partial\rho(x,r)}{\partial r}dr + \rho_1$$

gives the density itself. If $\Theta_x(x,z)$ and $\Theta_z(x,z)$ are measured in the entire flow field of structure $\rho(x,r)$ the density gradients are then determined in all the regions of interest.

Experimental Technique

The above defined theory shows the way for the experimental procedure: a laser beam is adjusted at a given x_e distance from one of the nozzle exits of the counter flow burner. The deviations $\Theta_x(x,z)$ and $\Theta_z(x,z)$ are measured as a function of z. "Axial and radial deviograms" are obtained to which the Abel inversion code can be applied to extract the density gradients. The density can than be calculated in the x_e plan. This procedure is subsequently used at different x_e values in order to cover the entire reaction region.

The principle of the experimental set-up is sketched in Figure 20. The counter flow diffusion flame is produced by two opposed nozzles supplied by fuel and oxygen as it was discussed in the previous sections (see Fig. 5). Deviations of the laser beam were measured to validate the procedure. The measurements were obtained with a position detection sensor placed at a distance of 600 mm from the axis of the flow. The laser beam diameter was reduced in the crossing region by a focusing lens (f = 200 mm) to improve the spatial resolution of the technique. The burner system can be moved in the O_x and O_y directions by a step by step translation system. The burner position (x,z) and the signals of the position sensor are recorded by a PDP-11/23 microcomputer. The Abel inversion can therefore be applied directly, i.e. without storing the raw data.

Experimental Results

For the present experiments, the experimental conditions were chosen arbitrarily as follows: burner separation H = 16 mm, nozzle diameter D = 20 mm, theoretical flow strain rate $a = 24\,a^{-1}$; the equivalence ratio of the hydrogen-air diffusion flame was equal to unity. The sensitivity of the position sensor in radial and axial deviations was respectively δ_x = 1.0 volt/mm and δ_z = 3.6 volt/mm. With this settin the entire flow field was investigated.

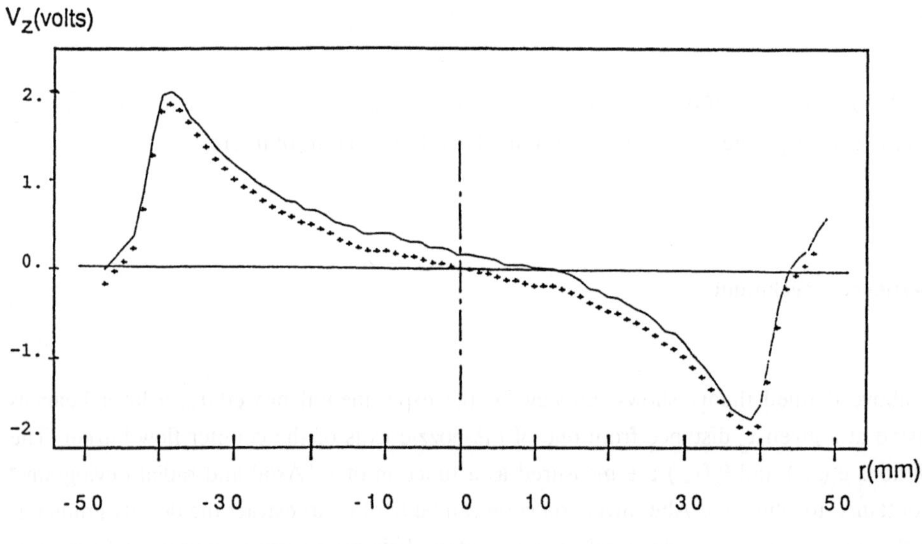

Fig. 23: ___ Radial laser beam deviation (radial "deviogram")
+++ Fitted and symmetrized radial deviogram (sensitivity of the position sensor δ_x = 1 volt/mm)

As an example the results corresponding to the radial and axial deviograms are reported measured at the flame surface X_F = 1.55 mm. Fig. 23 shows the radial component of the laser beam deviation, where (----) correspond to actual measured values and (++++) to the fitted and symmetrized ones. Fig. 24 is the result of the Abel inversion of the fitted radial deviogram. The flat central part of this curve shows the high temperature flame surface. Fig.

25 is the plot of the axial deviogram. Fig. 26 shows the Abel inversion of the measured (----) and the fitted (+ +) values of axial deviogram; it can be seen that noise error produced in the surrounding region of the axisymmetric flow is accumulated in the central part of the Abel inversion profile, because the Abel inversion is an integral transform. This is why the measured deviograms has to be fitted and symmetrized.

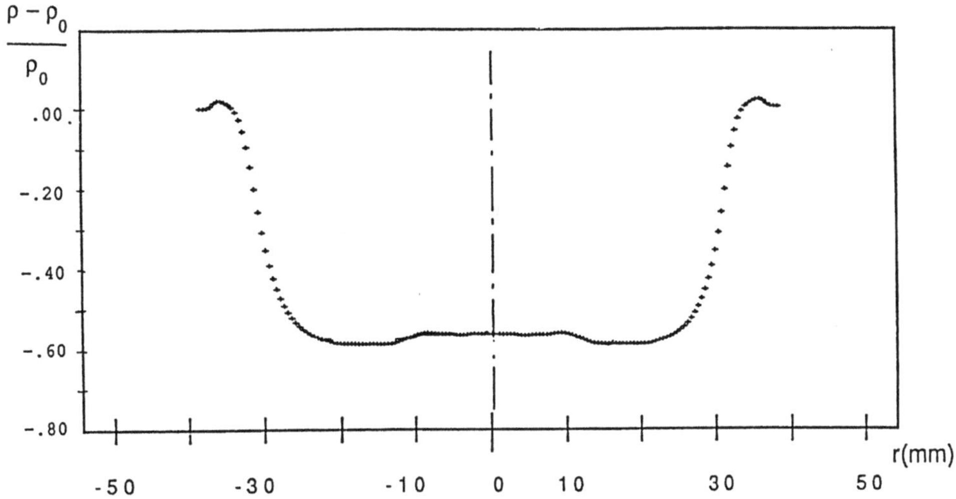

Fig. 24: + + Density calculated by Abel inversion of fitted radial deviogram

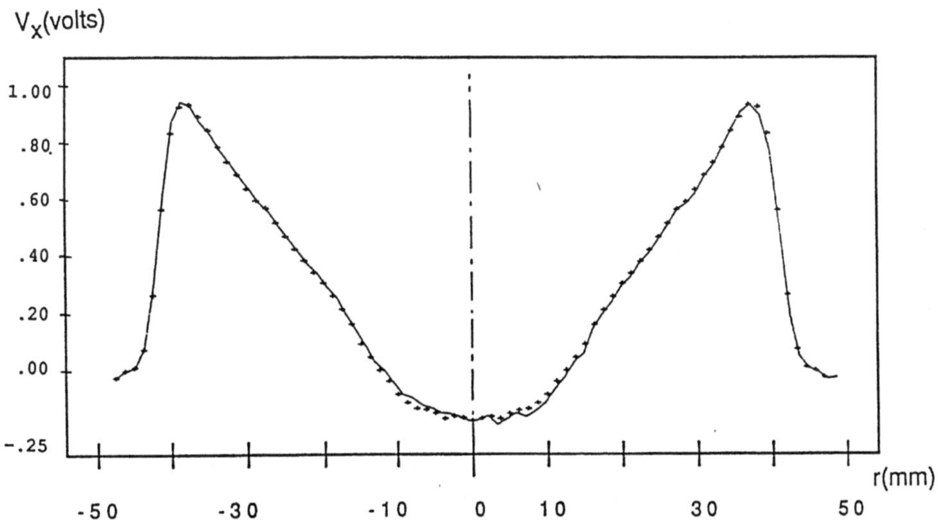

Fig. 25: ___ Axial laser beam deviation (axial "deviogram")

+ + Fitted and symmetrized axial deviogram (sensitivity of the position sensor δ_z = 3.6 volt/mm)

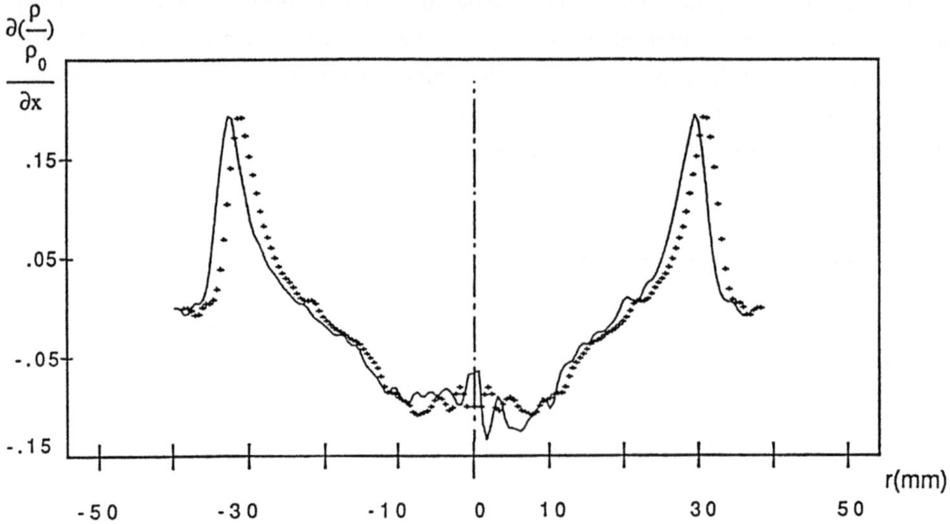

Fig. 26: ___ Density gradient calculated from the measured values by Abel inversion of the axial deviogram

+ + + Calculated by Abel inversion from fitted and symmetrized deviogram

CONCLUSIONS

In this paper, the potential flow conditions as used in flat counter flows flames (see Darahiba, Candel and Marble (1986)) were extensively investigated, and velocity measurements in a flat counter flow diffusion flame are presented. Such flames are of great interest in combustion because they are well adapted to study complex chemistry features and constitute a base of "flamelet models" (see Rog, Behrent and Warnatz (1986), Giovangigli and Smooke (1987)). Nevertheless, very few experimental velocity data are available in these flames and accordingly theoretical models are generally based on the potential flow theory proposed by Spalding. The results show that this theory is too simple and that two dimensional effects, due to viscosity, have to be taken into account to allow valid comparisons.

The next step of this work will be the simultaneous measurements of the two velocity components u and v in the reaction region. To solve the difficulties encountered with the strong refractive index gradients in the reaction region, it is planned to deduce the density field from the laser beam deviation and to compute the local axial velocity v through the mass conservation equation and the simultaneously measured local density and radial component of the velocity.

ACKNOWLEDGEMENT

This work emerged from a collaboration between the authors' two lavoratories, through E.E.C.-contract ST2J-0273-C(EDB), the JOULE-Project JOUE-0056-C, dealing with combustion and the exchange of scientists received support through the PROCOPE-Programme. All three supports are thankfully acknowledged.

BIBLIOGRAPHY

AZINHEIRA, J.: Thèse de Docteur Ingénieur, Ecole Centrale des Artes et Manufactures, Paris, 1986

DARABIHA, N., CANDEL, S.M. and MARBLE, F.E.: The Effect of Strain Rate on a Premixed Laminar Flame. Combustion and Flame, Vol. 64, pp 203-217, 1986

FENDELL, F.E.: Ignition and Extinction in Combustion of Initially Unmixed Reactants. J. Fluid. Mech., Vol. 21, pp 281-303

GIOVANGIGLI, V. and SMOOKE, M.: Extinction of Strained Premixed Laminar Flames with Complex Chemistry. Combustion Science and Technology, Vol. 53, pp 23-49, 1987

HAHN, W.A., WENDT, J.O.L.: NO_X Formation in a Flat, Laminar Opposed Jet Methane-Air Diffusion Flame, 9th Symposium on Combustion, p 121, 1981

KRISHNAMURTY, L., WILLIAMS, F.A. and SESHADRI, K.: Asymptotic Theory of Diffusion-Flame Extinction in the Stagnation-Point Boundary Layer, Combustion and Flame, Vol. 26, pp 363-377, 1976

LECLERC, A.: Déviation d'un Jet Liwuide par une Plaque Normale à son Axe. La Hoille Blanche, N° 6, pp 3-8, 1950

LINAN, A.: The Asymptotic Structure of Counterflow Diffusion Flames for a Large Activation Energies. Acta Astronautics, Vol. 1, pp 1001-1009, 1974

MARBLE, F.E. and BROADWELL, J.E.: The Coherent Flame Model Turbulent Chemical Reactions. Technical Report TRW-9-PU Project Squid Headquarters, Chafee Hall, Purdue University, West Lafayette, Indiana, 1977

PANDYA, T.P. and WEINBERG, F.J.: The structure of Flat, Counterflow Diffusion Flames. Proc. Roy. Soc., A 279, pp 544-561, 1964

PETERS, N., Progress in Energy and Combustion Science, Vol. 10, p 319, 1984

POTTER Jr., A.E. and BUTLER, J.N.: A Novel Combustion Measurements Based on the Extinguishement of Diffusion Flames. ARS Journal, N° 29, pp 54-56, 1959

ROG, B., BEHRENDT and WARNATZ, J.: Modelling of Turbulent Methane-Air Diffusion Flames: The Laminar Flame Model. Paper presented at 85. Bunsentagung Heidelberg, May 8-10, Ber Bunsenges Phys. Chem., 1986

ROLON, J.C., MARTIN, J.-P. and DURST, F.: Counter Jet Stagnation Flows Yielding Flat Diffusion Flames. Third International Symposium on Applications of Laser Anemometry to Fluid Mechanics, Lisbon, Portugal, July 7-9, 1986

ROSENHEAD, L.: Laminar Boundary Layers. Oxford-Clarendon Press, page 155, 1966

SESHADRI, K. and PETERS, N.: The influence of Stretch on a Premixed Flame with Two Steps Kinetics. Combustion Science and Technology, Vol. 33, pp 35-63, 1983

SPALDING, D.B.: Theory of Mixing and Chemical Reaction in the Opposed-jet Diffusion Flame. J. Amer. Rocket Soc., Vol. 3, pp 763-771, 1961

TSUJI, H.: Counterflow Diffusion Flames. Prog. Energy Combustion Science, Vol. 8, pp 93-119, 1982

WEINBERG F.J.: Optics of Flames, Butterworths, London 1963

WILLIAMS, F.A.: A Review of Flame Extinction. Fire Safety Journal, Vol. 3, pp 163-175, 1981

Investigation of a Combustion Zone Behind a Wedge

P. Bruel, P. Blazart, B. Deshaies, M. Champion, S. Duplantier, L. Sramek.

Laboratoire d'Energétique et de Détonique
URA 193 CNRS
ENSMA - Rue Guillaume VII
86034 POITIERS
FRANCE

ABSTRACT

An experimental study of a turbulent combustion zone stabilized by a wedge is presented. Informations collected through temperature and velocity measurements along with CCD video-cameras visualizations allow for a qualitative description of the various combustion processes involved. More precisely, it is clearly evidenced that more than one model should be used to represent numerically the whole flowfield characteristics.

INTRODUCTION

In many combustors as these of Ramjet and Turbojet engines, a combustion zone is stabilized by a flow recirculation. In particular, this is the case when obstacles such as vee gutters or wedges are used. In this situation, the combustion process is controlled by numerous parameters related to local transport properties, turbulent mixing, flow geometry and chemical kinetics. Due to the complexity of such a reacting flow and in view of numerical calculations, there is a need for simplification which is usually satisfied by defining limiting regimes controlled by a smaller number of parameters.

The goal of this work is to characterize experimentally the different kinds of limiting combustion regimes that can be found in the flow downstream of a wedge. Clearly, it is expected that not only one limiting combustion regime is necessary to describe the entire flowfield where the combustion process takes place. In that respect, it should be noted that numerical models are usually mainly based on a single limiting combustion regime and, accordingly, are more adapted to describe only a part of the entire combustion process.

R. S. L.Lee · J. H. Whitelaw · T. S. Wung (Eds.)
Aerothermodynamics in Combustors
IUTAM Symposium Taipei, Taiwan, 1991

Another goal of this work is to provide a better knowledge of the exact flow structure thus i) to indicate the exact region where a given model should be applied, ii) to provide test points for numerical computation, iii) to emphasize the need for setting new models in some specific regions.

EXPERIMENTAL CONDITIONS

The experimental setup, described in figure 1, is made of a wind tunnel involving three parts:

- a mixing chamber where an homogeneous mixture of air and propane is obtained.
- a settling section made of grids and honeycomb structures followed by a converging channel insuring a 1 or 2 per cent turbulence level.
- a one meter long rectangular channel whose width is D=16 cm and height H=3 cm where the flow becomes a fully developped turbulent flow with a 2-D mean structure.
- the test section itself, at the beginning of which a stainless steel wedge is positionned.

Quartz windows are located on both sides of the channel upstream and downstream of the wedge.

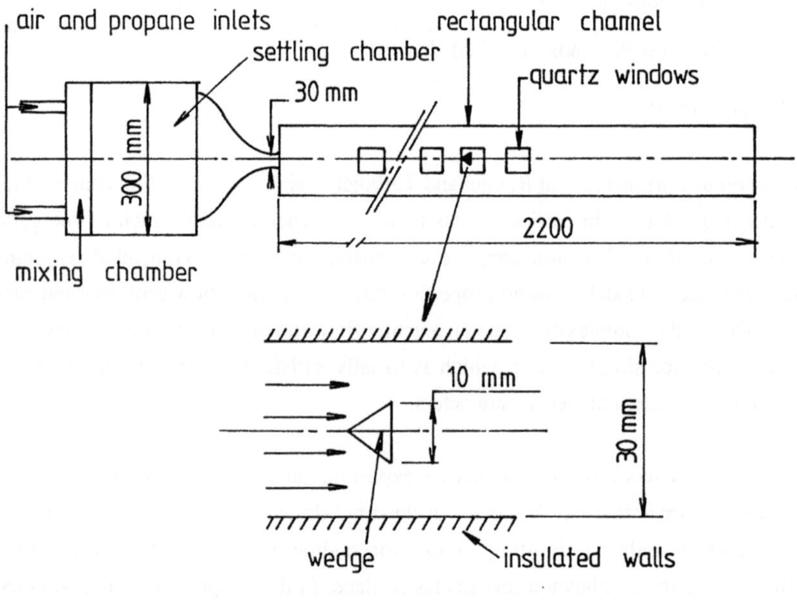

Fig.1. Experimental setup

The reacting flow is investigated using:

- a 2-components L.D.A. system,
- 20μm Pt/PtRh thermocouples,
- CCD video-cameras operating on visible and on the CO_2 wave lengths respectively.

The whole structure of the reacting flow stabilized with this experimental setup is schematically represented in figure 2 .

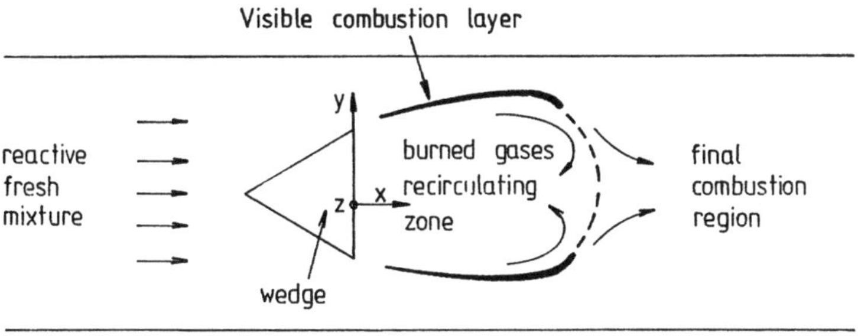

Fig.2. Mean structure of the reacting flow.

RESULTS AND DISCUSSION

The results presented in this paper have been obtained for a blockage ratio of 33%, an equivalence ratio of 0.67 which corresponds to the lowest practical value leading to a stable combustion and a total volumetric flow rate of 177 m^3 /hour at atmospheric pressure.
The compensation of the effect of the thermal inertia of the thermocouples has been performed using a variable time constant τ defined, in the case of cylindrical thermocouples by :

$$\tau = \rho\, d\, C / 4\, h$$

where ρ, d, C stand for the thermocouple density, diameter and specific thermal capacity respectively and h represents the convective heat transfert coefficient.This latter coefficient is determined using the following empirical relation between the local Nusselt and Reynolds numbers :

$$Nu = 0.891\ Re^{0.33}$$

The Reynolds number is obtained via the L.D.A. velocity measurements.Practically, τ lies between 2 ms and 7 ms. All the temperature probability density functions presented in this paper (thereafter referred to as p.d.f's) have been obtained using 300 intervals over the temperature range (0 K - 2000 K) and 4000 samples. They can be systematically compared to the corresponding uniformly distributed one.

Mean velocity field

The mean experimental 2D flow fields obtained under these conditions are given in fig.3 (mean velocity and streamwise turbulence intensity) concerning the flow upstream of the wedge and in fig.4 concerning the downstream flow in absence of combustion.

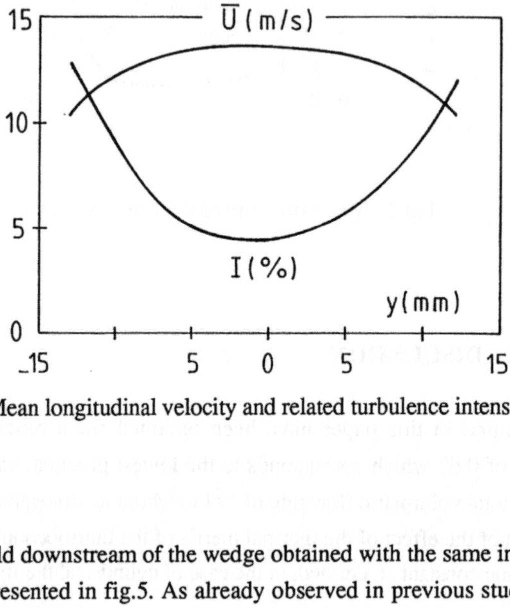

Fig.3. Mean longitudinal velocity and related turbulence intensity.

The mean velocity field downstream of the wedge obtained with the same inlet conditions but with combustion is presented in fig.5. As already observed in previous studies ([1], [2], [3]) the extension of the recirculation zone is longer in this latter case (approximately 3 times longer than the corresponding one without combustion). Though there is no simple explanation to such an extension which is in fact the result of a modification of the whole flow structure, this is an important feature which must be used to validate numerical models.

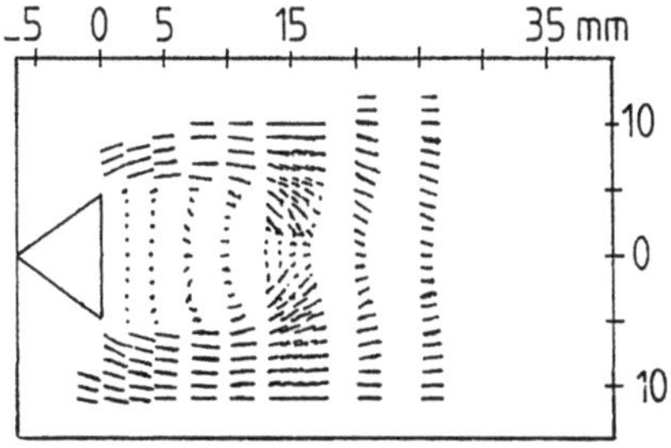

Fig.4. Mean velocity field (inert flow).

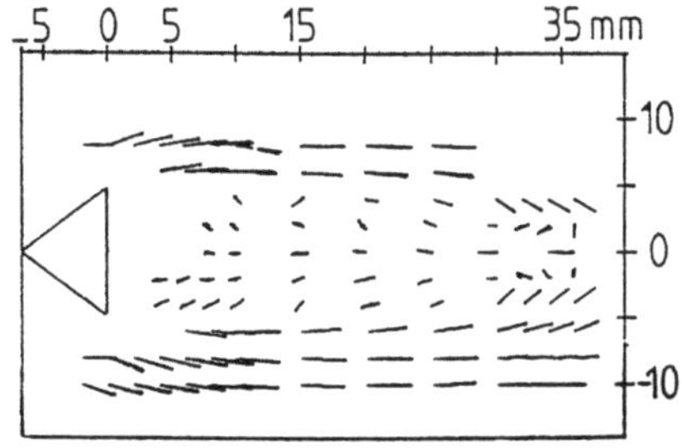

Fig.5. Mean velocity field (reacting flow).

Structure of the recirculation zone

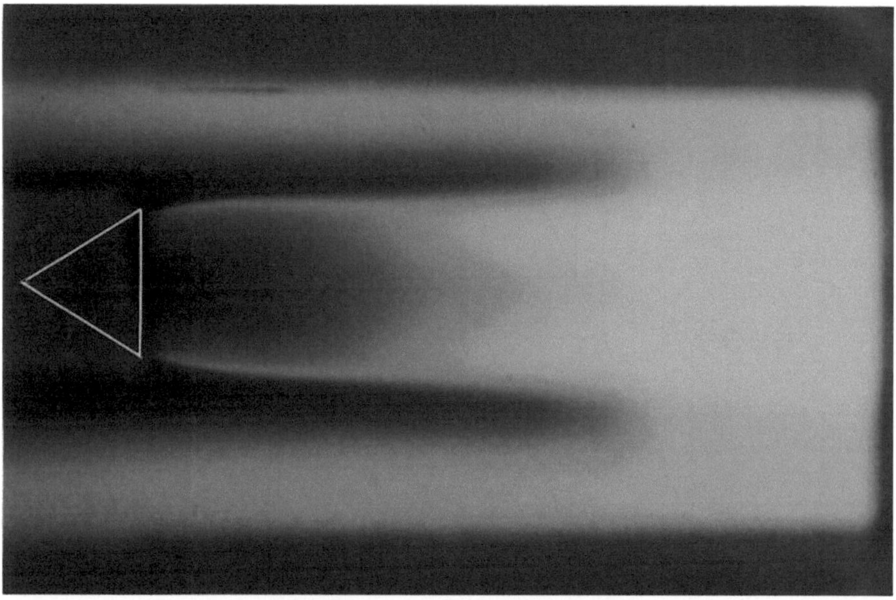

Fig.6. Reaction zone behind the wedge.

The picture of the reacting flow downstream of the wedge, given in figure 6, shows a quasi-

stationary visible front located in the peripherical part of the recirculating zone.
Instantaneous temperature measurements confirm the existence of a thin continuous reactive

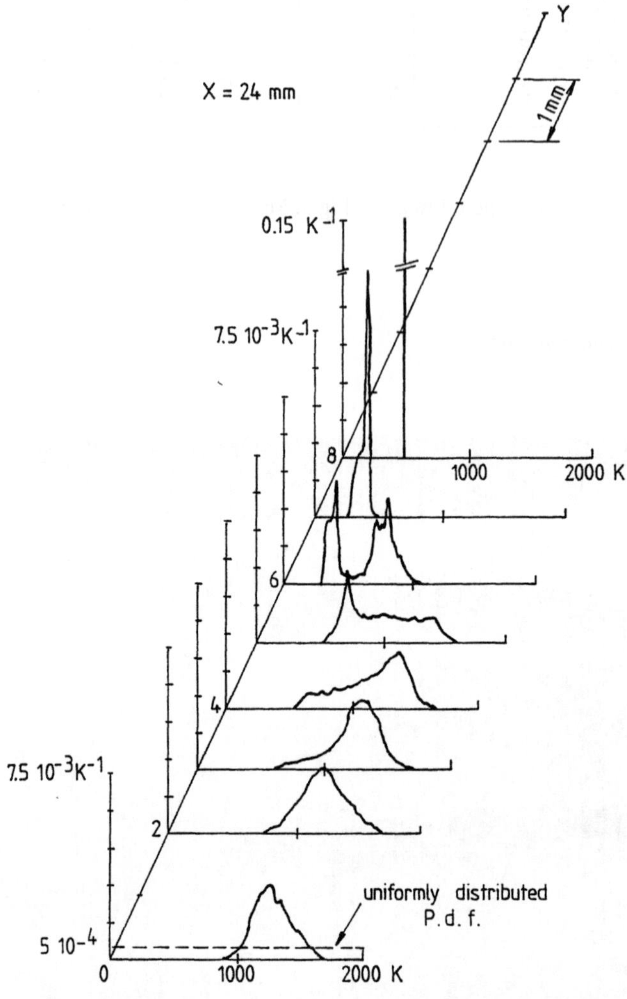

Fig.7. Transverse evolution of the temperature p.d.f. at x = 2.4 cm.

layer separating unburned reactants from burned products at least up to a streamwise distance from the wedge x=3cm.

This is illustrated by figure 7 where temperature p.d.f.'s obtained at x=2.4 cm are presented. Immediately on the unburned side of the reactive layer (y = 8 mm) the p.d.f. exhibits only small fluctuations centered around the temperature of the flow upstream of the wedge and is specific of a chemically free flow. On the plane of symetry of the channel (y = 0 mm) although the highest measured temperatures are not so far from the adiabatic value (Tad=1850K for the mixture under investigation), smaller values (i.e. T=900K) are also obtained. The p.d.f. exhibits a pseudo-gaussian shape and is centered around a value of 1300K. This indicates a recirculating kernel mostly, but not entirely, made of burned products thus involving a non negligible chemical rate. Specific measurements have to be performed to clarify this point and to investigate the exact nature of the mechanisms underlying the combustion process (homogeneous and / or flame fronts).

For streamwise abscissa x>3cm the continuous quasi stationnary reactive layer is progressively destroyed. This is illustrated in figure 8 where infrared video pictures of the recirculating kernel are presented. These pictures show that both pockets of hot products and cold reactants are formed alternatively at the end of the recirculating kernel. To understand these pictures, it should be noticed that, due to the exposure and sweeping times of the video system (40ms) an unburned or burned gas pocket appears as an oblique stripe. Going from blue to red indicates an increase in the degree of burning.

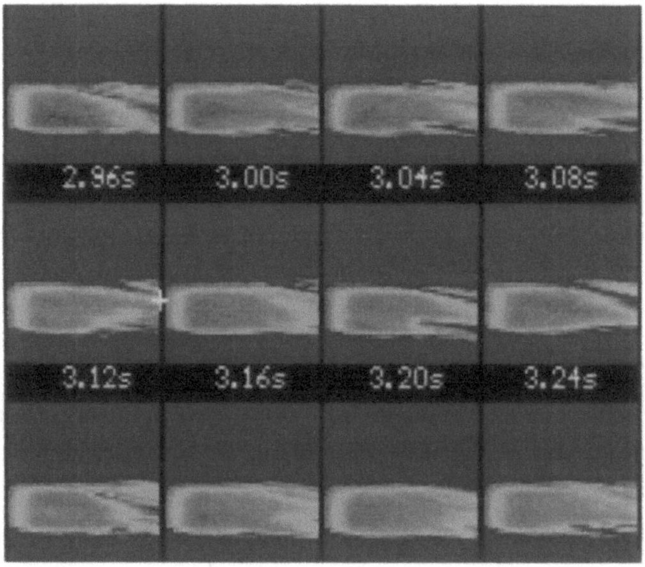

Fig.8. Instantaneous infrared video pictures of the recirculation zone.

The time average of these pictures, presented in figure 9, gives the mean shape of the recirculating zone. One can see that, in average, the highest temperatures are found in the central part of the kernel. A decrease of temperature is observed both in the vicinity of the wedge, due to heat exchange, and in the downstream part of the kernel, due to mixing between burned products and fresh reactants.

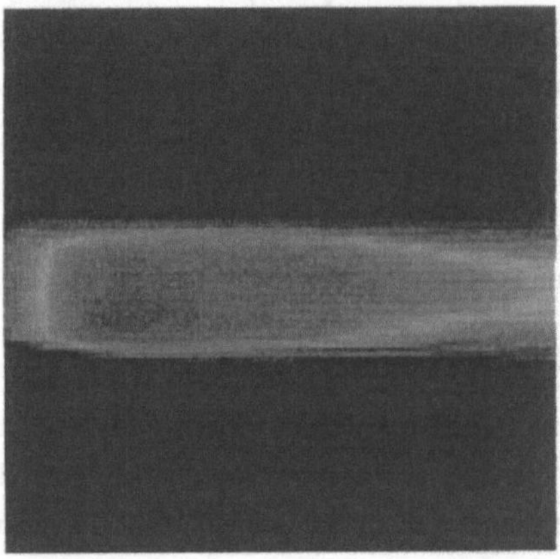

Fig.9. Time average of the instantaneous infrared video pictures of the recirculation zone.

Due to the destruction of the continuous reactive layer, part of the cold reactants are recirculated by the reverse flow and then burned in the wedge stabilized vortex. This contributes to the feeding of the recirculating zone with hot products.

Concerning the exact nature of the continuous reactive layer and on the basis of our present knowledge we may suppose a laminar flame like structure for this zone. Indeed the angle of the visible layer with the main flow is compatible with the laminar burning velocity (20cm/s) of the mixture under investigation. Detailed temperature and velocity measurements are required to confirm the existence and to understand the real flow kinematics leading to such a laminar flame front.

The region just near the wedge at the beginning of the visible front is of great practical importance for such a purpose. As expected in [4] and [5], the flame front can be the result of the development, just near the edge of the wedge, of a mixing layer involving both recirculated burned gases and cold reactants.

Structure of the region downstream of the kernel.

As aforementioned, the quasi stationary continuous combustion layer is destroyed, giving birth to a highly turbulent combustion region where pockets of hot and cold gases are convected and where in fact most of the fresh mixture is burned already.

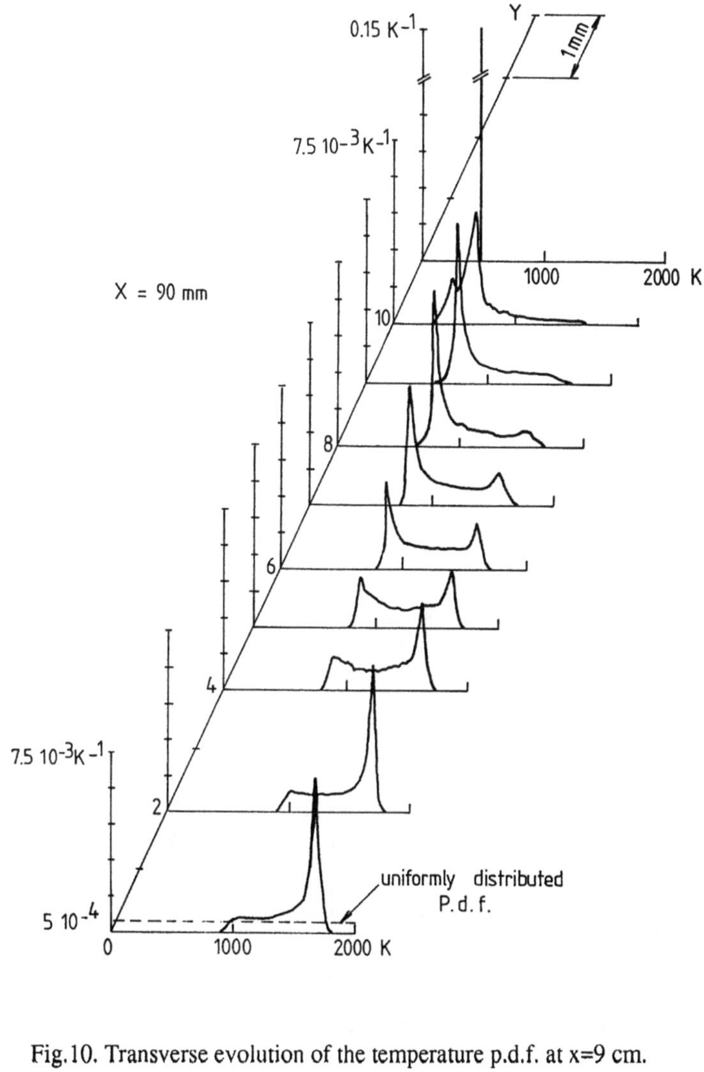

Fig.10. Transverse evolution of the temperature p.d.f. at x=9 cm.

In spite of the disparition of the continuous quasi steady combustion layer, the reacting flow can be still divided, in average, into two parts following the transverse direction:

- an internal part, around the plane of symmetry (y = 0 mm), where a pockets regime of combustion prevails. The proportion of burned gases is the largest at the plane of symmetry whereas unburned gases are more and more frequent when going towards the peripherical part of this central region.
- a peripherical layer, allowing a transition between the internal part, where intensive combustion prevails, and the wall, occupied mostly by fresh reactants.

This is illustrated by figure 10 which shows the evolution of the temperature p.d.f. in the transverse direction at x=9cm. On the plane of symetry (y = 0 mm), the p.d.f., centered around T=1700 K, indicates that the fluid is mostly made of hot products. When going from the plane of symmetry to the wall, bimodal temperature p.d.f.'s are obtained involving two maxima centered around the temperature of the reactants and of the hot products respectively. Inside the internal region described above, the maximun maximorum of the corresponding p.d.f. is positionned on the hot side of the temperature range.The reverse configuration is obtained when approching the wall, leading eventually to a p.d.f. centered around cold reactants temperature. It should be noticed that, for the p.d.f., the change from one configuration to the other occurs in a thin layer.

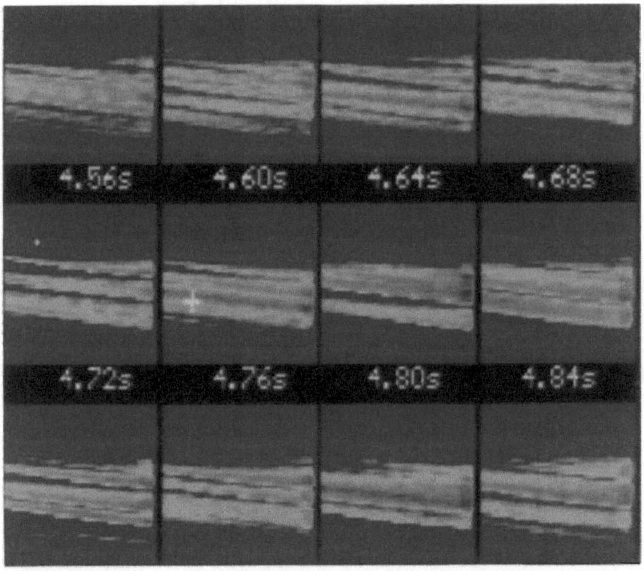

Fig.11. Instantaneous infrared video pictures of the flow downstream of the recirculation zone.

These results are confirmed by the infrared video pictures presented in figure 11 which clearly evidence the existence of hot and cold pockets in the central part of the flow whereas the flow near the wall is made of cold reactants. As in figure 8 going from blue to red corresponds to an increasing of both temperature and CO_2 mole fraction. Due to exposure and sweeping time of the video system, the pockets appear as oblique stripes. These pictures seem to indicate that, in the central part of the flow the convected pockets are not simply made of pure cold reactant or fully burned products separated by laminar flame fronts as assumed in basic statistical models of turbulent combustion. Within these pockets, temperature and CO_2 mole fraction are shown to increase in the streamwise direction thus indicating that each pockets acts as a small chemical reactor.

This behavior is confirmed by the temperature p.d.f. obtained on the plane of symmetry where the lowest measured temperatures correspond to partially burned mixture. On the contrary, at the edge of the internal region, temperature p.d.f. indicates the presence of pockets made of pure cold reactants.

The time average of these pictures, given in figure 12, shows the two regions described above and puts into evidence an increase of the degree of burning with the streamwise coordinate.

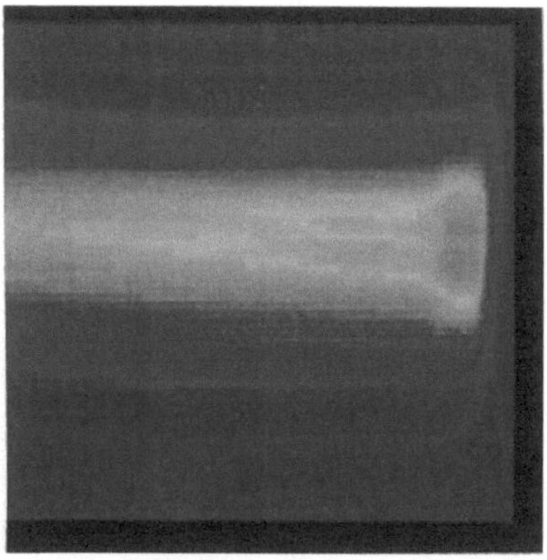

Fig. 12.Time average of the instantaneous infrared video pictures of the flow downstream of the recirculation zone.

The figure 13 puts into evidence, on the channel plane of symmetry, the progressive transition from the recirculation zone with a gaussian like shape for the p.d.f. to the region far downstream of the wedge with a single peak p.d.f..Mean temperature measurements show that the adiabatic temperature of 1850K is reached at a streamwise coordinate x=70cm onwards, indicating a complete burning of the incoming mixture.

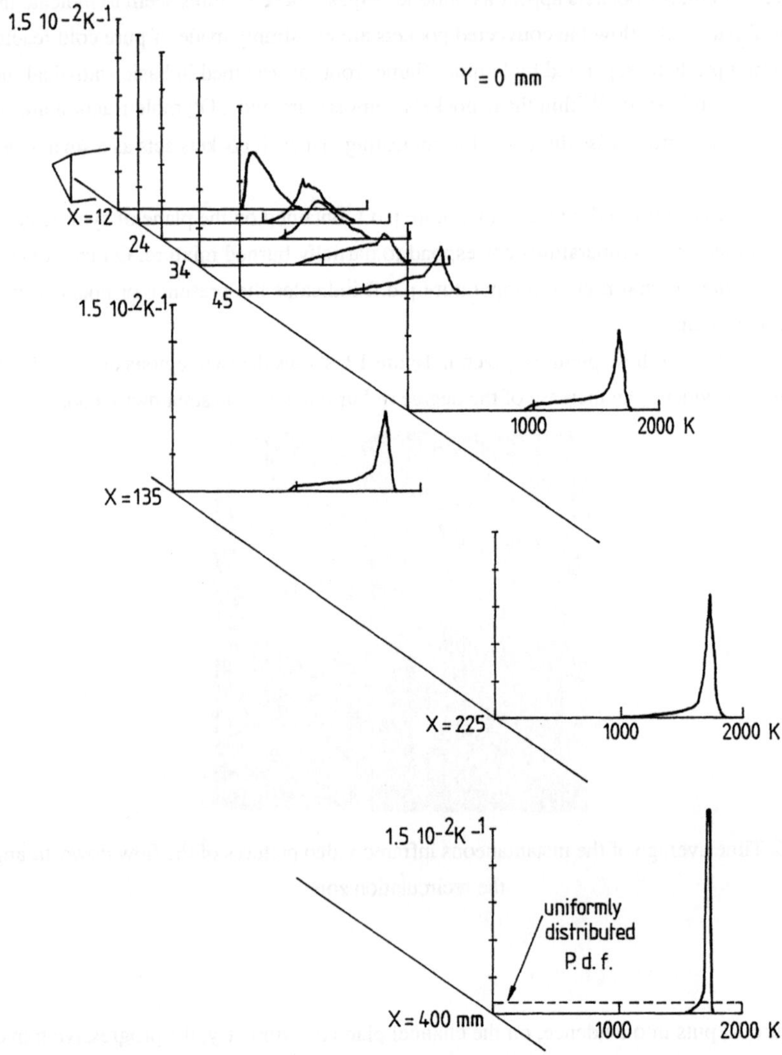

Fig.13. Streamwise evolution on the channel plane of symmetry of the temperature p.d.f.

CONCLUSION

To conclude it should be noticed that modelling precisely obstacle stabilized combustion would require a full set of models each of them being based on a different limiting regime:

- the stabilization process should involve a recirculating kernel with non negligible overall chemical rate at the edge of which a laminar like oblique flame front is expected.
- the transition region, located at the end of the recirculating zone where pockets are produced and which would require a non stationnary model. Production frequency of these pockets is of great importance since it determines the chemical rate in the terminal region,
- the terminal region where prevail both chemically controlled and flamelet regimes.

REFERENCES

1. Fujii, S, Gomi, M, Eguchi, K; Journal of Fluids Engineering, Vol. 100, p. 323 (1978).
2. Fujii, S, Eguchi, K; Journal of Fluids Engineering, Vol. 103, p. 328 (1981).
3. Heitor, M.V., Taylor, K.P.,Whitelaw, J.H., Combustion Sci. and Tech., Vol 62, (1988).
4. Marble, F.E., Anderson, T.C.; Jr. Jet Propulsion, Vol. 24, p. 85 (1954).
5. Williams, F.A.; Combustion Theory, The Benjamin / Cummings Publishing Company Inc. (1985).

The Influence of Swirl Generator Characteristics on Flow and Combustion in Turbulent Diffusion Flames

B. Mundus, Gaswärme-Institut e. V., Essen, Germany
H. Kremer, Ruhr-Universität Bochum, Bochum, Germany

Abstract

The present paper discusses experimental work to investigate the influence of swirl generator characteristics on the aerodynamic properties of unconfined turbulent swirl jets and the combustion behavior of unconfined swirl flames.

The distribution of the tangential velocity component at the burner exit characterizes different types of vortices typical of each swirl generator. These different vortices may be described by a modified solid body vortex approach. Swirl generators producing similar vortices generate swirl jets and swirl flames with similar properties.

Swirl intensity is described by means of a swirl number. The paper discusses the prediction of swirl numbers from swirl generator design data and the validation of these numbers by measurement.

1. Introduction

To generate turbulent swirl flames, a tangential velocity is normally introduced into combustion air flow. The burner component in which the turbulent velocity is produced is usually referred to as a swirl generator. Different types of swirl generators are employed in industry. They break down into two large groups. Swirl generators of the first group feature bladed screens to deflect combustion air flow. The arrangement of these screens may be radial or axial. The intensity of swirl is controlled by blade angle adjustment. In the case of the second type of swirl generator, combustion air flows are mixed. Flow rotation is produced by tangential air. To vary swirl intensity, the mixing ratio of the flows must be adjusted. Industrial burners use swirl generators of each of the types described or even combine both principles of producing swirl.

R. S. L. Lee · J. H. Whitelaw · T. S. Wung (Eds.)
Aerothermodynamics in Combustors
IUTAM Symposium Taipei, Taiwan, 1991

Swirl generator design has an influence on velocity component distribution at the burner exit and hence the flow field, the mixing properties and the reaction field of isothermal swirl jets and for swirl flames. The work presented in this paper was designed to investigate the behavior of different types of swirl generators for the swirl intensities encountered in practical applications.

2. Tests Set-up and Instrumentation

Four different swirl generators were tested. To compare the effectiveness of the different designs, the four generators had to be tested under the same conditions. For this reason, they were built to be housed in the same burner body. Fig. 1 shows the four different swirl generator designs.

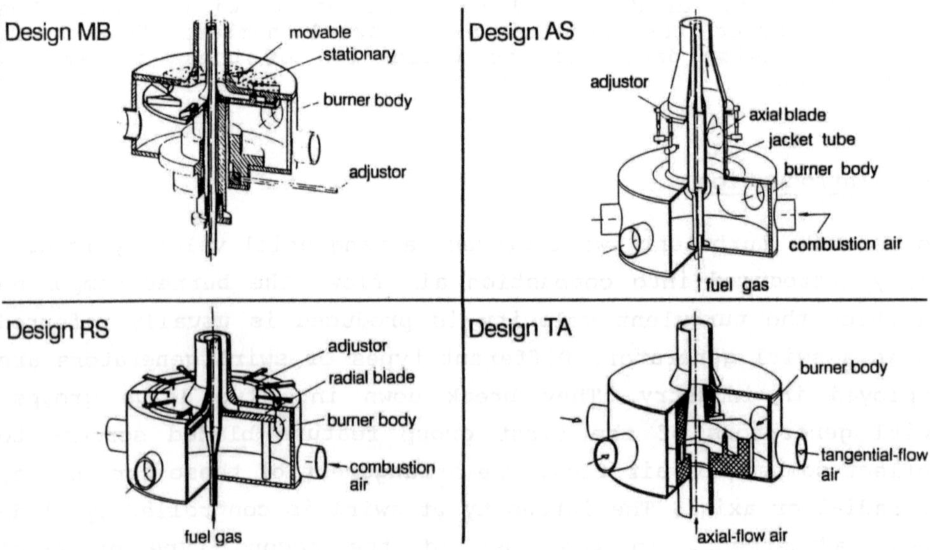

Fig. 1: Swirl Generators Tested

Apart from the movable block design (MB), a radial bladed-screen (RS) generator, an axial bladed-screen (AS) swirl generator and a tangential mixing (TA) generator were tested. The exit diameter of the cylindrical burner nozzle was D = 50 mm.

Fuel gas was supplied to the burner nozzle through a central injector. A large number of radial bores at the top end of the injector kept the momentum of the exiting gas low and prevented a change in the characteristics of the swirling air flow by the exiting gas. The gas injector diameter was r = 15 mm.

Velocity distributions were measured by an LDA system which allowed quasi-simultaneous two-component measurements by rapid Laser polarization change-over. Temperatures were measured by the two thermocouples-method. Gas analyzers were used to determine concentrations. The air ration was $\lambda = 1.1$ in all cases.

Supercritical swirl jets and swirl flames were measured. The intensity of swirl is characterized in this paper by the swirl number S_0

$$S_0 = \frac{\dot{D}}{R \cdot \dot{I}_0} \tag{1}$$

and the predicted swirl number $S_{0,th}$

$$S_{0,th} = \frac{\dot{D}_{th}}{R \cdot \dot{I}_0} \tag{2}$$

where the swirl number S_0 is obtained from the swirl momentum flux $\dot{D}$ calculated for measured values, the mean axial momentum flux at the burner exit $\dot{I}_0$ and the radius R of the burner exit plane. The predicted swirl number is obtained from the predicted swirl momentum flux $\dot{D}_{th}$ derived from swirl generator design /1/. As shown-below, the predicted and the measured swirl numbers are in good agreement.

3. Velocity Measurements

The velocity measurements provided distributions for the mean and the turbulent axial and tangential components and the turbulent shear stress component $\overline{u'w'}$. Transverse distributions in the jets and the flames was mainly measured near the burner in planes x/D = 0.1 and x/D = 1. Longitudinal distributions along the jet axes were also measured.

3.1 Tangential Velocity Distributions and Vortex Type

Fig. 2 plots the RS and TA swirl generator tangential velocity distributions at a distance of x/D = 0.1 from the burner as a function of the intensity of swirl. In the case of the RS swirl generator, the peak velocity moves towards the outside of the jet as the intensity of swirl increases and the linear increase in velocity near the jet axis is lost. For high intensities of swirl, the slope of the velocity distribution near the centerline is positive.

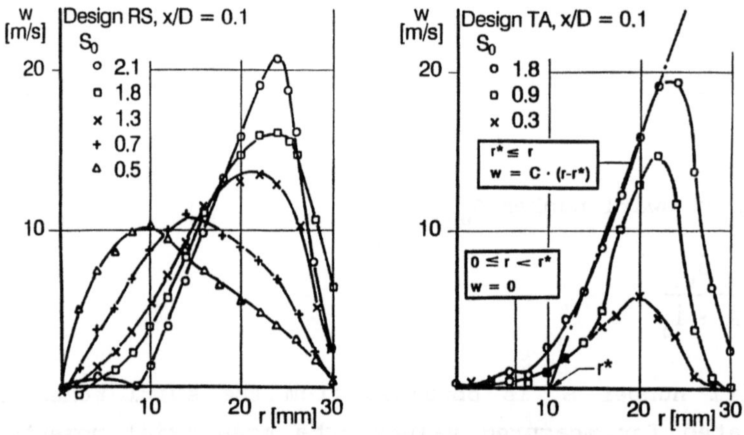

Fig. 2: Mean Tangential Velocity Distribution and Definition of the Modified Solid Body Vortex Approach

In the case of the TA swirl generator, on the other hand, the location of the peak velocity is nearly independent of the intensity of swirl. The correlation is at no swirl intensity linear and the slope is positive even for little swirl.

In the outerflow region, the differences between the tangential velocity distributions for the various swirl generators are not significant. The test showed, thus that the different swirl flows produced by the various swirl generators must be characterized mainly by the tangential velocity component distribution in the inner region of the jet. For all swirl generators and for all intensities of swirl, the change in the tangential velocity from the jet axis to the point of peak velocity may be described by the modified solid body vortex approach depicted in the right-hand side plot in Fig. 2:

$$0 \le r \le r^*: \qquad \bar{w} = 0$$
$$r^* \le r: \qquad \bar{w} = C \cdot (r - r^*) \tag{3}$$

This approach accounts for the positive slope of the profile in the vicinity of the jet axis by the inclusion of the radius r* term. The gradient C and the radius r* can be obtained from the velocity profile by linear regression. Fig. 3 plots C and r* for isothermal swirl jets as a function of the intensity of swirl.

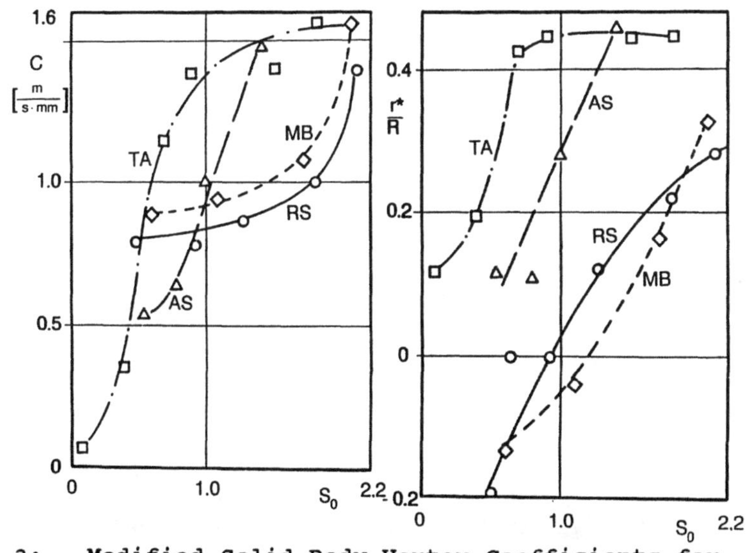

Fig. 3: Modified Solid Body Vortex Coefficients for Isothermal Swirl Jets

Qualitatively, the gradient C terms are the same for the radiant-flow swirl generators (RS and MB). For low intensities of swirl, C is nearly constant and then increases sharply for high swirl intensities. The change in C is nearly the opposite for the TA swirl generator. In the case of the AS design, the increase in C is practically linear. As for radius r*, the value increases for the RS, AS and MB designs, as the intensity of swirl increases. Regression produces negative r* values for low swirl intensities for the RS and MB designs. For high intensities of swirl, radius r* remains constant in the case of the TA design (see Fig. 2). In the case of the AS and TA design, radius r* remains positive even for low intensities of swirl implying that the slope of the tangential velocity distribution in the vicinity of the jet axis is positive for all intensities of swirl.

The vortices produced by the MB and RS swirl generators are very similar because of the common design features of these two generators. Uncoming flow and flow inside the generator is radial and the combustion air flow is then deflected from the horizontal to the vertical direction.

Fig. 4 plots the coefficients of the modified solid body vortex equations (3) for swirl flames. The chemical reaction does not have any influence on the qualitative differences between the swirl generators.

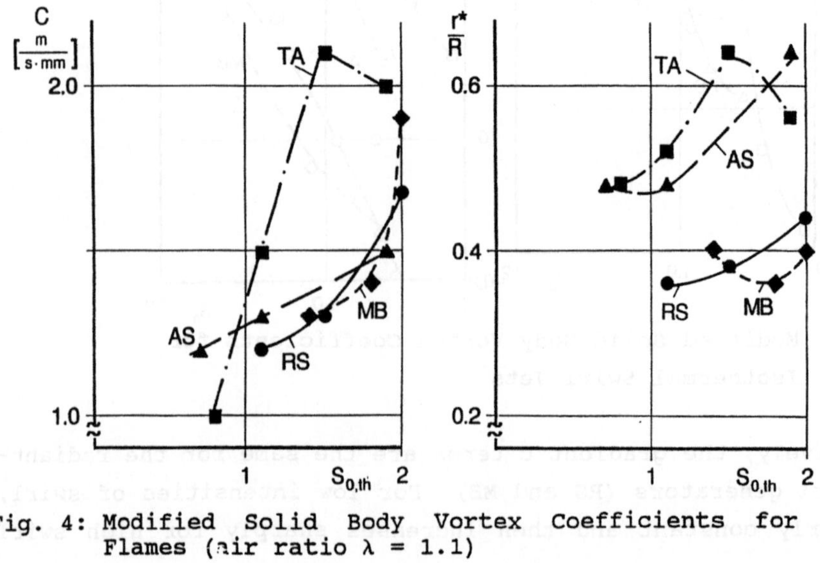

Fig. 4: Modified Solid Body Vortex Coefficients for Swirl Flames (air ratio λ = 1.1)

3.2 Axial Velocity Distribution

Axial velocity distributions provide information on the radial and the axial size of the inner recirculation region. Fig. 5 plots the radius of the recirculation region as a function of the swirl number for the x/D = 1 plane under isothermal conditions. The radius is referred to the burner exit radius. The recirculation region is largest for the radial-flow RS and MB designs. The AS and TA designs are very different from the other two swirl generators. If the curves are extrapolated for lower intensities of swirl, the intersections with the abscissa represent the intensity of swirl at which recirculation commences in this plane. This critical swirl intensity is quite apparently dependent upon the swirl generator design.

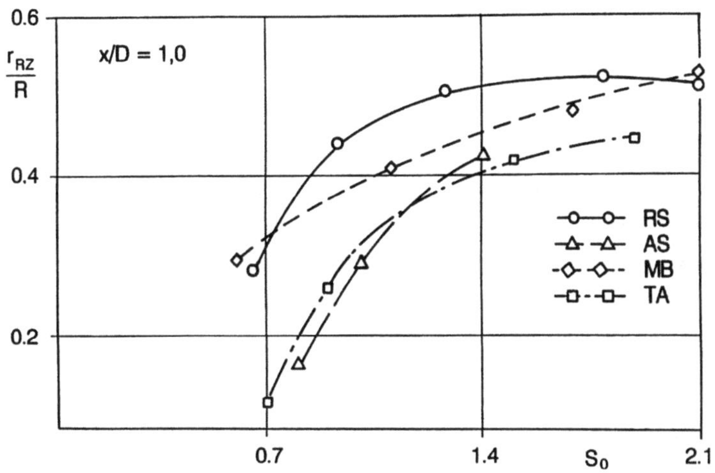

Fig. 5: Radial Size of Recirculation Region for Isothermal Jets

The design of the swirl generator also has an influence on the axial length of the recirculation region, as Fig. 6 demonstrates. The increase in the length of the recirculation region as a function of swirl intensity typical of each swirl generator design is, for isothermal conditions, in good agreement with the increase of the gradient C (see Fig. 3). Combustion reduces the length of the recirculation region, but the qualitative differences between the swirl generators remain the same.

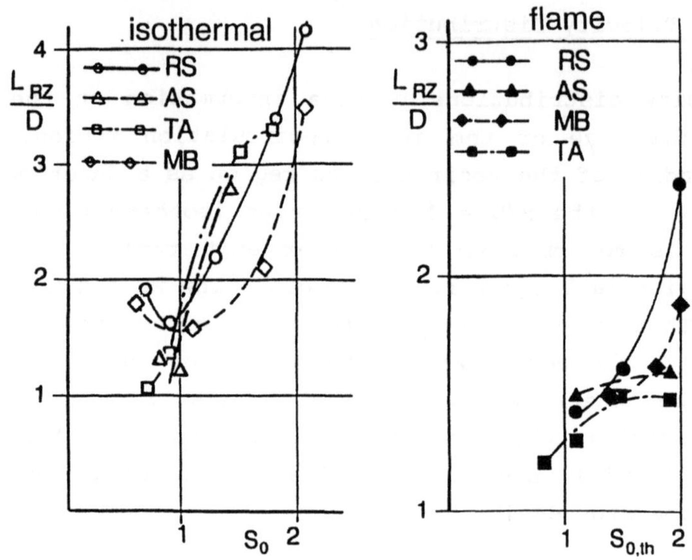

Fig. 6: Axial Length of Recirculation Region for Isothermal Swirl Jets and Swirl Flames

Swirl flow is highly turbulent. The design of the swirl generator also has an influence on the typ of swirl jet and swirl flame turbulence /2/. The correlation between the type of vortex and turbulence can be shown for normal turbulence stress in the tangential direction. Fig. 7 plots the maximum intensity of turbulence of the tangential component in the x/D = 0.1 plane for each swirl generator as a function of swirl intensity.

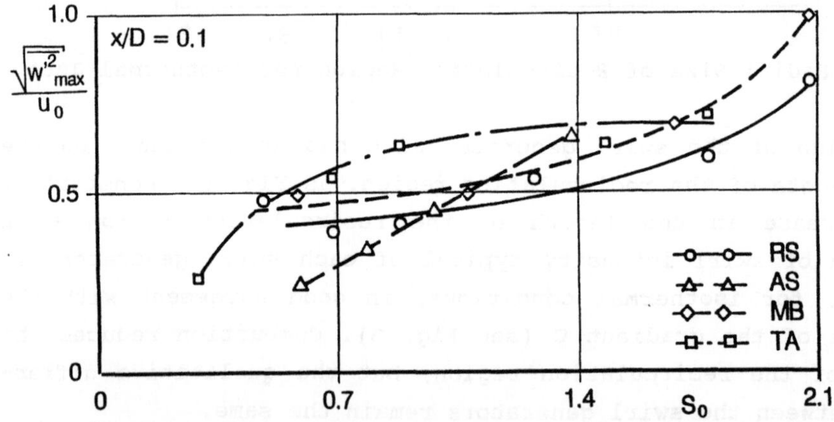

Fig. 7: Maximum Turbulence Intensity of Tangential Component

Maximum turbulence increases as swirl intensity increases, but the rate of increase depends on swirl generator design and coincides with the changes of gradient C in Fig. 3. The increase in maximum turbulence is non-linear for the MB and RS designs and linear for the AS design. Maximum turbulence tends to be constant for high swirl intensities in the case of the TA design.

4. Concentration

Fig. 8 shows the influence of swirl generator design on the concentration field. It plots the maximum methane concentrations

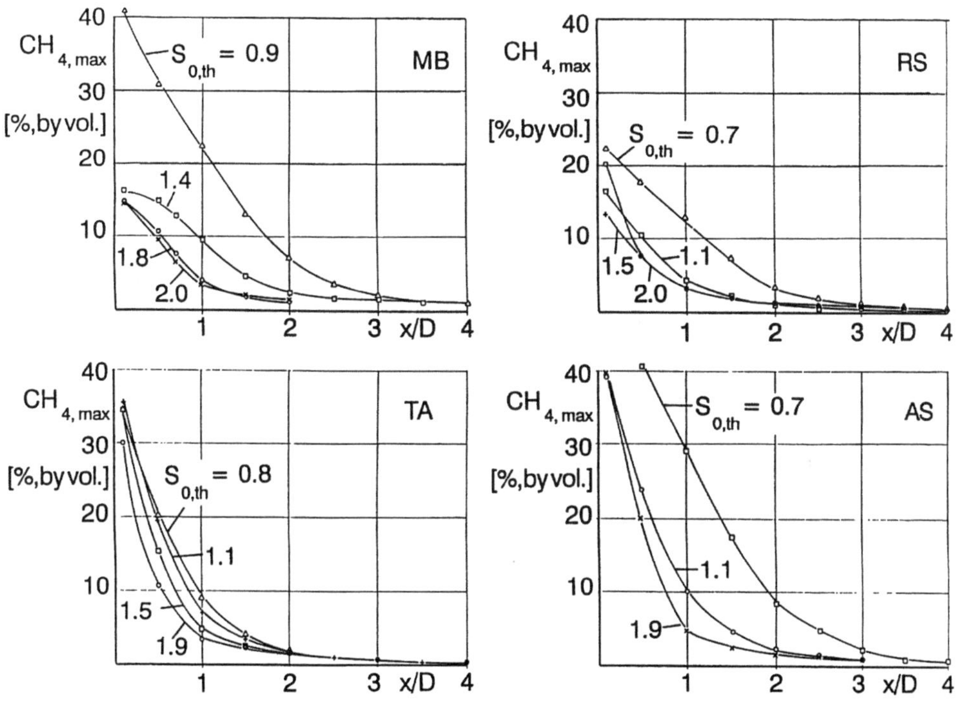

Fig. 8: Decrease in Peak Methane Concentration

measured for different swirl intensities as a function of distance from the burner exit plane. The pattern of the methane concentration decrease is the same for the MB and RS swirl generators. Excepting flame lift-off conditions (predicted swirl number of 0.9 and 0.7), the peak methane concentration is between 15 and 20 % (by volume) in the burner exit plane. The concentrations are about twice as high for the other two swirl generators. Initial mixture is better in the case of the MB and RS designs, because combustion air flow is diverted from the horizontal to the vertical direction /1/. In the case of the TA and AS swirl generators, though, the peak concentrations in the bottom region of the flame decrease more rapidly. At a distance of x/D = 2 from the burner exit, the maximum methane concentrations are about the same for all swirl generators.

5. Calculation of the Swirl Number

The calculation of a swirl number for a swirl generator assumes loss-free swirl momentum flux from the generator to the burner exit. Under these conditions, the swirl momentum flux produced by the swirl generator is equal to that of the swirl flow exiting from the burner. The equations used for calculating the swirl numbers for the swirl generator designs depicted in Fig. 1 are based on formulas given in /3,4/ and discussed in /2/. To validate these equations, the predicted swirl numbers were compared with swirl numbers calculated from velocity values measured by Laser-Doppler-Anemometer /1/ and by a five-hole probe /5/.

The RS Swirl Generator

The predicted swirl number for the RS swirl generator is given by

$$S_{0,th,RS} = \frac{\tan\alpha}{1 - \frac{n \cdot b}{2\pi r_Q}} \cdot \frac{R}{2h} \cdot (1 - \nu^2) \tag{4}$$

Where:

α = blade angle

n = number of blades

h = blade height

b = blade thickness

ν = r/R ratio

r_q = distance from blade trailing edge to burner axis

As Fig. 9 demonstrates, the predicted swirl number is in good agreement with the swirl number calculated from measured values.

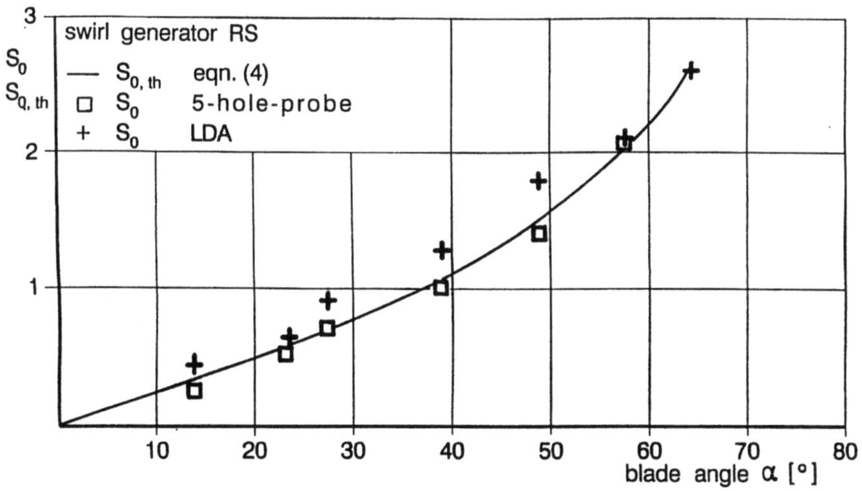

Fig. 9: Measured and Predicted Swirl Numbers for the RS Swirl Generator

The AS Swirl Generator

In the case of the AS swirl generator, the equation for the prediction of the swirl number includes the number of blades n, the blade thickness b, the blade angle α, the hub ratio at the burner mouth and the hub ratio at the blade axis $\nu_M = r_M/R$. The predicted swirl number is given by

$$S_{0,th,AS} = \frac{2}{3}\tan\alpha\,\frac{R\cdot(1-\nu_M^3)}{r_M(1-\nu_M^2)^2}\cdot\frac{(1-\nu^2)}{\left(1-\dfrac{n\cdot b}{\pi r_M(1+\nu_M)}\right)^2} \tag{5}$$

as /5/ does not publish any measured values for this screen, the intercomparison in Fig. 10 is limited to the LDA data.

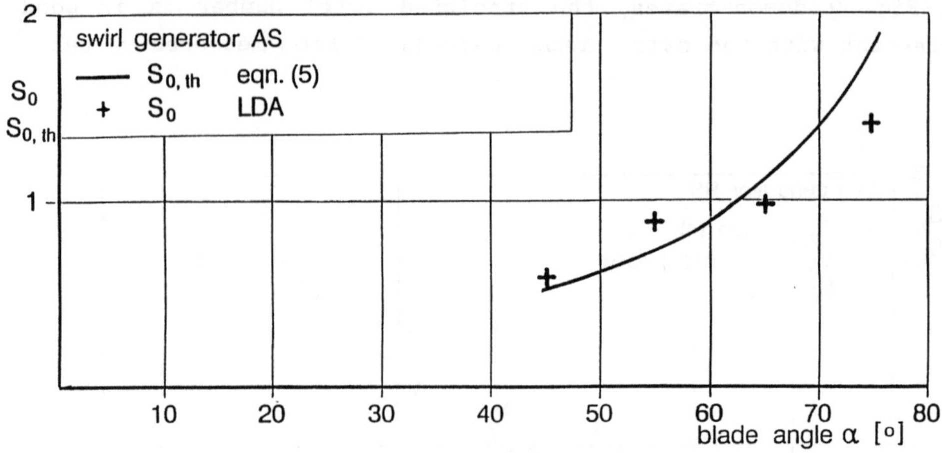

Fig. 10: Measured and Predicted Swirl Numbers for the AS Swirl Generator

The TA Swirl Generator

In the case of the TA swirl generator, the swirl intensity depends on the hub ratio at the burner mouth, the ratio between the tangential combustion air flow and the total combustion air flow, the number of openings for tangential combustion air n, the width d, and the height h of these openings and the eccentricity e of these openings relative to the burner axis. For the TA design swirl generator examined, the loss factor due to flow blending and tangential combustion air diversion is $f_V = 0.85$. The predicted swirl number is given by

$$S_{0,th,TA} = \left(f_V \frac{\dot{M}_{L,tan}}{\dot{M}_{L,tot}} \right)^2 \cdot \frac{\pi \cdot R \cdot e}{n \cdot h \cdot b} \cdot (1-\nu)^2 \qquad (6)$$

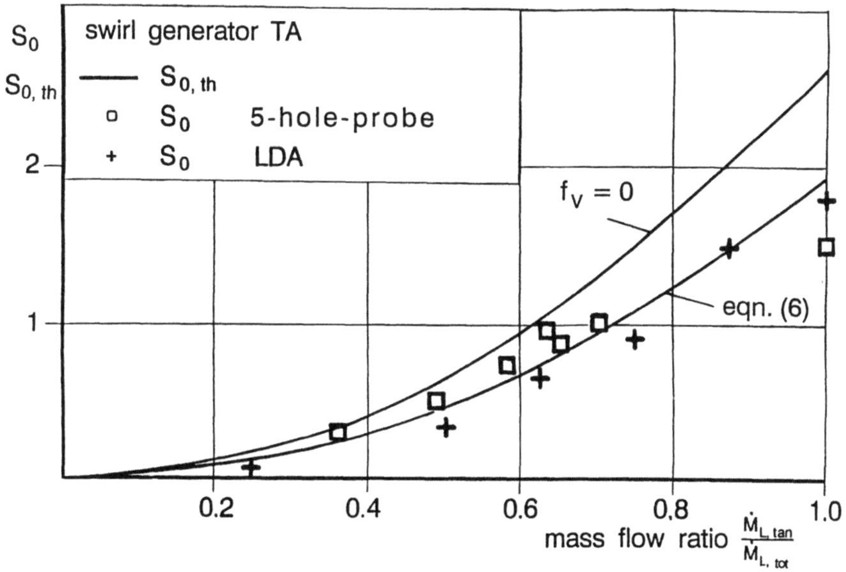

Fig. 11: Measured and Predicted Swirl Numbers for the TA Swirl Generator

The MB Swirl Generator

For the movable block design used in the tests, swirl numbers calculated by published equations were not in good agreement with the measured swirl numbers, because the height of the blocks was too low with respect to the inside diameter /1/. As Fig. 12 demonstrates, a modification of the equation published in /3/ and /4/ achieved good agreement with the swirl numbers obtained from measured data:

$$S_{0,th,MB} = \frac{\pi \cdot R \cdot \sin\alpha}{\xi \cdot n \cdot h} \cdot \frac{\cos\alpha \cdot (\xi/\xi_m) \cdot [1+\tan\alpha \tan(2\xi_m)] \cdot [1 - \nu^2]}{\left[1-(\xi/\xi_m)(1-\cos\alpha[1+\tan\alpha\tan(2\xi_m)])\right]^2} \quad (7)$$

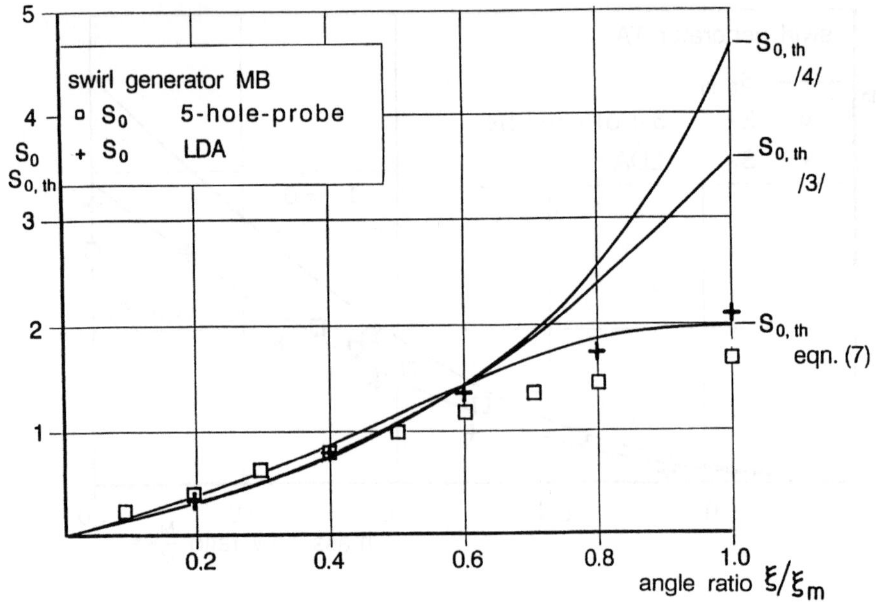

Fig. 12: Measured and Predicted Swirl Numbers for the MB Swirl Generator

Swirl intensity is a function of the block angle α, the hub ratio at the burner mouth ν, the number of block pairs n, the block height h, the angle ξ and the maximum angle ξ_m.

Literature:

/1/ Mundus, B.; Kremer, H.:
Untersuchung des Verbrennungsverlaufs verdrallter Diffusionsflammen in Abhängigkeit von der Art der Drallerzeugung. GWI report No. 8357, 1990

/2/ Mundus, B.:
Über den Einfluß der Drallerzeugerkonstruktion auf das Strömungs- und Reaktionsfeld turbulenter Diffusionsflammen. Thesis, Bochum University, 1990

/3/ Beer, J.M.; Chigier, N.A.:
Combustion Aerodynamics. Applied Science Publishers Ltd, 1972

/4/ Leuckel, W.:
Swirl intensities, swirl types and energy losses of different swirl generating devices. IFRF, Ijmuiden, Nov. 1967, G 02/a/16

/5/ Mundus, B.; Kremer, H.:
Untersuchung der strömungstechnischen Eigenschaften unterschiedlicher Bauformen von Drallerzeugern.
GWI report No. 7734, 1988

Flame Diagnostics by Pure Rotational CARS

A. Leipertz, E. Magens*, T. Seeger and H. Spiegel

Lehrstuhl für Technische Thermodynamik (LTT-Erlangen)
Friedrich-Alexander-Universität Erlangen-Nürnberg
Am Weichselgarten 9
D-8520 Erlangen
Federal Republic of Germany

* now with DLR, Institut für Antriebstechnik, D-5000 Köln 90, (FRG)

Summary

Pure rotational coherent anti-Stokes Raman scattering (CARS) has been applied successfully for flame diagnostics in simple premixed methane-air flames. Temperature determination is demonstrated as well in the high temperature region where only pure nitrogen was available as in the mixing zones of the flame with surrounding air, where also oxygen was present complicating the evaluation procedure. Here, also information on the concentration of both gas species was deducable. The influence of high pressure on the evaluation procedures is discussed in comparison to vibrational CARS. First results for the application of the technique to more complex nonlinear gas molecules are presented.

Introduction

Coherent anti-Stokes Raman scattering (CARS) has been used successfully since several years for the determination of temperature and concentration in combustion processes. For this, mostly the rotational-vibrational energy states of the molecules have been used for probing the population difference between these levels in order to determine the temperatures and number densities (vibrational CARS) [1-4]. Alternatively, the pure rotational energy levels within the vibrational ground state of the molecules can be used for the same purpose (pure rotational CARS). In particular combustion situations, e.g. high pressure applications, rotational CARS may provide some advantages over rotational CARS [5,6].

Figure 1 shows schematically the spectral appearance and some characteristics of both CARS techniques [5]. Some of the advantages and disadvantages of both techniques can already be deduced from these characteristics. The CARS frequency is in general identical to the difference frequency of two different irradiated laser frequencies (pump laser and Stokes laser). Thus, in general just a single gas species can be investigated by using vibrational CARS as the radiation of the Stokes dye laser is spectroscopically not broad enough to cover all the spectral range of interest for detecting several different molecules simultaneously. Recent approaches to overcome these limitations are given in Refs. [7-9].

R. S. L. Lee · J. H. Whitelaw · T. S. Wung (Eds.)
Aerothermodynamics in Combustors
IUTAM Symposium Taipei, Taiwan, 1991

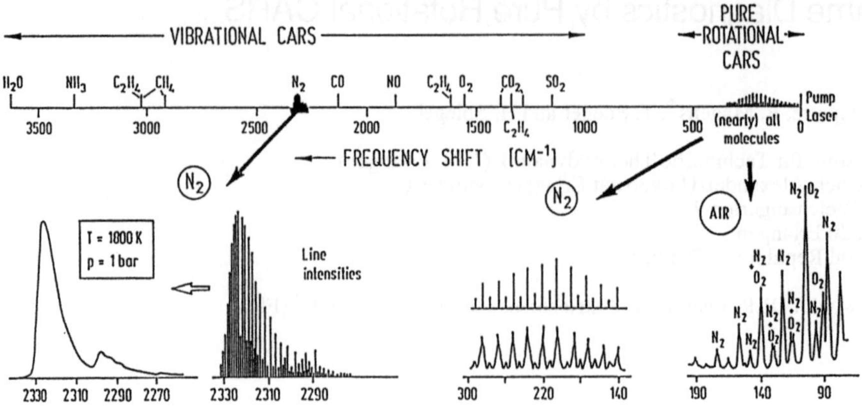

Fig. 1. Schematic of vibrational and pure rotational CARS spectra.

This is different for pure rotational CARS. Except for spherical top molecules (e.g., CH_4), every gas molecule gives contributions in the same spectral range of a few hundred wavenumbers. Thus, all present molecules can be investigated simultaneously using the same Stokes dye laser. Problems may arise from line overlap between the contributions of different molecules (see e.g., the air spectrum in Fig. 1) which in general is not the case for vibrational CARS. This problem is considered in detail elsewhere [10] and will not be treated here.

Temperature information can be obtained from the vibrational CARS spectra by a weighted fit of the theoretical and experimental spectra [2-4]. As the shape of the spectra is governed by the convolution of the single line intensities with the spectral resolution of the system, all quantities influencing this shape have to be considered very carefully: e.g., laser bandwidths [11] with cross-coherence effects [12,13], collisional narrowing at high pressures [14-16] and the scaling of the nonresonant CARS contributions [17,18]. Besides that, difficulties may occur from Stark shifts [19,20] and saturation effects [21,22] when using too high laser powers, leading in summary alltogether to inaccuracies and noise problems. The evaluation procedure is time consuming. In complicated cases a single fit may take up to a few hours. Quick-fitting techniques can sometimes be used to fasten the evaulation process [23].

This situation is somewhat less complicated when using the pure rotational CARS spectrum. Due to an improved line separation (see, e.g., N_2 in Fig. 1) a simplified modeling of the spectra is possible with faster evaluation procedures [24,25]. The Raman linewidths are more narrow and the cross sections larger than those of vibrational CARS resulting in larger signal intensities. Thus, pure rotational CARS gives extrem high accuracies for $T < 1000$ K (see, e.g., Ref. [24] for $T < 300$ K), suffers however - although possible [26] - at flame temperatures from low signal intensities. Problems with line overlap from different molecules when using the conventional line integration technique (LIT) [10] may be overcome by using the Fourier analysis technique (FAT) [25] which has been used here for signal analysis.

In comparison to vibrational CARS only few experimental results using pure rotational CARS are published so far. This may be due to problems with the suppression of the strong irratiated laser beams which occur in the spectral neighbourhood of the CARS beam and with the poor dye efficiency when using the third harmonic of the Nd:YAG pump laser for simultaneously pumping the dye laser. First pure rotational CARS measurements have been performed by Barrett [27] in hydrogen with large Raman shifts and by Beattie et al. [28] in air. In flames, first spectra have been recorded by Shirley et al. [29]. First temperature measurements using pure rotational CARS have been published by Murphy and Chang near room temperature [24] and by Fleming et al. [30] at higher temperatures. Its first application to flame diagnostics was reported by Zheng et al. [26]. During the last years several temperature measurements have been reported, e.g., Refs. [31,32], and different new approaches were introduced in order to improve the efficiency of CARS generation and the accuracy of the measurements, e.g., by using the dual broadband approach [33,34] and an new evaluation technique based on Fourier analysis [10,25].

In this paper first results are presented using this technique for the evaluation of experimentally obtained rotational CARS spectra in high temperature regions of premixed methane-air flames where only nitrogen is available for the evaluation process, and for concentration and temperature determination in the mixing region of partially premixed methane-air flames with surrounding air where besides nitrogen also oxygen can be used. The advantage of pure rotational CARS over vibrational CARS in high pressure situations is considered theoretically and first results are given for the investigation of more complex nonlinear gas molecules which may appear in real technical combustion.

Application of the Fourier analysis technique

Basic information on pure rotational CARS is given elsewhere (see e.g., Refs. [10,32]) and will not be treated here. For the evaluation procedure two different approaches may be used. The conventional line integration technique (LIT) has been introduced by Murphy [24,35]. Here, the total integrated peak areas of the individual rotational lines are used in order to find the temperature value from a least square fit between the experimental data and the calculated spectral intensity distribution. This approach does not exploit the specific periodic structure of the rotational spectra. Using Fourier analysis for the evaluation of the experimentally obtained spectra, the periodicity of the rotational CARS spectra offers advantageous possibilities for gas temperature and concentration evaluation over LIT, which in particular is true in gas mixtures [10,25].

Additionally a data reduction procedure is possible when using FAT. While for the purpose of recording a typical CARS spectrum several thousands datapoints are necessary for the description, there are only a few Fourier coefficients necessary (as much as rotational transition lines are observed) which have to be stored without loss of any information. Furthermore, not all of these coefficients must be considered for evaluating the temperature. Only a small subset of a few coefficients are necessary and this subset is reduced with increasing temperature. In this way a faster temperature evaluation is possible. This can be done by directly using the Fourier space (real or imaginary parts of the Fourier coefficients) or by reconstructing the envelope of the rotational CARS spectrum from the low frequency Fourier coefficients by an inverse Fourier transformation. This envelope yields the temperature information in the position of its maximum peak and in its FWHM-value. A very fast temperature evalation is possible by separating these low frequency Fourier coefficients experimentally with a boxcar filter [10].

Experimental spectra normally are noisy due to laser and detector noise and background, e.g., combustion luminosity. In general, this noise is frequency dependent, perturbing only some of the Fourier coefficients. By selecting some undisturbed coefficients the envelope may still be reconstructed. This finally results in an additional S/N enhancement.

In order to proof the applicability of FAT to rotational CARS spectra, first tests have been performed with simulated spectra. Figure 2 shows the calculated rotational CARS spectrum as it may occur in real combustion systems, assuming for the calculation a gas mixture of 60% N_2, 6% O_2, 6% CO_2, 5% CO, 3% H_2 and 20% H_2O and a gas temperature of 1500 K. Additional random noise was taken into account as it appears in optical measurement techniques as shot noise, detector noise etc., following Poisson statistics. The noise level was in the same order of magnitude as the signal contributions of the smaller concentrations. An independent evaluation was done for nitrogen, oxygen and carbon dioxide. Table 1 shows the results for temperature and concentration for 50 such randomly calculated spectra indicating very good agreement between the input data and the calculated temperature and concentration values.

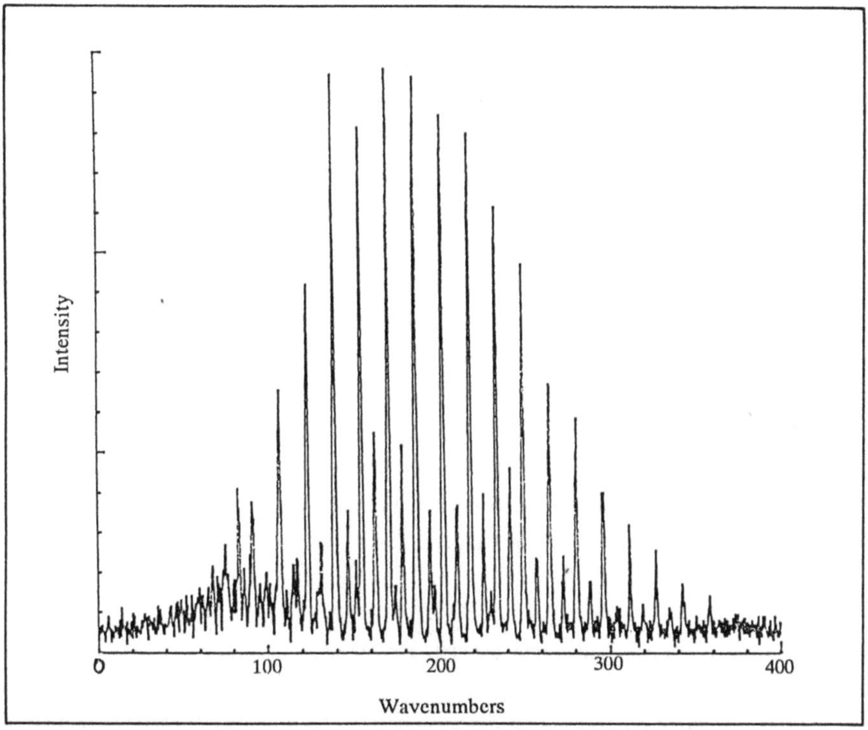

Fig. 2. Calculated pure rotational CARS spectrum for a gas mixture of 60% N_2, 6% O_2, 6% CO_2, 5% CO, 3% H_2 and 20% H_2O. Gas temperature: 1500 K. Additional contributions by random noise (Poisson statistics).

Gas	$T \pm \Delta T$ [K]	$c \pm \Delta c$ [%]
N_2	1508 ± 25	60.00 ± 0.60
O_2	1507 ± 190	6.11 ± 1.01
CO_2	1492 ± 110	5.95 ± 0.46

Table 1. Results from 50 randomly generated simulated pure rotational CARS spectra for a flame gas mixture of 60% N_2, 6% O_2, 6% CO_2, 5% CO, 3% H_2, 20% H_2O and T = 1500 K. Evaluation has been done independently for the N_2, the O_2 and the CO_2 molecule.

Results in simple flames

The FAT technique has been used for the evaluation of conventionally obtained experimental rotational CARS spectra. Details on the experimental setup and on the evaluation technique used cannot be given here. The experimental setup used was similar to that described in Ref. [26]. The evaluation technique is based on details given in Refs. [10,25] which is extended in some aspects [36]. Experimental spectra have been taken in the post flame region of a premixed methane-air flame generated by a flat flame burner of 50 millimeters diameter. The burner was driven in two different ways. In order to generate a well established high temperature zone downstream the flame front the burner gas was premixed to near stoichiometric conditions. In a second different approach the air content to the fuel gas was decreased to a more diffusion-like behaviour generating a fluctuating flame body.

Figure 3 shows an example of the results obtained in the high temperature region of the premixed methane-air flame. It shows the comparison of the experimentally obtained pure rotational CARS spectra for two different temperatures in the upper traces with the calculated ones in the lower traces.

Here only nitrogen has been taken into account. The agreement is quite well. Several of those measurements have been used for temperature evaluation for a particular combustion situation. For three different burning conditions as varied by the flow rate of methane and air the results are given in Table 2. The rotational CARS measurements has been compared with ra-

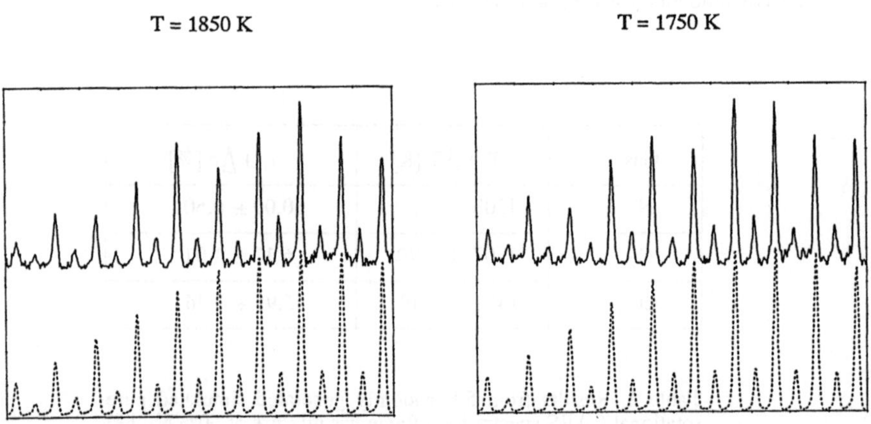

Fig. 3. For two different gas temperatures comparison of experimentally obtained pure rotational CARS spectra taken in the high temperature region of a premixed methane-air flame (upper traces) with calculated spectra (lower traces) for pure nitrogen.

Air ratio	Height over Burner	$T_{equ.}$ [K]	Thermo-couple [K]	CARS FAT [K]	CARS LIT [K]
0.94	0.7 d	2228	1780	1811 ± 75	1743 ± 132
1.17	0.7 d	2074	1865	1832 ± 35	1768 ± 119
1.17	0.1 d	2074	1970	1969 ± 110	1951 ± 181

Table 2. Rotational CARS gas temperature results obtained in the high temperature region of premixed methane-air flames in comparison with calculated equilibrium values and radiation corrected thermocouple readings. For both evaluation techniques (FAT, LIT) the results displayed are mean and rms values from sets of 10-30 single temperature measurements.

diation corrected thermocouple readings. Each of the CARS results indicated are given as mean and rms values from sets of 10 - 30 single temperature measurements. The experimental CARS spectra were evaluated by both techniques (LIT, FAT). The FAT technique gives results which are closer to the thermocouple readings providing also smaller standard deviations. The calculated equilibrium temperature values indicated differ clearly from the thermocouple readings and from the CARS results as well. This may be due to different input parameters for the calculation and for the experiment, taking not into account in the calculation the high liquid water content of the fuel gas. This, however, will result in lower temperatures for the measurements.

The situation is much more complicated in the mixing region of a diffusion-like flame body with surrounding air. Figure 4 shows the comparison between experimental and calculated rotational CARS spectra in those mixing regions. These spectra were taken and calculated for two different burning conditions. The upper part show spectra related to nearly preheated air, the lower part spectra taken in highly fluctuating regions further downstream. The indicated temperature values are not corrected which seems to be necessary due to the different temperatures which appear during the measurement period in the same gas sample when accumulating the spectra and due to "different" temperatures for the oxygen and the nitrogen molecule (oxygen being closer related to the combustion process). This results in a shift of the shape form and of the maximum position of the rotational CARS spectra. Additionally the different concentration values occuring simultaneously at the same measurement position have to be taken into account. The procedure in treating those problems will be discussed in much detail in a forthcoming paper [37]. Table 3 gives the very first results we have obtained in the mixing zone for three different situations which are related to preheated air and to fluctuation

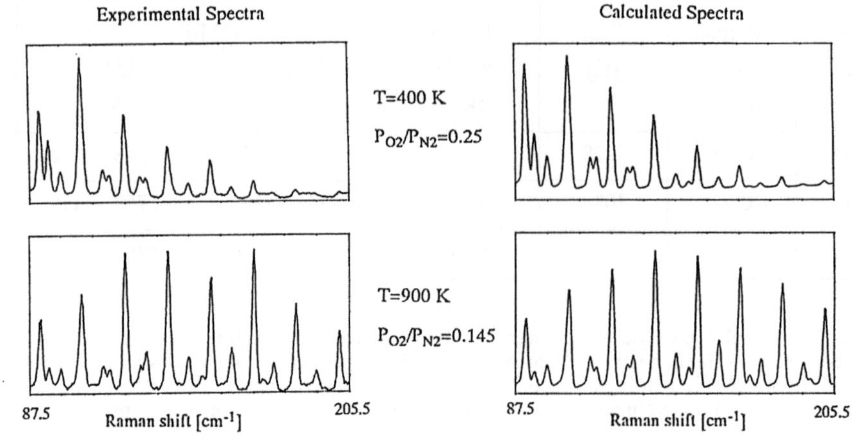

Fig. 4. Comparison of experimentally obtained pure rotational CARS spectra which have been taken in the mixing region of a partially premixed flame with surrounding air for two different flame conditions (left) with calculated ones (right). In the calculation procedure, temperature and ratio of the partial pressures of O_2 and N_2 have to be adjusted simultaneously. Indicated temperatures are not corrected. Corrected data are given in Table 3 for both the temperature and the pressure ratio values.

CARS temperature T/[K]	371 ± 8	448 ± 22	820 ± 70
CARS pressure ratio P_{O2}/P_{N2}	0.24 ± 0.02	0.23 ± 0.02	0.14 ± 0.02

Table 3. Temperature and O_2/N_2 partial pressure ratio results obtained from pure rotational CARS spectra which were taken in the mixing zones of a fluctuating partially premixed methane-air flame with surrounding air. The shown values are mean and rms values from sets of 10-30 single measurements.

regions. Both temperature and concentration values are calculated again in form of mean and rms values for sets of 10 - 30 measurements. There are no reference measurements available for the indicated ratio of the partial pressure of oxygen and nitrogen. Thermocouple readings were close to the measured values given in the table for the lowest and highest temperatures.

The results so far obtained give a clear manifestation for the usability of pure rotational CARS for flame diagnostics in high temperature regions and for the mixing region as well. Further tests are going to be performed in well defined combustion situations.

High pressure considerations

We are now starting to investigate the influence of high pressure on the evaluation procedures of pure rotational CARS in comparison to vibrational CARS. The expected advantages of pure rotational CARS over vibrational CARS in high pressure situations will be considered theoretically as well as experimentally. Line separation in rotational CARS (8 cm^{-1} for N_2) is larger compared with the line separation in vibrational CARS (0.4 cm^{-1} for N_2). That means that the line narrowing effects, which are well known in vibrational CARS, are not so dramatically influencing the spectral shape of rotational CARS, even at higher pressures. Figure 5 shows experimental vibrational CARS spectra of N_2 for high pressure conditions and the effect of line-narrowing which appears already at pressures of a few bar [38]. In contrast to that the calculated spectra in Figure 6 indicate that this effect is less severe in pure rotational CARS due to the larger line separations. The lines are only individually broadened due to the large pressure [39]. A similar indication can be found for experimental spectra [6].

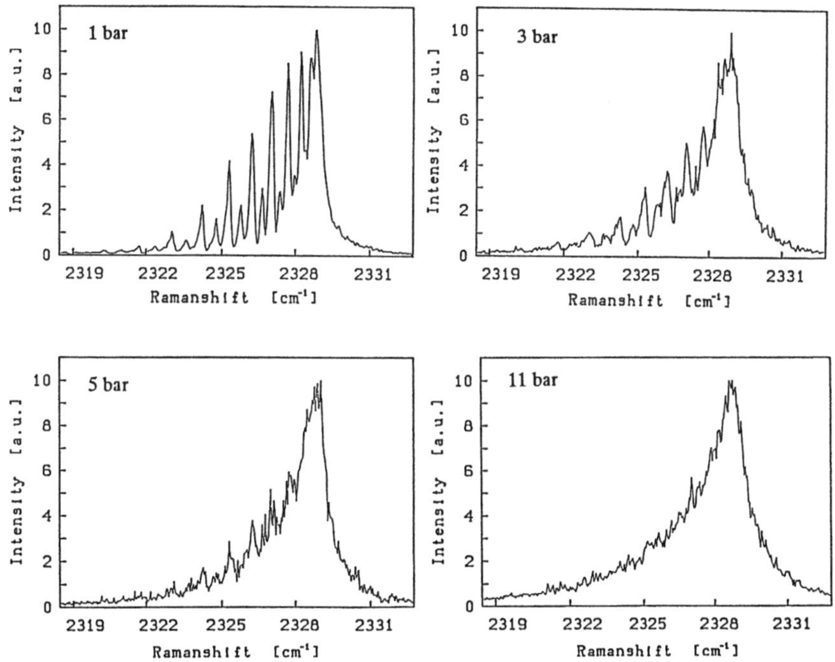

Fig. 5. Scanning vibrational CARS spectra of N_2:
Pressure dependance (line-narrowing effect) [38].

One of our future steps is the experimental test of various rotational relaxation models having been developed for describing the line-narrowing effects in vibrational CARS (MEG, XMEG, DEG, PEG) [3,40]. Furthermore, we are interested in more detail in the influence of high pressure on the accuracy of temperature and concentration measurements by pure rotational CARS evaluated with both LIT and FAT.

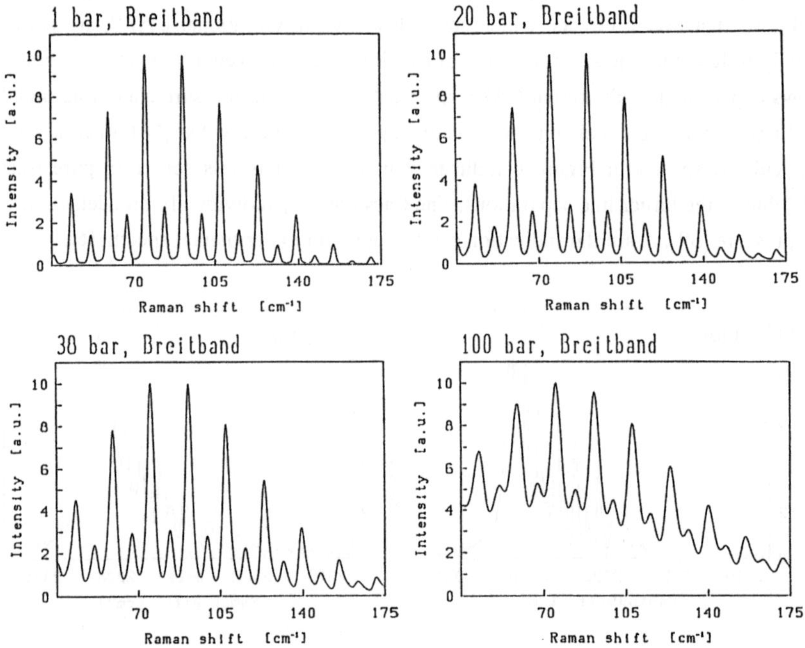

Fig. 6. Pure rotational CARS spectra calculated for different pressures ($T = 300$ K, FWHM = 4.8 cm^{-1}) [39].

Nonlinear gas molecules

More complex nonlinear gas molecules may appear in real technical combustion situations. The calculation of such spectra is a very difficult task and has, to the best of our knowledge, so far not been done successfully in general for all different types of gas molecules. We developed a computer code [36] which has to be tested by experimental spectra in different aspects. First results are presented here. As an example, Figure 7 shows the measured and calculated pure rotational CARS spectra of C_2H_4 at room temperature. With the complexity of a gas molecule, not only the number of transitions and therefore the number of lines is growing (partially due to the cancelation of degeneracy of the energy states), but also the line spacing between individual lines becomes more narrow. Because of that the line narrowing effects at high pressure are expected to play a more important role even also for pure rotational CARS when taking the spectra of complex nonlinear gas molecules.

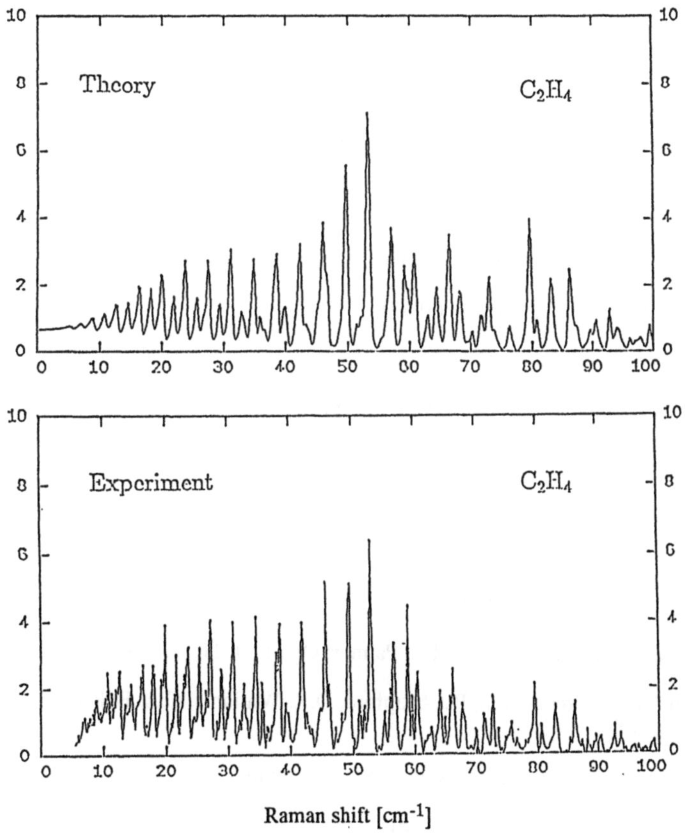

Fig. 7. Calculated and measured pure rotational CARS spectra of C_2H_4 at room temperature.

Acknowledgement

The authors gratefully acknowledge financial support for the work by the Volkswagen-Stiftung and by the Deutsche Forschungsgemeinschaft.

References

1. P.R. Regnier, J.P.E. Taran: Appl. Phys. Lett. **23**, 240 (1973)

2. R.J. Hall, A.C. Eckbreth: "CARS: Application to Combustion Diagnostics", in "Laser Application", Vol 5 (Eds. J.F. Ready, R.K. Erf), Academic Press, New York 1984, p. 214

3. D.A. Greenhalgh: "Quantitative CARS spectroscopy", in "Advances in Non-Linear Spectroscopy", (Eds. R.J.H. Clark, R.E. Hester), Wiley, New York 1988, p. 193

4. T. Parameswaran, D.R. Snelling: Nat. Res. Council Canada, Report No. TR-GD-013 (1988)

5. A. Leipertz: "Raman processes and their applications", in "Instrumentation for Combustion and Flow in Engines", (Eds. D.F.G. Durao, J.M. Whitelaw, P.O. Witze), Kluwer, Dordrecht 1989, p.107

6. S. Kröll, P.-E. Bengtsson, M. Alden, D. Nilsson: Appl. Phys. **B 51**, 25 (1990)

7. A.C. Eckbreth, T.J. Anderson: Appl. Opt. **24**, 2731 (1986)

8. A.C. Eckbreth: SPIE Vol. **742**, 34 (1987)

9. R.P. Lucht: Opt. Lett. **12**, 78 (1987)

10. A. Leipertz, E. Magen, T. Lasser: AIAA-Preprint No. 85-1569 (1985)

11. M.A. Yuratich: Mol. Phys. **38**, 625 (1979)

12. R.E. Teets: Opt. Lett. **9**, 226 (1984)

13. D.A. Greenhalgh, R.J. Hall: Opt. Commun. **57**, 125 (1986)

14. R.J. Hall: Opt. Eng. **22**, 322 (1983)

15. M.L.Koszykowski, R.L. Farrow, R.E. Palmer: Opt. Lett. **10**, 478 (1985)

16. F.M. Porter, D.A. Greenhalgh, P.J. Stopford, D.R. Williams, C.A. Baker: Appl. Phys. **B 51**, 31 (1990)

17. G.J. Rosasco, W.S. Hurst: Phys. Rev. **A 32**, 281 (1985)

18. W. Kreutner, W. Meier, T. Plath, W. Stricker, M. Woyde: Laser und Optoelektronik **22**, 67 (1990)

19. R.L. Farrow, L.A. Rahn: Phys. Rev. Lett. **48**, 395 (1982)

20. R.L. Farrow, R.P. Lucht, G.L. Clark, R.E. Palmer: Appl. Opt. **24**, 2241 (1985)

21. H. Weil, P.W. Schreiber: Appl. Opt. **21**, 941 (1982)

22. A. Gierulski, M. Noda, T. Yamamoto, G. Marowsky, A. Slenczka: Opt. Lett. **12**, 608 (1987)

23. A.C. Eckbreth, G.M. Dobbs, J.H. Stufflebeam, P.A. Tellex: Appl. Opt. **23**, 1328 (1984)

24. D.V. Murphy, R.K. Chang: Opt. Lett. **6**, 233 (1981)

25. T. Lasser, E. Magens, A. Leipertz: Opt. Lett. **10**, 535 (1985)

26. J.B. Zheng, J.B. Snow, D.V. Murphy, A. Leipertz, R.K. Chang, R.L. Farrow: Opt. Lett. **9**, 341 (1984)

27. J.J. Barrett: Appl. Phys. Lett. **29**, 722 (1976)

28. I.R. Beattie, T.R. Gilson, D.A. Greenhalgh: Nature **276**, 378 (1978)

29. J.A. Shirley, R.J. Hall, J.F. Verdieck, A.C. Eckbreth: AIAA-Preprint No. 80-1542 (1980)

30. J.W. Fleming, A.B. Harvey, W.T. Barnes: "Pure rotational CARS thermometry", in "Temperature: It's Measurement and Control in Science and Industry", Vol. V, (Ed. J.F. Schooley), American Inst. Physics, New York, 1982, p. 589

31. B. Dick, A. Gierulski: Appl. Phys. **B 40**, 1 (1986)

32. M. Alden, P.-E. Bengtsson, H. Edner, S. Kröll, D. Nilsson: Appl. Opt. **28**, 3206 (1989)

33. A.C. Eckbreth, T.J. Anderson: Opt. Lett. **11**, 496 (1986)

34. M. Alden, T.J. Anderson, H. Edner: Appl. Opt. **25**, 4493 (1986)

35. D.V. Murphy: Ph. D. Dissertation, YALE University 1981

36. E. Magens: Doctoral Thesis (in preparation)

37. A. Leipertz, E. Magens: paper in preparation

38. J. Woellert: Diploma Thesis, LTT-Erlangen 1991 (Report No. LTT-DA 91-1)

39. B. Hirschmann: Diploma Thesis, LTT-Erlangen 1991 (Report No. LTT-DA 91-2)

40. F.M. Porter, D.A. Greenhalgh, P.J. Stepford, D.R. Williams, C.A. Baker: Appl. Phys. **B 51**, 31 (1990)

Thermometry Inside a Swirling Turbulent Flame, CARS Advantages and Limitations

CLAUDE CAHEN, PASCALE BELLAICHE, DANIELE GARRETON*

Direction des Etudes et Recherches - Electricité de France
REME - 25 allée privée - 93206 Saint Denis, France.
*LNH - 6 quai Watier - 78400 Chatou, France.

1. Introduction

The objective of this research is to study the detailed structure of turbulent diffusion flames. The interaction between the aerodynamical, thermal and chemical effects render their study very complex. In particular, high spatial and temporal resolution measurement methods are needed to increase the understanding of the microscopic processes which are believed to drive the macroscopic behaviour of these reacting turbulent flows.

Most practical combustion systems fall in the category of turbulent diffusion flames. In these flames, the mixing of the fuel and the oxydizer is slow compared to the reaction rates, such that the mixing controls the burning rate. A particular configuration of these combustion systems is the swirl stabilized diffusion flame. This configuration is being largely used to reduce the dimension of the flame and to enlarge the reaction zones but is seldom investigated due to its complexity.

In order to keep as close as possible to the industrial applications and, though it is rather challenging, we have selected a 20 kW swirler where the fuel was axially introduced and the oxydizer tangentially. The two flows were substantially separated to get a better understanding of the mixing processes. The swirling air created a central depression which, in turn, produced a recirculation zone fueling the burning zone with fresh air and maintaining the flame hung at the burner nozzle.

Due to its high spatial and temporal resolutions Coherent Anti-Stokes Raman Scattering (CARS) spectroscopy has been selected as the method of choice for temperature field measurements.

Hereby, we present the main advantages and the encountered limitations while using this technique. The temporal and spatial resolutions are investigated. The subsequent trade-offs are examined. Some measurement limitations are emphasized and shown to provide an undirect method for other flow parameter measurements. The confrontation of CARS results with thermocouple and velocity measurements are also discussed. Finally, a

R. S. L.Lee · J. H. Whitelaw · T. S. Wung (Eds.)
Aerothermodynamics in Combustors
IUTAM Symposium Taipei, Taiwan, 1991

methodology for the comparison of the experimental and numerical results is presented. The consequences seem to validate the numerical model which is being developed as an engineering design tools.

2. CARS temperature measurements

2.1. temperature measurements

To restore the temperature field, two techniques were being used: thermocouple measurements for total monitoring and Coherent Anti-Stokes Raman Scattering (CARS) for local and instantaneous investigations. Despite their intrusive features, the thermocouple measurements allowed an easy total exploration of the flame. In figures 1 and 2, the burner geometry and a temperature cartography are shown, the thermocouple temperature measurements have been corrected for conduction and convection effects while the radiation effects have been estimated to be negligible. The temperature field was symetric and confirmed the correct design of the burner. However, since the thermocouple time response was flow dependent, both the turbulence and the recirculation zone precluded the reliability of the instantaneous measurements. In contrario, the non-intrusive, instantaneous CARS measurements could be used.

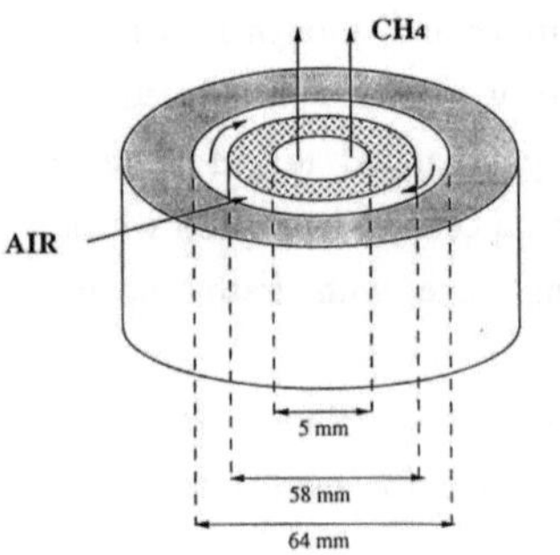

Fig. 1. The burner geometry

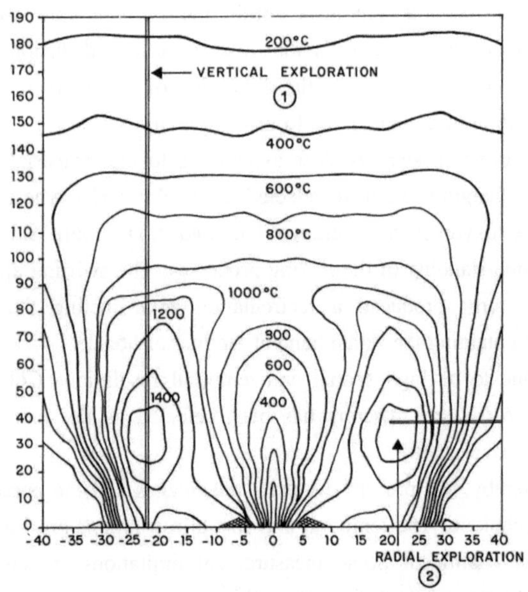

Fig.2. Temperature cartogaphy

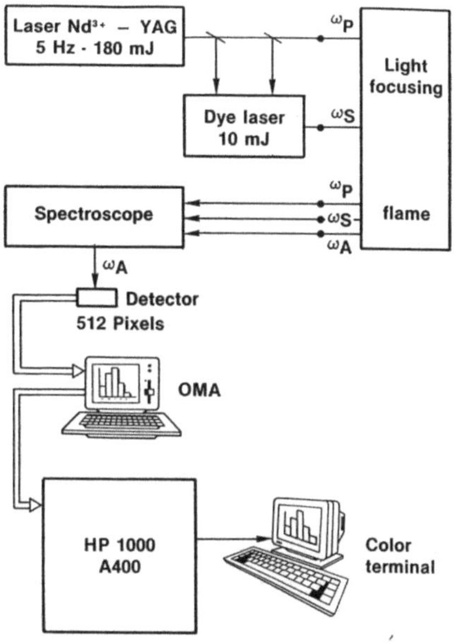

Fig. 3. The experimental set-up

2.1. CARS basic principles

CARS spectroscopy is now one of the most popular techniques for the temperature field measurements and basic principles are known though its limitations are seldom emphasized. CARS is a four wave mixing process which occurs when two laser beams whose frequency difference is tuned to match exactly a molecular vibration frequency, are focused inside the probe volume. In practice, one laser frequency is fixed while the other is tuned to the probe molecule resonance. Moreover, if the tunable laser is operating in a broadband emission configuration, the beat frequencies could overlap the different ground state levels of the molecule and the coherently driven dipole will radiate a third field which is a laser like signal whose spectrum is a weakly distorted anti-Stokes Raman spectrum of the molecule under study. This distortion makes the synthesis of the spectrum more complicated but still analytically amenable. Unfortunately, these distortions can be increased by high turbulence and spatial inhomogeneity of the species concentration like in our experiment. Therefore, the CARS data reduction had to be carefully processed. This is one of the basic reasons why CARS spectroscopy is still seldom used in high turbulence flows despite its intrinsic high temporal (pulsed lasers) and spatial (phase matching) resolutions.

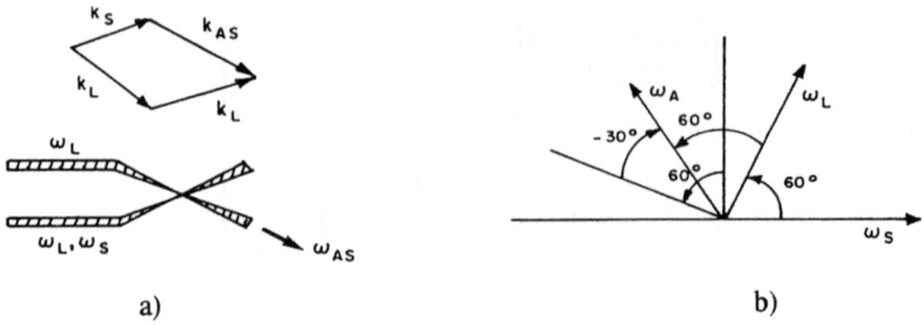

Fig.4. BOXCARS a) and cross polarization b) configurations

2.3. experimental set-up

Our CARS system was a commercially available bench upgraded for narrowband/broadband configurations. The laser source consisted of the frequency doubled output of a Nd^{3+}-YAG laser, TEM_{00} single mode, split into two beams. One was directly focused into the flame while the other one pumped the Stokes dye laser, in a broadband configuration (150 cm^{-1} emission width). The dye laser emitted wavelength was tuned onto a vibrational Raman resonance of the nitrogen N_2 molecule. The spatial resolution was theoretically estimated to be 0.5 cm. A further increase of the spatial resolution was limited by the possible gas breakdown if the focal lengths were decreased. The performance of the laser system is summarized in table I.

	Energy	wavelength	pulse rep. rate
Pump laser	100 mJ	532 nm	5 Hz
Stokes laser	10 mJ	607 nm	5 Hz

Table I : Lasers performance

Actually, to ensure a good overlap between the N_2 Q-branch and the broadband emission, a Fabry-Perot etalon was inserted into the laser oscillator. The signal was sent into a grating spectroscope (2100 g/mm) and recorded with a 512 pixel diode array controlled by an optical multichannel console where the multiplex spectra were stored prior to their ultimate transfer to the microcomputer (Fig.3). The ideal measurement configuration for the weakest distortions [1] is the so-called BOX(USED)CARS and cross polarisation configuration (Fig. 4). This increases both the spatial resolution and the characteristic features of the recorded spectra. Unfortunately, our experimental situation did not always allow this optimal choice and the subsequent trade-offs which were required, are hereby presented.

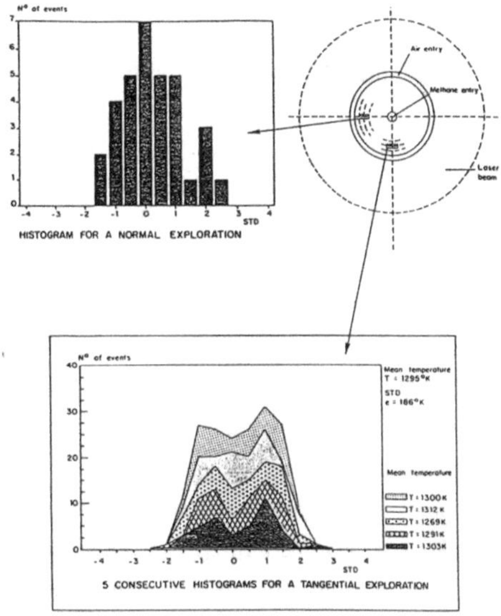

Fig.5. recorded histograms for two symetric positions. The normal exploration smooths the double peak structure

2.4. BOXCARS advantages and limitations

The BOXCARS configuration takes advantage of the two initially well-separated pump and Stokes laser beams to clearly localize the probe volume at the beams intersection. Simple geometrical considerations make it possible to size this probe volume (typically 5 mm * 100 μm for our experiment). We checked the actual spatial resolution by recording temperature histograms for two equivalent positions with respect to the flow symmetry (Fig. 5). The largest dimension of the probe volume was either perpendicular or parallel to the steep temperature gradients. The difference between the two recorded histograms (double narrow peak distribution for parallel and broad single peak distribution for perpendicular) clearly indicated a spatial integration of the measurement over at least 10 mm which was greater than the expected 5 mm resolution. However, this procedure was a true in-situ check of the actual spatial resolution and the unexpected behaviour was probably due to the non-linear generation of the CARS signal combined with the effects of the turbulence. Finally, the BOXCARS configuration had to be selected to preserve a reasonable spatial resolution of the temperature measurements and, depending on the probe volume position inside the flame, either parallel or perpendicular measurements were retained.

2.5. Cross polarization advantages and limitations

The problem of the cross polarisation configuration was more complex. First, this configuration was adopted to cancel the non-resonant contribution of all the species present inside the probe volume (methane, by-products, oxygen or non-resonant nitrogen). Due to the adopted burner geometry, the air and methane flows were well seperated and the methane was anticipated to strongly contribute to the non-resonant background. If the corresponding spectra were better resolved, the signal intensity was decreased by at least two orders of magnitude compared to the generally adopted parallel polarisation configuration. We have checked that these two configurations led to comparable temperature measurements inside a steady medium [2]. Consequently, to ensure a good signal to noise ratio (SNR) at any position, i.e., to deal with expected weak signals, we could select either parallel (numerical account of the non-resonant background) or cross (physical account of the non-resonant background) polarisation configurations. This choice was particularly important in the intermediate temperature range (900 - 1200 K) where some detectivity considerations have compelled the parallel polarisation configuration (Fig. 6).

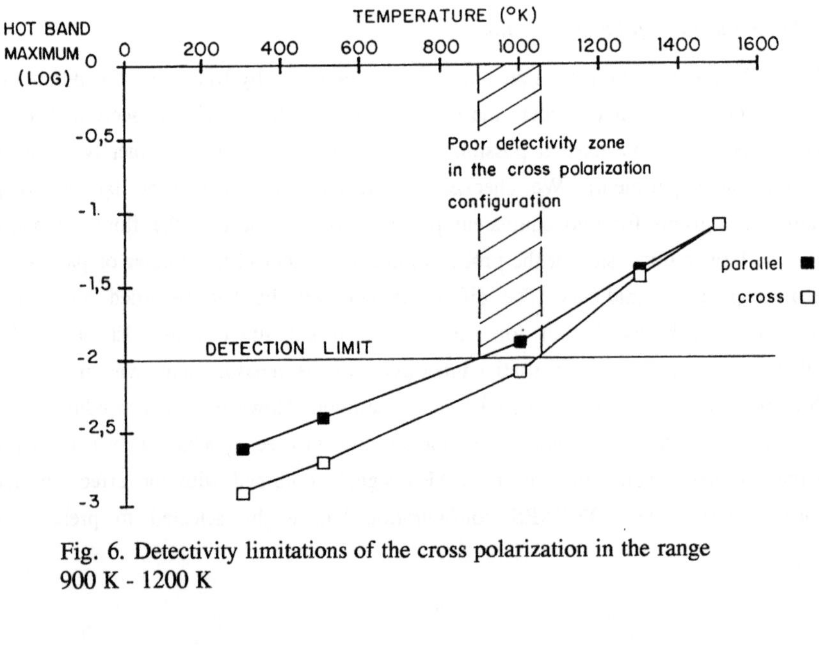

Fig. 6. Detectivity limitations of the cross polarization in the range 900 K - 1200 K

3. Comparison of CARS and thermocouple measurements

As indicated in figure 2, both vertical and radial explorations have been performed. Using a measurement setup including either parallel or cross polarisation configuration and for the two explorations, the measurements have compared favourably (Fig. 7,8). However, for the radial exploration and when crossing the temperature steepest gradients (100°C/mm), the instantaneous CARS temperature measurements have exhibited large fluctuations (rms = 300°C) and the comparison was weakly informative.

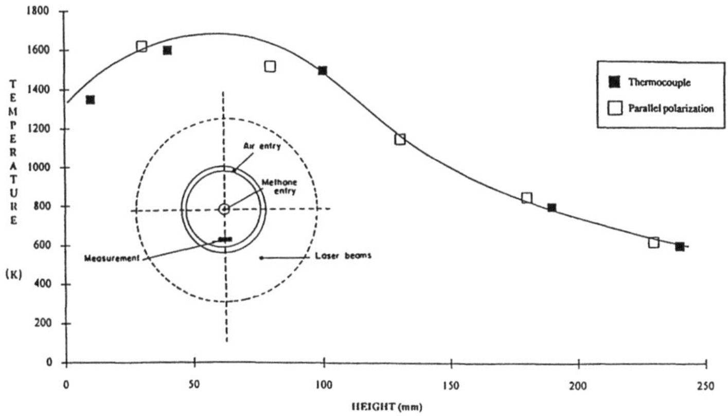

Fig. 7. CARS and thermocouple comparison for the vertical exploration shown in fig. 2

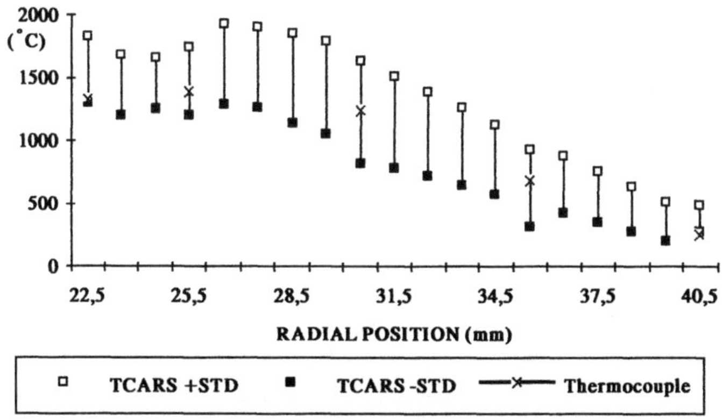

Fig. 8. CARS and thermocouple comparison for the radial exploration shown in fig. 2. The temperature fluctuations as measured with CARS are represented

4. Methodology for the comparison of the model predictions and the experimental results

4.1. model Probability Density Function

The main idea was to establish a pertinent methodology to compare the instantaneous CARS temperature measurements with the model predictions. To account for the chemical processes, a combustion model, called Probabilistic EUlerian Lagrangian (P.EU.L.), was combined with a usual k-ε model. This combustion model accounted for one finite chemical reaction rate [3]:

$$CH_4 + 7/4\,O_2 \longrightarrow 1/2(CO + CO_2) + 2\,H_2O \qquad (1)$$

Each cell was considered as an homogeneous reactor where the mean concentrations could be computed using a Probabilty Density Function (PDF) averaging operator which was presumed as a β function. This PDF described the passive scalar carbon atom mass fraction, f. The β function coefficients depend directly on the mean and the fluctuations of f. The shape of the distribution pictured the balance between the mean and the fluctuations of f :

$$p(f) = f^{(a-1)}.(1-f)^{(b-1)} \qquad (2)$$

$$a = f_{mean}.[f_{mean}(1-f_{mean}) - f'^2]/f'^2 \qquad (3)$$

$$a/b = f_{mean}/(1-f_{mean}) \qquad (4)$$

Figure 9 shows an example of a predicted β distribution for a given position inside the flamme (radius = 24.5 mm, height = 40 mm). The mean value of f was 0.13 and the rms 0.013. The rich and lean zones fast equilibrium diagram is also reported in figure 9, the observed values corresponded to a rich zone with large fluctuations (10%).

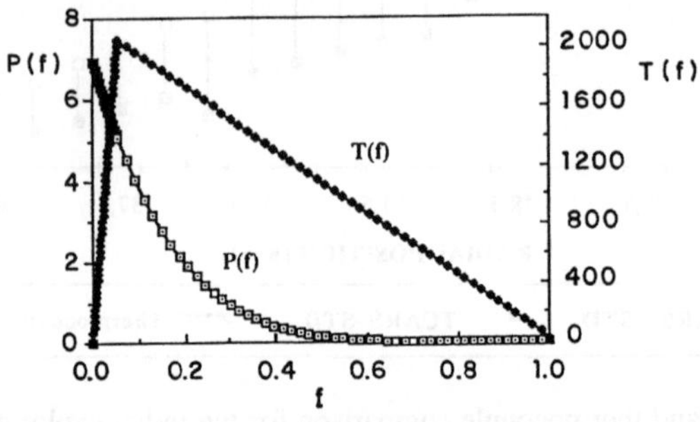

Fig. 9. β function computed for a given point inside the flamme (z = 40 mm, r = 24.5 mm). The fast equilibrium diagram is also reported

4.2. analytical representation of the measured histograms

According to the β function PDF used for the modeling, it seemed advantageous to represent also the CARS temperature histograms as a β function whose coefficients were directly deduced from the recorded mean temperature and the associated rms. As an example, in figure 10 and for the same position inside the flame, the corresponding β function associated to the locally recorded histogram is shown. As expected, the representation was good due the analytical properties of this distribution.

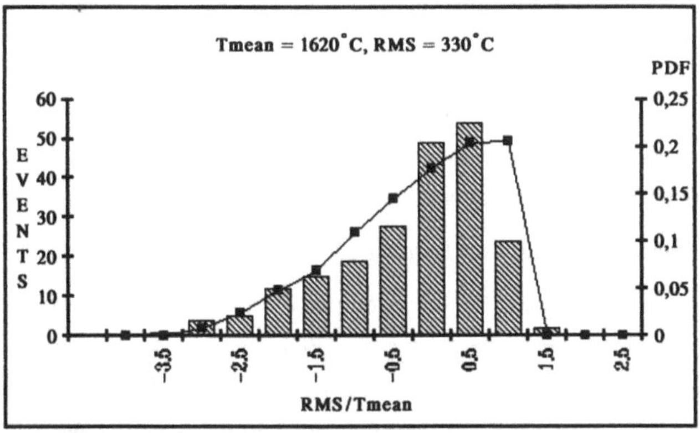

Fig. 10. CARS measured histogram for the same point of fig. 9. The deduced β function is also shown

However, this representation could be questionnable in the case of a double peak histogram as shown previously (see Fig. 5). This double peak temperature distribution has been often observed inside turbulent reacting flows [4] and has been associated to the alternate crossings of hot and fresh gas pockets inside the probe volume. The crossing time has been related to the largest eddy typical time (30 ms in our flame) and was fully compatible with this observed alternate crossing. This phenomenon caused a fluctuation of the flamefront position and therefore refraction index variations disturbing the laser beam propagations. For these measurements, these propagation effects have been investigated by correlating the CARS spectrum maximum position on the photodiode array and the corresponding temperature (correlation >0.7). Assuming that the double peak structure was not generated by an experimental bias, it was possible to separate this distribution into cold and hot distributions for which a corresponding β function could be computed (Fig. 11). The sum of these two distributions represented fairly the directly observed temperatures histogram (Fig. 12).

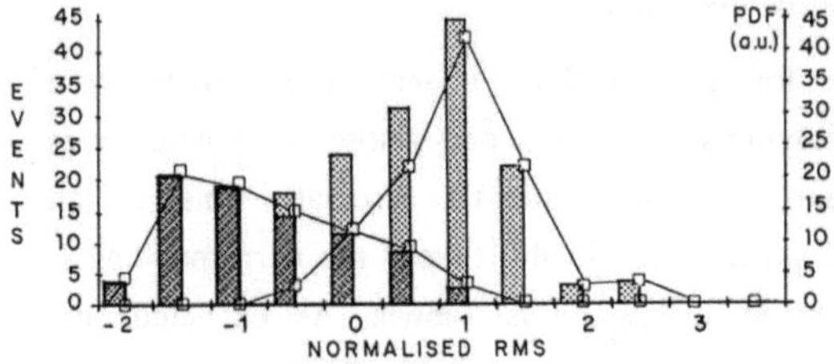

Fig. 11. Separation of the distribution into cold and hot populations

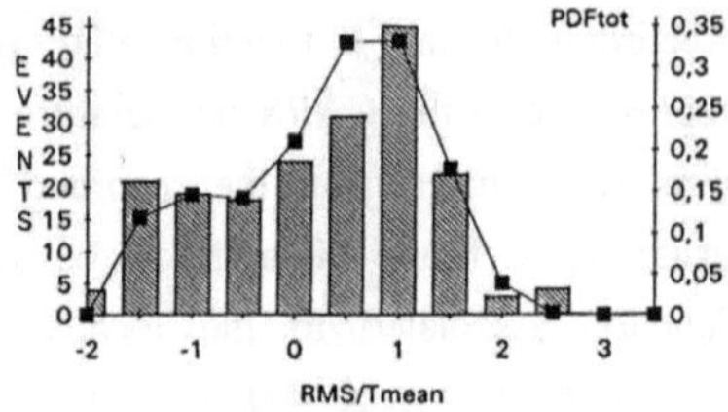

Fig. 12. The total distribution as the sum of the two β functions

Therefore, the analytical representation of the experimental data was considered as a reasonnable approach for further analysis.

4.3. comparing model and experiment temperature distribution

According to the above analysis, an easy comparison of the model predictions and the experimental results could be made. Using the PDF and the fast chemical equilibrium diagram (see Fig. 9), we have computed a local temperature distribution (which is not anymore but still close to a β function) and compared to the experimentally determined β function (Fig. 13).

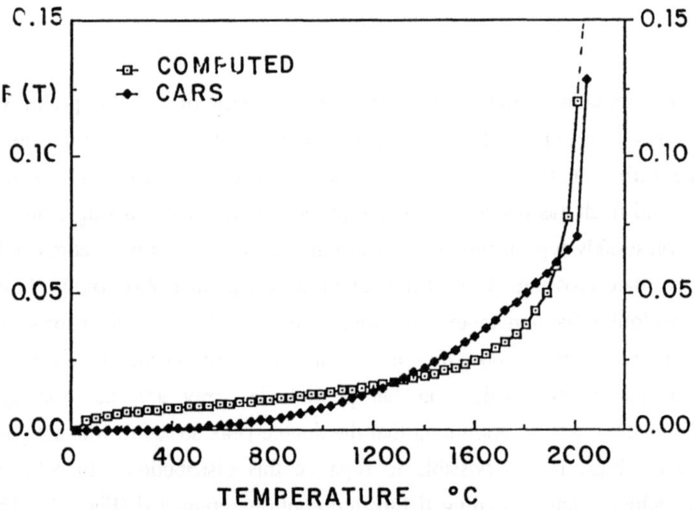

Fig. 13. Comparison of computed and observed PDF for the same point (z = 40 mm, r = 24.5 mm). The agreement is correct

The comparison was fully satisfactory. The mean experimental and computed temperatures (resp. 1893 K and 1835 K) fairly compared while the rms (resp. 330 K and 526 K) exhibited a weak discrepancy. The model seemed to privilege the cold temperatures. This encouraging result was only regarded as an estimate since the model distribution was a Favre average distribution. We supposed that the density variations had a weak influence on the distribution shape and, in parallel, the CARS signals were intrinsically weighted by the local density. This approach clearly needed further investigations however the methodology for the comparison seemed correct.

5. The observed flame structure

5.1. flamefront position

A key point to describe the flame was the observation of the recirculation zone. Consequently, the most relevant observation was the velocity field. This was done with a Doppler anemometer (LDA). The complete measurements are not presented here but only the most relevant result concerning the flame structure is hereby presented . Plotting the vertical velocity versus the radius at different heights (Fig.14), it clearly has appeared that a vertical cylinder of radius 24.5 mm was the null velocity locus.

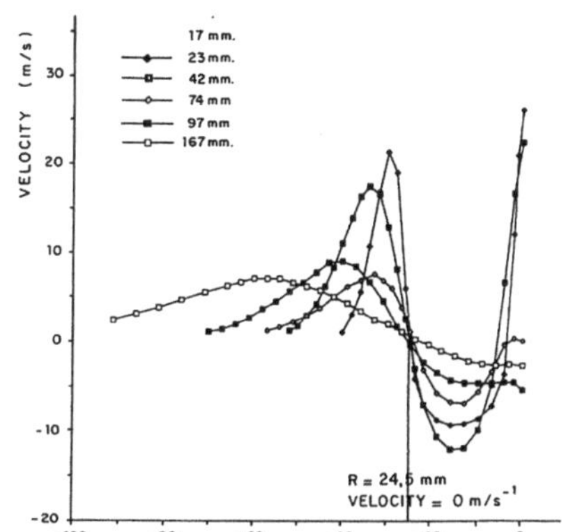

Fig. 14. LDA velocity measurements at different heights

The flame was rather a cylinder than a bowl flame as expected from the burner geometry. Using the CARS measurements and the analyical representation mentionned above, we have plotted the a exponent of the β function (see Eq.3) versus the radius at a height z = 40 mm (Fig.15). A peak value was retrieved at r = 24.5 mm. A peak value corresponded to high temperatures and low fluctuations. When connected to the velocity measurements and the

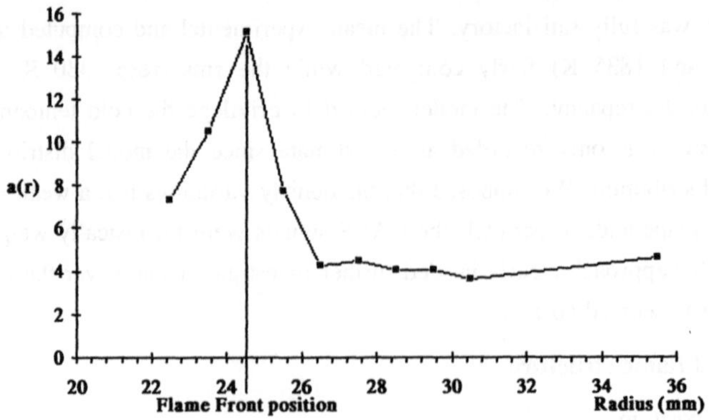

Fig. 15. a exponent of the β function vs radius

velocity fluctuations (Fig. 16) which reached a maximum at r = 24.5 mm, this last result was regarded as the flamefront position and its thickness was about 3 mm corresponding to the expected increase in turbulent flows. Furthermore, the swirl has strongly structured the flame in the radial direction. This was confirmed by recording the vertical a coefficient evolution which exhibited a smooth profile.

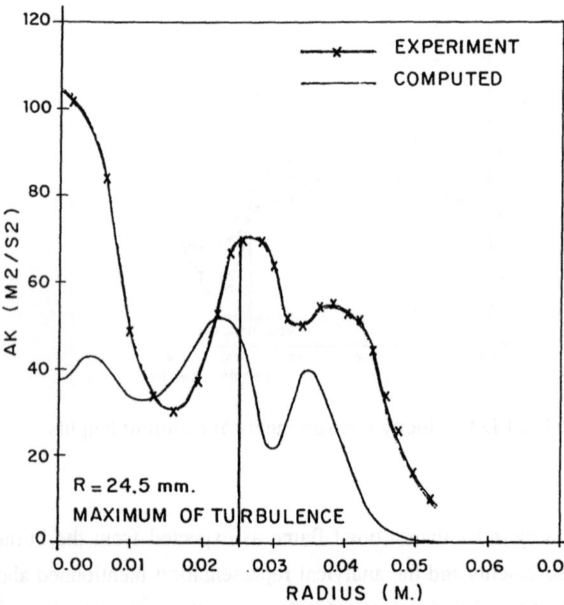

Fig. 16. Measured and computed turbulent intensity at z = 40 mm

5.2. vertical methane profile at the burner center

At the immediate vicinity of the burner exit, the high methane concentration have precluded the temperature measurements even in the cross polarisation configuration as anticipated. However, this important non-resonant background could be used to estimate the relative methane to air ratio. The observed semi-quantitative methane profile, recorded in the parallel polarization configuration to enhance the methane contribution, agreed very well with the predicted profile (Fig.17).

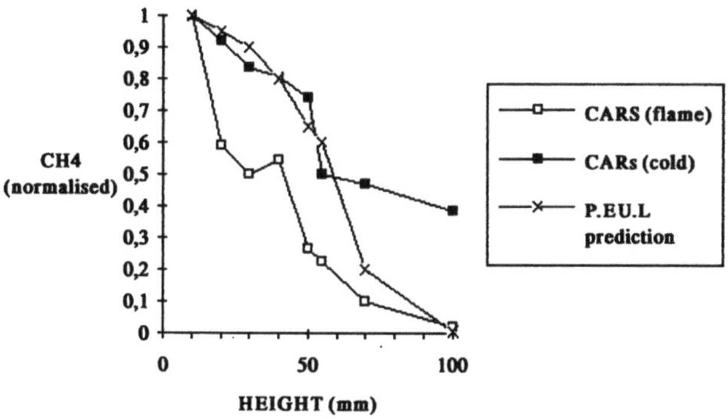

Fig. 17. Semi-quantitative CH_4 profiles at the center of the burner

An other profile was also recorded inside the unburned flow and is also shown in figure 17. According to the chemical reaction (Eq. 1), the P.EU.L. approach accounted for a total reaction and the computed methane profile represented the profile of all the carbon species.

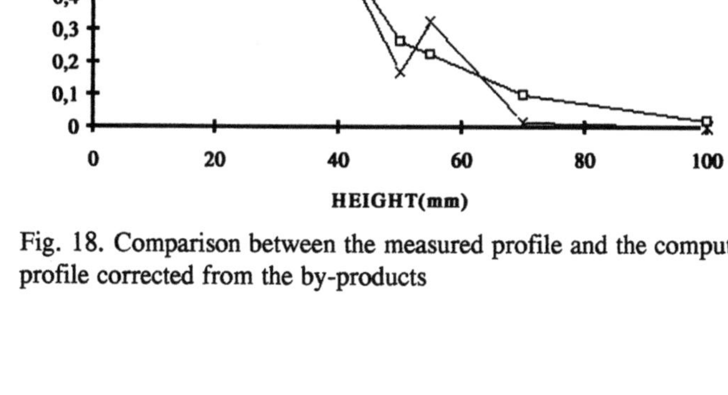

Fig. 18. Comparison between the measured profile and the computed profile corrected from the by-products

Consequently, the difference bewtween the two CARS profiles could be considered as a by-products profile which could correct the model profile. The comparison was very satisfactory (Fig. 18) and confirmed the validity of this approach even though a reduced kinetics scheme could be a major improvement.

6. Conclusion

Though the CARS technique could not allow any systematic exploration of the flame, it brought an inestimable physical insight of the flame behaviour. Although if some limitations could appear, they could often be turned into advantages. A variety of measurements is possible and, if careful data proceesing is done, the results could be largely informative. We have suggested a methodology to compare the experimental observations and the model predictions which seemed to be a promising approach. Finally, the complementarity of different techniques was clearly required to assess the consistency of a measurement set and helped to validate the computationnal model.

[1] Eckbreth A. : " Laser diagnostics for combustion temperature and species", Gupta K. and Lilley D. Editors, Abacus press, Cambridge USA, 1988.

[2] Bachmann M., Bellaïche P., Cahen C., Garréton D., Vervisch L.,"Modeling and thermometric study of a swirl stabilized turbulent diffusion flame" 23rd Combustion Institute symposium, Orléans, 1990.

[3] Garréton D., Vervisch L., "Numerical study of a swirling turbulent flame" Aerothermodynamics in combustors, Taipei, June 3 -5 1991.

[4] see for example Linuma K., Asanuma T., Ohsawa T., Doi J. Editors "Laser diagnostics and modeling of combustion" Springer-Verlag, Paris, 1987.

A Unified Model of Mean Reaction Rate in Turbulent Premixed Flames

M. KATSUKI, Y. MIZUTANI, T.YOSHIDA and T. YASUDA

Osaka University
Department of Mechanical Engineering
2-1 Yamada-oka, Suita, Osaka 565 Japan

Summary

We developed a model of the mean reaction rate in a turbulent premixed flame to predict time-averaged profiles of the velocity, temperature and species concentrations in practical combustion systems. The fundamental features of the model consists of chemistry- and mixing-controlled reaction rates which vary with the local Damköhler number. The comparison of numerical predictions with experiments demonstrated the applicability of the concept to the flames of wide range of Damköhler number.

Introduction

Recent progress in numerical fluid dynamics has been expanding into the direct simulation of three-dimensional, time-dependent turbulent reacting flows. Nevertheless, the simulation of this kind for practical combustors and furnaces of complex geometry seems yet unrealistic due to the insufficient capacity of existing computers. To satisfy the practical requirements in combustion engineering, therefore, we are obliged to restrict ourselves at present to solving the time-averaged balance equations for fluctuating properties associated with proper boundary conditions. In order to close the time-averaged conservation equations of the first moment for turbulent reacting flows, not only a turbulence model but also a turbulent reaction model which gives a mean reaction rate in turbulence is necessary.

R. S. L.Lee · J. H. Whitelaw · T. S. Wung (Eds.)
Aerothermodynamics in Combustors
IUTAM Symposium Taipei, Taiwan, 1991

The reaction models based on Bray-Moss-Libby's concept are widely accepted recently [1,2]. They should be applied to flames, the thickness of which is far thinner than the scale of dissipating eddies. In extremely intense turbulence, however, the eddy motion breaks a flame-sheet into reacting gas particles of comparable to the laminar flame thickness. Thus, we need a new reaction model widely applicable to the flames in intense turbulence of small scale as well as laminar flamelets.

Some basic concepts of the fine structure of turbulent premixed flames are roughly classified into two kinds in terms of Damköhler number, Da, defined as the ratio of the characteristic time for turbulent mixing τ_m to that for chemical reaction τ_c. The so-called *wrinkled laminar flame* or *laminar flamelets* appears when Da is large and the dissipating eddy-scale is larger than the laminar flame thickness, whereas the *distributed reaction zone* appears when Da is small and the scale of dissipating eddies is smaller than the laminar flame thickness. The reaction rates at these two extremes are dominated by chemistry and mixing, respectively. The objective of the present study is to develop a unified model of mean reaction rate in turbulent premixed flames applicable over a wide range of Damköhler number.

Experiments

In our previous study [3,4], we observed confined turbulent premixed flames to elucidate the local fine structure of flames and its influence on mixing processes. Figure 1 shows a combustor geometry, and Table 1 summarizes the experimental conditions. The observation section was a 300-mm-long square cross section duct, 50-mm-wide. A pilot burner with a 1-mm-wide port between two 1-mm-thick plates was centered at the bottom of the observation section. The pilot flame was a hydrogen-air premixed flame with an equivalence ratio, ϕ_p, of 0.6. Mixtures of city gas, which is 88 % methane, and air having the same equivalence ratio, ϕ_m, with nominal velocities U_1 and U_2, respectively, flowed on either side of the centered pilot burner. The initial turbulence intensities, u_1' and u_2' were varied by inserting a turbulence generating grid 30 mm upstream of the pilot port. More details are found in Ref.[3].

Figure 2 shows the schlieren photographs of the observed flames. The weakly turbulent approaching flow exhibits a

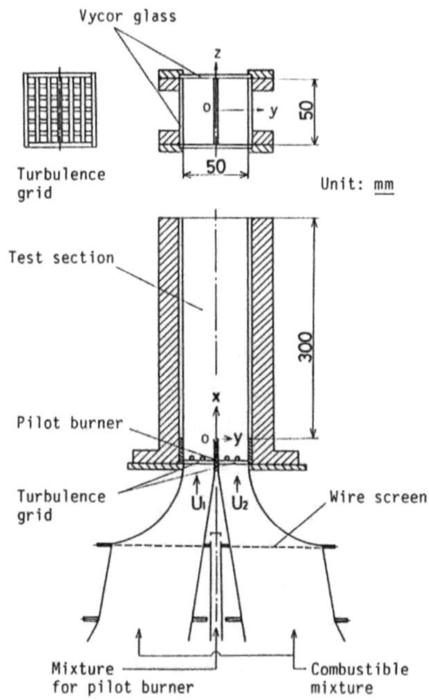

Fig. 1 Combustor geometry

Table 1 Experimental conditions

Flame	U_1 m/s	U_2 m/s	u_1' m/s	u_2' m/s	ϕ_m	U_p m/s	ϕ_p
I	10	20	0.15	0.4	0.65	10	0.6
II	10	20	2.2	4.2	0.65	10	0.6

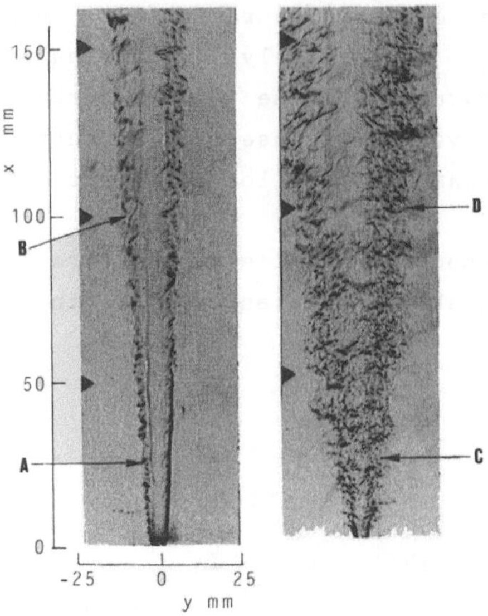

(a) Flame I (b) Flame II

flame structure resembles that of the so-called *wrinkled laminar flame*. In contrast, the strongly turbulent mixing layer exhibits a structure likely to be of the *distributed reaction zone* type. We discussed these fine flame structures based on the measured turbulence characteristics in Ref. [4]. Here, we briefly show the obtained characteristic features essential for the present objective, that is, the modelling of turbulent premixed combustion.

We examined the probability density functions (PDF's) of fluctuating temperatures, and some PDF's observed at the positions pointed in Fig.2 are shown in Fig.3. Because the fast-chemistry, hence large *Da*, makes a PDF bimodal, such a distribution observed at point B, corresponding to the structure of *laminar flamelets*, appears in many turbulent flames. A plateaulike distribution observed at point C, on the other hand, is ascribed to the interrupted reaction by the extremely intense turbulence of the smaller scale than the

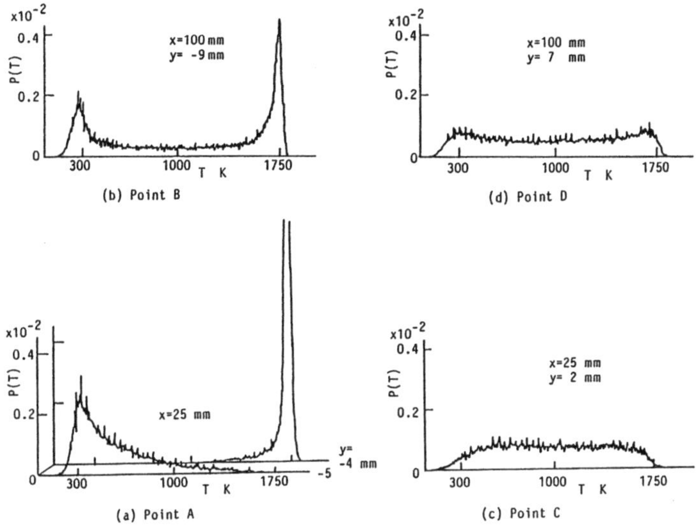

Fig. 3 PDF of temperature fluctuations

laminar flame thickness, which may correspond to the structure of *distributed reaction zone*. The probabilities of intermediate temperatures decrease with the increase in *Da* indicated in each caption of Fig.3.

We concluded that the shape of PDF varies with the degree of the interaction between reaction and turbulence, that is, the local Damköhler number in flames. In turn, the local Damköhler number can be used as the index of turbulent flame structure.

Governing Equations

The present model has been developed to predict time-averaged profiles of velocities, temperature and species concentrations in combustors, and to be incorporated into the time-averaged version of conservation equations for mass, momentum, enthalpy, and species concentrations. Additional equations for turbulence energy k and its dissipation rate ε are solved in the present model to express turbulent transport.

The Favre-averaging (density weighted averaging) is adopted to take the effect of variable density into account. For a steady two-dimensional flow, the governing equations for the above variables can be cast into the following general form using the general dependent variable ϕ:

$$\frac{\partial}{\partial x}(\bar{\rho}\tilde{u}\phi)+\frac{\partial}{\partial y}(\bar{\rho}\tilde{v}\phi)-\frac{\partial}{\partial x}(\Gamma\frac{\partial\phi}{\partial x})-\frac{\partial}{\partial y}(\Gamma\frac{\partial\phi}{\partial y})=S \quad (1)$$

The dependent variable, the exchange coefficient and the source term for each conservation equation are tabulated in Table 2 [5]. The empirical constants appearing in k-ε turbulence model and turbulent Prandtl numbers for the dependent variables are shown in Table 3 [6].

Because the governing equations are discretized in the numerical procedure, the source term, R_{fu}, of fuel mass fraction should represent the spatially- and temporally-

Table 2 Exchange coefficient and source term in Equation (1)

ϕ	Γ	S
1	0	0
u	μ_{eff}	$\frac{\partial}{\partial x}(\mu_{eff}\frac{\partial u}{\partial x})+\frac{\partial}{\partial y}(\mu_{eff}\frac{\partial v}{\partial x})-\frac{\partial}{\partial x}\{p+\frac{2}{3}(\rho k+\mu_{eff}\,\mathrm{div}\,\vec{U})\}$
v	μ_{eff}	$\frac{\partial}{\partial x}(\mu_{eff}\frac{\partial u}{\partial y})+\frac{\partial}{\partial y}(\mu_{eff}\frac{\partial v}{\partial y})-\frac{\partial}{\partial y}\{p+\frac{2}{3}(\rho k+\mu_{eff}\,\mathrm{div}\,\vec{U})\}$
k	μ_{eff}/σ_k	$G_t^*-\rho\varepsilon$
ε	$\mu_{eff}/\sigma_\varepsilon$	$c_1 G_t^*\varepsilon/k-c_2\rho\varepsilon^2/k$
h	μ_{eff}/σ_h	0
m_{fu}	μ_{eff}/σ_m	R_{fu}

$G_t=\mu_{eff}[2\{(\frac{\partial u}{\partial x})^2+(\frac{\partial v}{\partial y})^2\}+(\frac{\partial u}{\partial y}+\frac{\partial v}{\partial x})^2]$

$G_t^*=G_t-\frac{2}{3}\,\mathrm{div}\,\vec{U}(\mu_t\,\mathrm{div}\,\vec{U}+\rho k)-\frac{\mu_t}{\rho^2}(\frac{\partial\rho}{\partial x}\frac{\partial p}{\partial x}+\frac{\partial\rho}{\partial y}\frac{\partial p}{\partial y})-\frac{9}{55}\rho k\,\mathrm{div}\,\vec{U}$

$\mathrm{div}\,\vec{U}=\frac{\partial u}{\partial x}+\frac{\partial v}{\partial y}$

Table 3 Empirical constants

c_D	c_1	c_2	σ_k	σ_ε	σ_h	σ_m
0.09	1.44	1.92	1.0	1.3	0.7	0.7

integrated reaction rate over the finite volume of a computational grid.

Combustion Model

Visual length of flames and mixing processes in a combustor are primarily governed by the large scale motions of turbulence, but combustion reaction is held to be closely related to the fine structure of turbulence. The turbulent combustion model, therefore, cannot be embodied without taking the fine structure of turbulence into account.

As described previously, a local mean reaction rate depends on the interaction of chemistry with turbulence, the degree of which, the flame structure in other words, can be represented by the local Damköhler number. We assume, therefore, that the local reaction rate is the sum of chemistry- and mixing-controlled rates. Their two extremes corresponds to the structure of *wrinkled laminar flame* and of *distributed reaction zone*, respectively. Thus, the local mean reaction rate, R_{fu}, appears in the conservation equation of fuel is expressed in terms of Da as follows.

$$R_{fu} = R_c \{1 - \exp(-k_1 Da)\} + R_m \exp(-k_1 Da) \qquad (2)$$

where k_1 is an empirical constant of the order of unity, R_c and R_m are the chemistry- and the mixing-controlled reaction rates, respectively. The local Damköhler number, Da, is expressed as

$$Da = \tau_m / \tau_c = (\ell/u')/(\delta_L / S_L) \qquad (3)$$

where δ_L and S_L are the laminar flame thickness and the laminar burning velocity, respectively. The length scale ℓ and the turbulence intensity u' are evaluated from the local turbulent quantities, k and ε. The contribution of chemistry- and mixing-controlled rates varies with Da, which situation is shown in Fig.4 assuming k_1 equals unity.

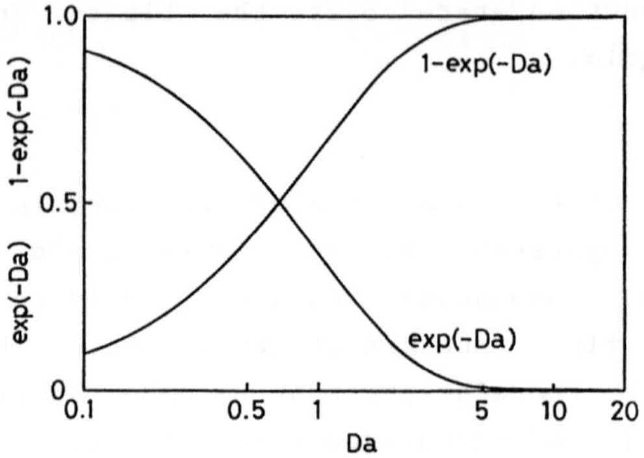

Fig. 4 Change in the contribution of chemistry- and mixing-controlled reaction rates

The chemistry-controlled reaction rate, R_c, is evaluated using Arrhenius rate law taking account of the influence of fluctuations in temperature and concentrations as follows.

$$R_c = -F\,\bar{\rho}^2\,\tilde{m}_{fu}\,\tilde{m}_{ox}\,\exp(-E/R\tilde{T})\,\{1+X\} \tag{4}$$

where X denotes the expansion terms of fluctuating components [7] and it is represented in the following form.

$$X = a_0\,\frac{\overline{m''_{fu}\,m''_{ox}}}{\tilde{m}_{fu}\,\tilde{m}_{ox}} + a_1\left(\frac{\sqrt{\overline{T''^2}}}{\tilde{T}}\right)^2 + a_2\left(\frac{\overline{T''\,m''_{ox}}}{\tilde{T}\,\tilde{m}_{ox}} + \frac{\overline{T''\,m''_{fu}}}{\tilde{T}\,\tilde{m}_{fu}}\right) + \cdots \tag{5}$$

$$a_0 = 1, \quad a_1 = \frac{1}{2}(E/R\tilde{T})^2 - E/R\tilde{T}, \quad a_2 = E/R\tilde{T}$$

Following the eddy-break-up concept [8,9], the mixing-controlled reaction rate, R_m, is related to the dissipation rate of turbulence as well as the probability of the contact of burnt gas with unburned reactants.

$$R_m = -A\,\bar{\rho}\,\{V_b(1-V_b)\}^{1/2}\,\varepsilon/k \tag{6}$$

where A denotes an empirical constant and V_b is the volume fraction of burnt gas, which is estimated from the reactedness, $\overline{\zeta}$.

$$V_b = \frac{m_{fu,u}(\rho_u - \overline{\rho}) + (m_{fu,u} - m_{fu,b})\,\overline{\rho}\,\overline{\zeta}}{m_{fu,u}\,\rho_u - m_{fu,b}\,\rho_b} \tag{7}$$

The local reactedness $\overline{\zeta}$ is defined as follows using the local mass fraction of fuel $\tilde{m}_{fu}$.

$$\overline{\zeta} = (m_{fu,u} - \tilde{m}_{fu})/(m_{fu,u} - m_{fu,b}) \tag{8}$$

where $m_{fu,b}$ and $m_{fu,u}$ are the fuel mass fractions in the fully burnt and unburned reactants, respectively. When we use a leaner combustible mixture than the stoichiometric, V_b in Eq.(7) is simplified as follows because $m_{fu,b} \equiv 0$,

$$V_b = 1 - (\overline{\rho}\tilde{m}_{fu})/(\rho_u m_{fu,u}) \tag{9}$$

Numerical Prediction

In order to examine the validity of the proposed model, some numerical calculations were carried out with the inlet conditions equivalent to those of flames observed in the experiment. However, for the first approximation the influence of fluctuation in temperature and concentration in Eq.(4) was neglected in the present calculation.

The predicted profiles of mean axial velocity component, turbulence intensity and mean temperature for weakly turbulent Flame I are shown in Fig.5 together with the experimental data. The global agreement is excellent, which suggests that the present model works well when the wrinkled laminar flame structure dominates the whole flame. In contrast, the results for strongly-turbulent Flame II are shown in Fig.6. The agreement between prediction and experiment is deteriorated in regard to the mean temperature on high velocity (right) side and the turbulence intensity in downstream cross sections.

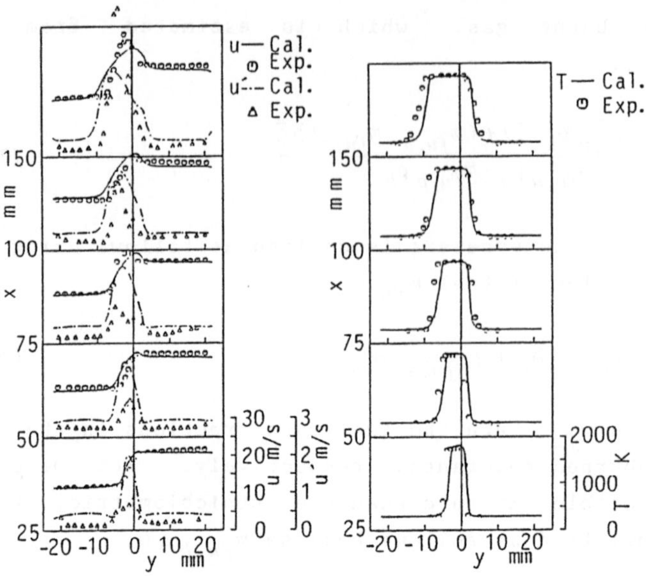

Fig. 5 Comparison between experiment and calculation for Flame I

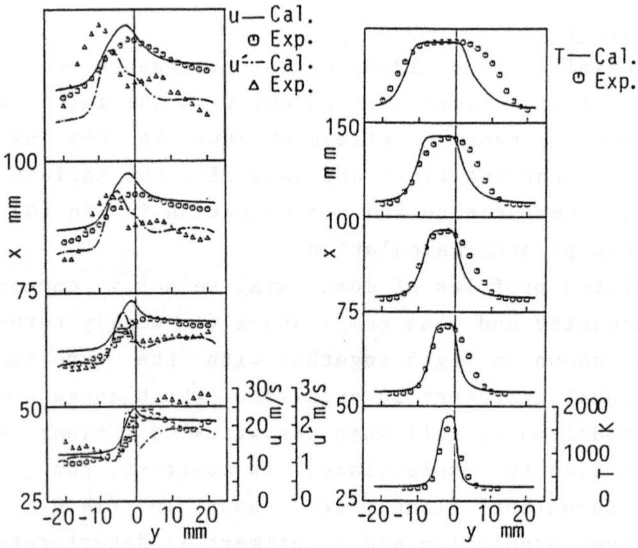

Fig. 6 Comparison between experiment and calculation for Flame II

Since the combustion model is intimately related with the local Damköhler number, it is essential to predict the turbulence field correctly. Therefore, the difference between prediction and experiment is caused probably by the fact that the actual anisotropy of turbulence and the unknown flame-turbulence interaction cannot be well predicted by the present k-ε turbulence model. The further refinement of the reaction model including the effect of turbulence model is scheduled.

Conclusion

In formulating the time-averaged conservation equations for turbulent reacting flows, we have discussed the expression of the time-mean reaction rate applicable to various flames over a wide range of Damköhler number. The rate was expressed by the contributions of chemistry- and mixing-controlled rates which depend upon the local flame structure, hence the local Damköhler number.

Due to our limited knowledge of the interaction between reaction and turbulence, some modifications of the present model may be needed later in connection with the turbulence model. In the present study, however, the numerical predictions have demonstrated the capability of the concept and its applicability to turbulent premixed flames.

References

1 Liew, S.K., Bray, K.N.C. and Moss, J.B., Combustion Flame, 56(1984), 199-213.

2 Peters, N., Twenty-first Symp. (Int.) on Combustion: The Combustion Institute, (1986), 1231-1250.

3 Katsuki, M., Mizutani, Y., Yasuda, T., Kurosawa, Y., Kobayashi, K. and Takahashi, T., Combustion Flame, 79 (1988), 9-18.

4 Katsuki, M., Mizutani, Y., Yasuda, T., Kurosawa, Y., Kobayashi, K. and Takahashi, T., Combustion Flame, 82 (1990), 93-105.

5 Bradley, D., Kwa, L.K., Lau, A.K.C., Missaghi, M. and Chin, S.B., Combustion Flame, 71(1988), 109-122.

6 Khalil, E.E., Spalding, D.B. and Whitelaw, J.H., Int, J. Heat Mass Transfer, 18(1975), 775-791.

7 Khalil, E.E., Modelling of Furnaces and Combustors: Abacus Press, 1976.

8 Spalding, D.B., Thirteenth Symp. (Int.) on Combustion: The Combustion Institute, (1971), 649-657.

9 Lockwood, F.C., Combustion Flame, 29(1977), 111-122.

A Simple Reaction Mechanism for Soot Formation in Non-Premixed Flames

R. P. Lindstedt

Fluids Section, Imperial College of Science, Technology and Medicine, Exhibition Road, London SW7 2BX

Summary

A reaction mechanism for soot formation in non-premixed flames is proposed and tested for different flame configurations and fuels. The mechanism is combined with detailed and simplified gas phase chemistry for ethylene and methane flames respectively. The cases considered include counterflow ethylene flames with oxygen enriched and depleted oxidant streams, a buoyant axi-symmetric methane-air flame and a methane-air flame on a Wolfhard-Parker burner. It is shown that for all these configurations excellent agreement can be obtained for the predictions of soot volume fractions and particle characteristics.

Introduction

Accurate prediction of soot formation in laminar and turbulent combustion is a formidable problem. However, the topic is of great practical importance particularly for non-premixed combustion in applications ranging from turbulent jet flames to internal combustion engines and gas turbine combustion chambers. While most practical applications are strongly turbulent in nature a fundamental requirement of any model is the successful prediction of soot characteristics under a wide range of laminar conditions. Unfortunately, even this problem is one of great complexity in particular as any deduced reaction mechanism must be sufficiently compact to be implemented into turbulent flame predictions for example via an extended laminar flamelet presumed *pdf* approach or via a transported *pdf* approach.

The present paper outlines a global reaction mechanism for the formation, growth and combustion of soot particles in laminar non-premixed flames. The model is combined with detailed and global chemistry descriptions for the gas phase. The reaction mechanism [1] involves nucleation, surface growth, particle coagulation and combustion steps and has been created for application to the prediction of turbulent flames. The soot nucleation and surface growth reactions are linked to the gas phase by presuming that soot formation is dependent upon the breakdown path of the fuel and that the presence of pyrolysis products, such as acetylene and poly-unsaturated cyclical hydrocarbons, are of primary importance to the soot formation process. In the present work acetylene is used as the indicative critical specie. This is naturally a considerable simplification, particularly for the particle formation step, but the introduction of a more

R. S. L. Lee · J. H. Whitelaw · T. S. Wung (Eds.)
Aerothermodynamics in Combustors
IUTAM Symposium Taipei, Taiwan, 1991

appropriate indicative specie for particle formation such as benzene adds significant uncertainties to the gas phase chemistry.

The deduced reaction mechanism is applied to the coflow methane-air flames measured by Garo et al [2] and Syed [11] and to counterflow ethylene/oxygen/nitrogen flames measured by Vandsburger et al [3]. For the methane flame a global reaction mechanism [4] with added reaction steps for acetylene formation and destruction have been used to minimize computational effort. For the ethylene flames a carefully constructed detailed mechanism with 36 species and 111 forward reaction steps using the latest available rate constant data has been used [5].

Soot Formation Mechanism

The soot formation mechanism used in this paper is based on the presumption that the fuel breakdown path is critical in determining sooting characteristics. Specifically it is assumed that the amount of soot formed is proportional to the amount of acetylene generated during the breakdown of the fuel. The model involves the solution of two additional conservation equations for the soot mass fraction and the particle number density. The model for soot volume fraction, particle number density and size is closed by the assumption of a spherical particle shape. The formation of soot mass is assumed to occur in two parts. Firstly, the formation of incipient soot particles or 'nucleation' - in the present model these particles are assumed to be formed from acetylene as:

$$C_2H_2 \quad \text{-->} \quad 2\,C(s) + H_2 \qquad (1)$$

The notation C(s) is not strictly correct as young soot contains significant amounts of hydrogen on a molar basis, but is used in the absence of more suitable alternatives. The reaction rate constant for this step has been determined by Leung et al [1] by the use of counterflow ethylene-air flames. It has been formulated to account for the formation of incipient particles as well as early surface growth. The latter is important since experimental studies [3] indicate that newly formed particles are significantly (up to an order of magnitude) more reactive than older particles. Here the assumption made is that the minimum particle size consists of 100 carbon atoms giving a particle size of 1.24 nm. The rate expression derived from these computations was:

$$R_1 = 0.1\,10^5\,e^{-21100/T}\,[C_2H_2] \quad (\text{kmol/m}^3/\text{s})$$

The second reaction responsible for the increase in soot mass is assumed to be surface growth due to adsorption of acetylene on the surface of the particles:

$$C_2H_2 + n\,C(s) \text{ --> } (n+2)\,C(s) + H_2 \qquad (2)$$

The use of acetylene as the specie responsible for soot mass-growth is strongly supported by many studies e.g. [1]. However, there is considerable doubt concerning the role of the available surface area as an appropriate parameter in the formulation of reaction rate expressions for soot mass growth. Past studies have resulted in suggestions that the soot mass-growth is proportional to the surface area e.g. Harris and Weiner [8] as well as independent of surface area e.g. Bockhorn et al [9]. The reaction rate expression for mass addition is for the simplified model used here assumed to be a *function of the surface area* and to have a linear dependence on the concentration of acetylene.

$$R_2 = k_2(T)\ f(S)\ [C_2H_2] \qquad (kmol/m^3/s)$$

It can further [1] be shown that an adequate expression for the function f(S) can be found if it is assumed that the number of active sites present locally in the flame are proportional to the square root of the total surface area available. The resulting expression may be written as:

$$R_2 = k_2(T)\ [C_2H_2]\ [C(s)]^{1/3}\ [\rho N]^{1/6}$$

$$k_2(T) = 0.60\ 10^4 \sqrt{\pi \left(\frac{6M_{c(s)}}{\pi \rho_{c(s)}}\right)^{\frac{2}{3}}}\ e^{-12100/T}$$

Where N is the number of particles/kg-mixture, $M_{c(s)}$ molecular weight of carbon (12.011 kg/kmol) and $\rho_{c(s)}$ the soot density (2000 Kg-soot/m^3-soot).

The soot oxidation step used in this work is based on the work by Lee, Thring and Beer [6]. However, the pre-exponential factor has been adjusted by a factor 14 to give a similar surface-specific peak burnout rate to that observed by Garo et al [2] in their experiments on coflow methane-air flames. Given the present uncertainties in determining an appropriate rate constant or even the major oxidizing specie, which may differ depending on location in the flame, the current arrangement is viewed as sufficiently accurate for the present purpose. It is further assumed that at flame temperatures soot is oxidised to carbon monoxide as:

$$C(s) + 1/2\ O_2 \longrightarrow CO \qquad (3)$$

The resulting rate expression can be written as

$$R_3 = k_3(T) \frac{\pi}{M_{c(s)}} \left(\frac{6M_{c(s)}}{\pi \rho_{c(s)}}\right)^{\frac{2}{3}} [O_2]\ [C(s)]^{2/3}\ [\rho N]^{1/3}$$

$$k_3(T) = 0.125\ 10^6\ T^{1/2}\ e^{-19680/T}$$

The soot nucleation step outlined above also gives rise to the source term in the number density equation. The decrease in particle numbers is simply assumed to correspond to agglomeration given by the normal square dependence. The reaction may be written schematically as:

$$n\,C(s) \;\text{-->}\; C_n(s) \qquad (4)$$

with the resulting rate expression [1] given by:

$$R_4 = 2N_A R_1/C_{min} - k_4(T)\,[C(s)]^{1/6}\,[\rho N]^{11/6}$$

$$k_4(T) = 2\,C_a \left(\frac{6M_{c(s)}}{\pi\rho_{c(s)}}\right)^{\frac{1}{6}} \left(\frac{6\kappa T}{\rho_{c(s)}}\right)^{\frac{1}{2}}$$

where N_A is Avogadros number 6.022 10^{26} particles/kmol, C_{min} is the number of carbon atoms in the incipient soot particle (100), κ the Boltzmann constant (1.38 10^{-23} J/K), and C_a the agglomeration constant, assigned the value 9.

Gas Phase Reaction Mechanism

The gas phase reaction mechanism used for the ethylene flames is based on the work of the CEC data evaluation group [7] and further developed to improve prediction of diffusion flames [1,5].

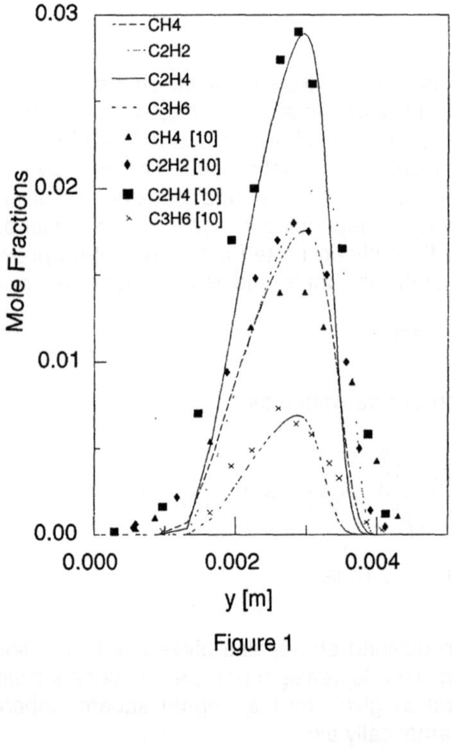

Figure 1

Of particular relevance to the present study is the accuracy of the prediction of pyrolysis products which can be shown to be excellent with the deduced mechanism as can be seen in Fig. 1 above. For the coflowing methane-air flames [2,11] an extended version of the global reaction mechanism derived by Jones and Lindstedt [4] was used. The modifications amounted to including steps for the formation and destruction of acetylene:

$$CH_4 + CH_4 \longrightarrow C_2H_2 + 3H_2 \quad (5)$$

$$C_2H_2 + 3/2\,O_2 \longrightarrow CO + CO_2 + H_2 \quad (6)$$

$$C_2H_2 + 2H_2O \longrightarrow 2CO + 3H_2 \quad (7)$$

with reaction rates (in kmol/m^3/s),

$$R_5 = 0.12\ 10^{13}\ e^{-42778/T}\ [CH_4]$$

$$R_6 = 0.15\ 10^{12}\ e^{-15098/T}\ [C_2H_2]^{1/2}\ [O_2]^{5/4}$$

$$R_7 = 0.30\ 10^{9}\ e^{-15098/T}\ [C_2H_2]\ [H_2O]$$

While the underlying global reaction mechanism [4] is valid for a wide range of flames the additional reactions are valid only for unstrained and weakly strained diffusion flames with a strain rate 'a' < 40 /s.

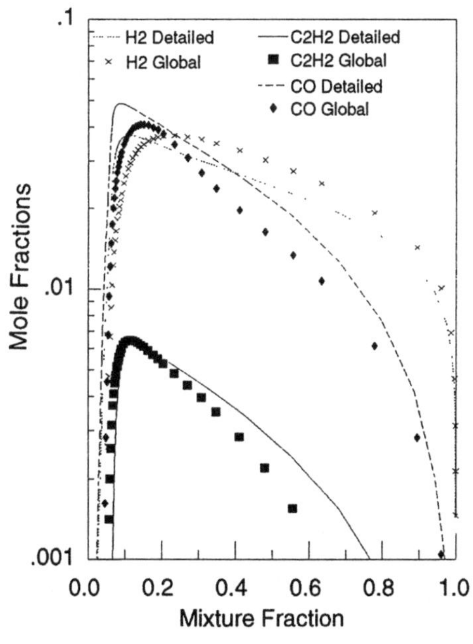

Figure 2

However, for such flames the agreement between the global and detailed mechanisms is very acceptable with typical errors less then 20% as can be seen from Fig. 2 above where predictions for a counterflow methane-air flame at a strain-rate of 15/s are compared.

Results and Discussion

The first set of computations were of three counterflow ethylene-air flames burning against oxygen enriched and depleted air streams investigated experimentally by Vandsburger et al [3]. These flames are well characterised with respect to soot levels with measurements of soot volume fractions, particle number densities and particle sizes. The velocity gradient estimated from the experiments was 63/s which was applied without modification in the numerical computations along with a fuel injection velocity of 0.10 m/s and 109 distributed nodes. The oxidant streams consisted oxygen/nitrogen ratios of 18/82, 22/78 and 28/72 respectively. The comparison of measured and predicted soot volume fractions can be seen from Figure 3 below.

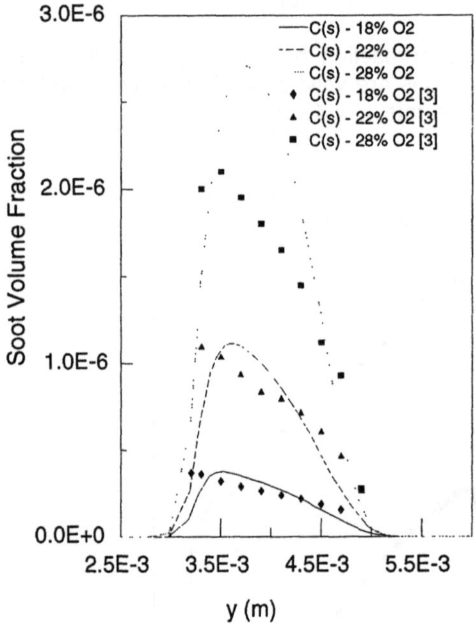

Figure 3

The agreement between predicted and measured soot volume fractions is generally very satisfactory. The predictions of particle number densities are also satisfactory, see Figure 4 below, but young particles close to the main reaction zone appears to agglomerate more quickly than older particles residing in colder parts of the flame. It is thus clear that as expected the accuracy of the above agglomeration expression is poor in the particle inception zone as well as at low temperatures close to the stagnation point.

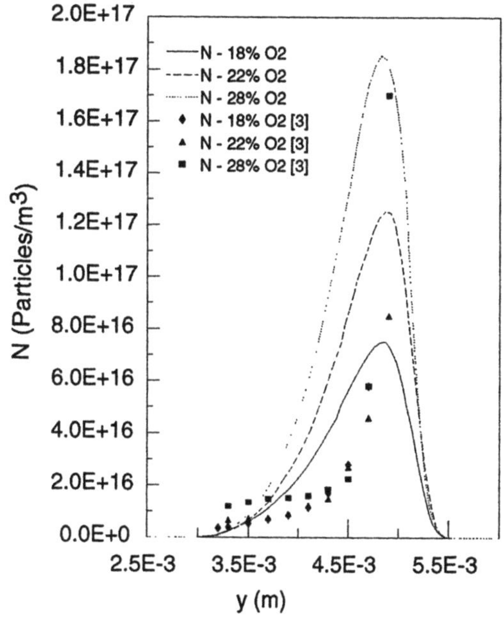

Figure 4

The area of the region of particle formation can also influence the particle number density and size in the early part of the flame. However, the predicted width of this region is around 1 mm or less which appears in good agreement with that observed experimentally [3]. An interesting feature of these flames is that the location in mixture fraction space of the peak growth rates of soot particles is shifted from 0.144 for the case of 18% oxygen in the oxidant stream to 0.187 for the case of 28% oxygen. This indicates that the increased temperature leads to soot formation being maintained into the more fuel rich parts of the flame.

The axi-symmetric coflow methane-air flame measured by Garo et al [2] was computed with 197 distributed nodes and the soot model unchanged from that applied to the counterflow ethylene flames. This flame provides an interesting comparison as not only are the fuel and oxidant streams changed, but the flame structure is also significantly different. The prediction of the gas phase using the global reaction mechanism for this flame is excellent [4] provided that computations begin at the first measuring station (12.5 mm downstream) and not at the burner exit. This poses a problem as soot nuclei start to form early in the flame and as no soot profiles are available at the first measuring station. Consequently, a preliminary solution was obtained from the burner exit to the first

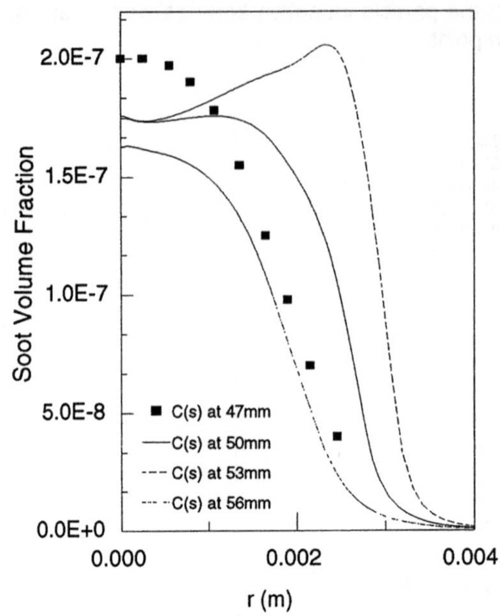

Figure 5

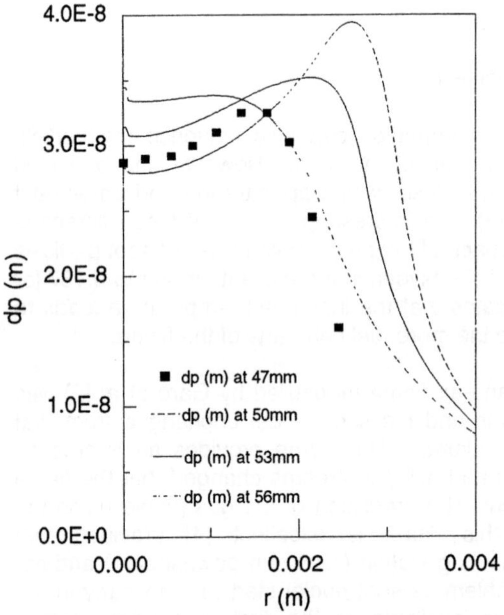

Figure 6

measuring station where the soot data were re-mapped according to the location in mixture fraction space. This procedure is approximate but considered more realistic than assigning zero values to the soot volume fractions and particle number densities at the first measuring station. The predicted soot volume fractions and particle sizes along the radius (r) at various distances downstream can be seen in Figures 5 and 6 respectively, where the measurements are from the radial profile at 47 mm downstream. The discrepancies between computed and measured profiles is probably to a large extent due to the approximate nature of the boundary conditions imposed at the first measuring station. This would appear to be given further weight by consideration of the soot volume fraction along the centreline of the flame which displays a similar shift with peak concentrations obtained at 47 mm downstream in the experiments compared to 51.5 mm in the computations as can be seen from Figure 7 below.

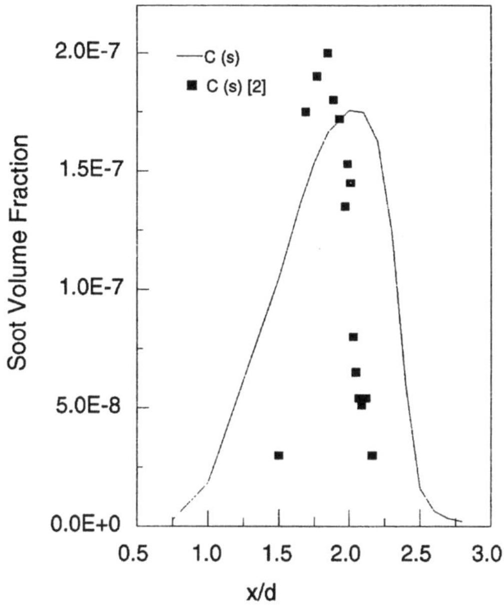

Figure 7

It can, however, not be ruled out that further improvements in predictions may be conditional upon the introduction of detailed gas phase chemistry and the introduction of an oxidation step dependent on OH radicals. However, the slope of the burnout of the soot along the centreline is well described by the current oxidation step as also shown in Figure 7.

To further investigate coflowing flames data obtained by Syed [11] on a Wolfhard-Parker burner was used. This data set contains measurements of soot volume fraction, mixture fraction and temperature profiles at several measuring

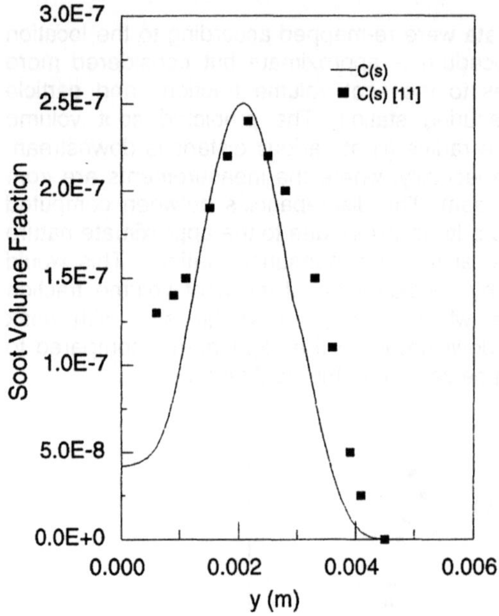

Figure 8

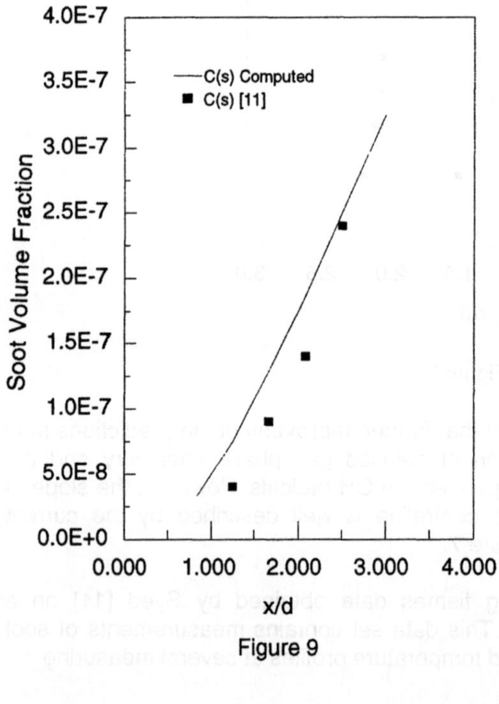

Figure 9

stations on a burner configuration less susceptible to producing difficulties with burner exit boundary conditions. Computations for this flame were consequently started at the burner exit and predictions are shown above for the final measuring station in Fig. 8 at x/d = 2.50. The parameter 'd' is here defined as the half-width of the Parker-Wolfhard burner. The agreement must be classified as excellent. However, there are differences lower in the flame as can be seen in Fig. 9 where the predicted and measured peak soot volume fractions in the flame are compared as a function of downstream distance. The over-prediction of the soot volume fraction at low heights is likely to be dependent primarily upon two factors. (i) The absence of an oxidation step involving the OH radical and (ii) the simplified nucleation expression. It is well known, e.g. Smyth et al [12], that the downstream evolution of aromatic hydrocarbons such as benzene differs from that of acetylene in flames of this kind. As a consequence earlier soot nucleation will occur with the present simplified reaction scheme.

Conclusions

The results obtained by the soot model presented here show very promising qualitative and quantitative agreement with measured data for a range of flames. In particular, predictions of soot volume fractions using the model appear to be close to the uncertainties associated with experiments and the reaction mechanisms used for the predictions of the gas phase. The predictions of other properties such as particle number densities and aggregate sizes are also satisfactory particularly if the uncertainties in the measurements of these properties are considered. However, further development work is required to generalise the model. This is particularly the case with respect to nucleation and oxidation steps as well as for the description of the evolution of particle number densities.

Acknowledgement

The financial support of the Ministry of Defence at RAE Pyestock and the efforts of Mr. K.M. Leung in the modelling of the gas phase chemistry are gratefully acknowledged.

References

1. Leung, K.M., Lindstedt, R.P. and Jones, W.P., 'A Simplified Reaction Mechanism for Soot Formation in Non-Premixed Flames', To appear Comb.and Flame

2. Garo, A., Prado, G. and Lahaye, J., Comb. and Flame 79, pp 226-233 (1990)

3. Vandsburger, U, Kennedy, I. and Glassman, I. Comb. Sci. and Tech. 39, pp. 263-285 (1984)

4. Jones, W.P. and Lindstedt, R.P. Comb. and Flame 73, p. 233 (1988)

5. Leung, K.M. and Lindstedt, R.P. 'Detailed Kinetic Modelling of Non-Premixed C_1 to C_3 Hydrocarbon Combustion', Imperial College Fluid Section Rept. FS/90/21, December 1990.

6. Lee, K.B., Thring, M.W. and Beer, J.M., Comb. and Flame 6, p137 (1962)

7. Baulch, D.L. and Co-workers in the CEC Kinetic Data Evaluation Group, 'Summary of Preferred Rate Data for Combustion Modeling', Final Rept. Sept. 1990

8. Harris, S.J. and Weiner, A.M. Comb. Sci. and Tech. 33, pp. 267-275 (1983)

9. Bockhorn, H., Moser, A., Wenz, H. and Warnatz, J., 19th Symp. (Int.) on Comb., The Comb. Inst., p.197, (1982)

10. Tsuji, H. and Yamaoka, I., 12th Symp. (Int.) on Comb., The Comb. Inst. pp 997-1005 (1969)

11. Syed, K. Ph.D. Thesis, Cranfield Institute of Technology, 1990.

12. Smyth, K.C., Miller, J.H., Dorfman, R.C., Mallard, W.G. and Santoro, R.J., Comb. and Flame 62, pp. 157-181 (1985)

Part III

Two Phase Flows

Droplet Vaporization Law in Non-Dilute Sprays

H.H. CHIU

INSTITUTE OF AERONAUTICS AND ASTRONAUTICS NATIONAL CHENG-KUNG UNIVERSITY TAINAN, TAIWAN, R.O.C.

SUMMARY

The vaporization of a droplet, interacting with its neighbors in a non-dilute spray environment is examined and a vaporization scaling law established. The interacting droplet is modeled with a centrally located droplet and its vapor bubble which is surrounded by a cloud of droplets. The distribution of the droplets and the size of cloud are characterized by a pair-distribution function. The analysis reveals that the vaporization of a droplet is retarded by the collective thermal quenching, vapor concentration accumulated in the outer sphere, and by the limited percolative passages for mass, momentum and energy fluxes. The retardation is scaled by the local collective interaction parameters group combustion number of a droplet cloud, droplet spacing, renormalization number and the local ambient conditions. The numerical results of a selected case study reveal that the vaporization correction factor falls from unity monotonically as the group combustion number increases, and saturation is likely to occur when the group combustion number reaches 35-40 with interdroplet spacing of 7.5 diameters and the environment temperature of 500 K. The scaling law suggests that dense sprays can be classified into: (1) a "Diffusively Dense" cloud characterized by uniform thermal quenching in the cloud, (2) a " Stratified Dense" cloud characterized by a radial stratification in temperature by the differential thermal quenching of the cloud, or (3) a "Sharply Dense" cloud marked by a dense cloud structure and the magnitude of the correction factor that depends on the collective parameters, pair-distribution function, group combustion number and stratification factor of the cloud.

R. S. L.Lee · J. H. Whitelaw · T. S. Wung (Eds.)
Aerothermodynamics in Combustors
IUTAM Symposium Taipei, Taiwan, 1991

INTRODUCTION

Increasing theoretical findings[1−13] and experimental evidence[14−18] attest to the widely held belief that the short-range collective interaction[2−6,11−13] among the neighboring droplets and the long-range interaction[1,7,9,10] with the droplets at a distance, on a hydrodynamic scale, have profound impact on the state of a droplet, i.e. the states of saturation, vaporization, ignition, combustion and extinction, as well as the droplet interfacial process rates in non-dilute cloud or spray environments. These collective interactions, produced by local hydrodynamic and transport processes in a complex topological environment, serve to control the percolative passage for dispersing mass, momentum energy fluxes and the effective interfacial area for the property exchange processes. These interactions result in the collective thermal quenching, the accumulation of vaporizing species and the tendency for stagnating the microscale local Stefan flow and mean flow through dynamic equilibration between the two phases. Although various models of droplet vaporization have been proposed to this date, a universal theory of short-range interaction and a comprehensive droplet vaporization law required for the non-dilute spray calculation have not been established. The objectives of this paper are to present the basic concepts, theoretical approaches and the results of Renormalized Droplet (RND) theory to establish vaporization laws for droplets in a stationary collisionless non-dilute cloud environment.

RENORMALIZED DROPLET (RND)

Structure of Model

The present model termed renormalized droplet (RND) consists of a test droplet interacting with its neighboring field droplets as shown in Fig.1. Two principal structural elements are (1) a Droplet-In-Bubble (DIB) and (2) a cloud of non-uniformly smeared droplets which functions as an external clothing for the DIB. The distribution of the neighbouring droplets is described by the pair-distribution function representing the joint probability of finding a test droplet and its neighbor at a particular separation distance. The pair-distribution function vanishes in the immediate vincinity of the bubble, representing the evacuated volume effect of droplets, and then increases rapidly to a value greater than unity at the radial distance comparable to a mean droplet spacing.

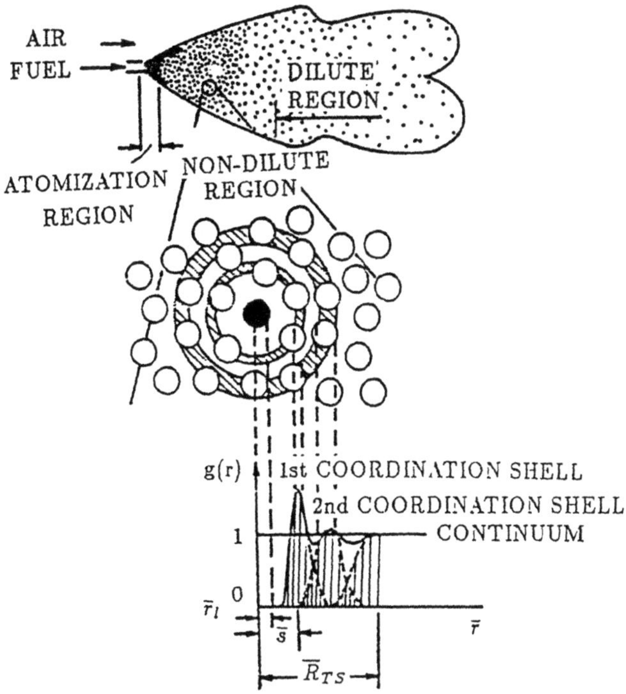

Fig.1 Structure of renormalized droplet in non-dilute spray and the schematic of quasi-droplet cloud in transition sphere.

This first high droplet density region populated by the nearest neighbor droplets is termed the "first coordination shell." The population peaks of the "second and higher order coordination shells" lose their sharpness as they merge with one another and are ultimately lost in a continuous environment where the pair distribution function approaches to unity. The size of the transition sphere, R_{ts}, defined as the radial location where the pair distribution function is 0.99, depends on the droplet size, spacing and arrangement.

The ratio of the size of the transition sphere, R_{ts}, to the characteristic hydrodynamic scale, L, for example, the radius of a cloud, is typically much smaller than unity, and thus the sphere degenerates into a point in the limit when R_{ts}/L vanishes. Accordingly, the average properties of the gas over the transition sphere approach to the local properties of the continuum flow as the limit R_{ts}/L goes to zero. This "correspondence hypothesis" and the interlinking transition sphere constitute the two key factors for the determination of the droplet process rate by the local gas properties of the continuum flow field.

Mathematical Analysis

Non-dimensional equations governing RND are formulated separately for the inner DIB region and an external droplet cloud, respectively, as follows.

$$\frac{1}{\eta^2}\frac{d}{d\eta}\left(\eta^2\psi_i\right) = 0 \qquad 1 \le \eta \le \eta_{co} \tag{1}$$

and

$$\frac{1}{\eta^2}\frac{d}{d\eta}\left(\eta^2\hat{\psi}_i\right) = G_s g \mu \hat{B}_i \qquad \eta_{co} \le \eta \le \eta_{ts} \tag{2}$$

with

$$\psi_m = \rho v \tag{3}$$

$$\begin{pmatrix}\psi_F\\ \psi_T\end{pmatrix} = \psi_m\begin{pmatrix}\alpha_F\\ \alpha_T\end{pmatrix} - \sigma\frac{d}{d\eta}\begin{pmatrix}\alpha_F\\ \alpha_T\end{pmatrix} \tag{4}$$

where $\eta = r/r_l(0)$, ρ and v are non-dimensionalized by the gas density, and velocity on the droplet surface, respectively, $Gs = 4\pi n r_{lo}^3$, g is a pair distribution function, μ is the vaporization shape factor defined by (12), ψ_i are the fluxes of various properties; mass for i=m, fuel vapor for i=F and thermal energy for i=T, and $\wedge$ refers to the properties pertaining to an outer region. The definitions of properties α_i and constants B_i are summarized in Table 1.

Table 1

Schvab-Zeldovichs Variables and Constants

i	α_i	$\hat{\alpha}_i$	$\hat{B}_i$
F	$-(y_F/W_f\nu_F)$	$(-\hat{y}_F/W_F v_F)$	$\varepsilon_F = -(W_F\nu_F)^{-1}$
T	$\int_{T_b}^{T} C_p dT/q$	$\int_{T_b}^{T} \hat{C}_p \hat{T} dT/q$	$\gamma_F = L/q$

where y is a mass fraction, W is a molecular weight, v is a stoichiometric coefficient, C_p is a specific heat, L is the latent heat of vaporization (per unit mass) of the fuel at the boiling temperature, q is the heat of combustion.

The system of Eqs. (1)-(4) is integrated by a repeated quadrature. The constants of integration are determined from the conditions of the impermeability of the gaseous species on the droplet surface and the balance of the heat conducted to the droplet with the latent heat of the vaporization. The integration gives the following vaporization laws

$$\dot{m}(o) = \frac{4\pi\rho D r_\ell(o)\ \ell n(1 + \frac{\alpha_{T_{co}}}{\gamma_F})}{1 - \varsigma_{co}} \tag{5}$$

and

$$\dot{m}(o) = \frac{r\pi\rho D r_\ell(o)\ \ell n(1 + \frac{\alpha_{F_{co}} - \alpha_{F\ell}}{\alpha_{F\ell} - \varepsilon_F})}{1 - \varsigma_{co}} \tag{6}$$

where

$$\varsigma_{co} = \eta_{co}^{-1}$$

These laws, which agree with those obtained by Tishkoff[4], are valid for any value of ς_{co}, provided the bubble contains no droplet other than the test droplet. This standard law will be termed "canonical vaporization law," the spherical bubble surface will be identified by "canonical environment" and the bubble temperature by "canonical temperature." In contrast to what is described above, an alternative law in which the vaporization rate is determined by the gas temperature on the surface of the transition sphere, Fig.1, will be called "renormalized vaporization law," the surface of transition sphere is "renormalized environment" and temperature on the surface is the "renormalized temperature" which is numerically equal to the local gas temperature of the continuum flow field.

Renormalized Droplet Law

Since the continuum theory of sprays predicts the local gas temperature but not the canonical temperature, the canonical vaporization laws (5) and (6) cannot be used in the spray calculation as previously described. The alternative proposed in this paper is to use a renormalized vaporization law. The analytical procedure adopted for the development of the renormalized droplet law is described below. First of all, the conservation equations governing the fluxes ψ_m, ψ_F and ψ_T for inner region, Eq.(1), and the corresponding fluxes $\hat{\psi}_m, \hat{\psi}_F$ and $\hat{\psi}_T$ for outer region, Eq.(2), are solved simulatneously with appropriate boundary conditions. The inner temperature, α_T, as described in 2,1 satisfies the conservation of energy on the surface of the test droplet. Specifically, the net heat conducted on the droplet surface is sufficient to vaporize the fuel leaving the test droplet. The fuel vapor concentration satisfies the mass conservation on the surface of the test droplet. The non-dimensional temperature $\hat{\alpha}_T$ and fuel vapor concentration $\hat{\alpha}_F$ on the outer surface of the transition sphere are equal to the prescribed values, $\hat{\alpha}_{Tts}$ and $\hat{\alpha}_{Fts}$, respectively. Secondly, these inner and outer solutions $\alpha_T, \hat{\alpha}_T, \alpha_F$ and $\hat{\alpha}_F$ are matched on the surface of the canonical bubble. This last step serves to determine the integration constants that appear in the inner and outer solutions uniquely. The resulting solutions are given by

$$\frac{\hat{\alpha}_T + \gamma_F}{\hat{\alpha}_{T_{ts}} + \gamma_F} = \frac{\hat{\alpha}_F - \varepsilon_F}{\hat{\alpha}_{F_{ts}} - \varepsilon_F} = \exp\left\{ -\frac{1}{\sigma} \int_{\eta}^{\eta_{ts}} \frac{1}{\eta'^2} \left(1 + G_s \int_{\eta_{co}}^{\xi^2} g\mu\xi^2 d\xi \right) d\eta' \right\} \tag{7}$$

In order to obtain the renormalization laws, one first determines $\hat{\alpha}_{Tco}$ and $\hat{\alpha}_{Fco}$ by replacing the η appearing in the lower limit of the integration of Eq.(7) by η_{co}. Subsequently, by substituting the resulting expressions of $\hat{\alpha}_{Tco}$ and $\hat{\alpha}_{Fco}$ in Eq.(5) and (6) one arrives at the following renormalized vaporization laws:

$$\dot{m}(o) = 4\pi\rho D r_\ell(o) C_v ln\left(1 + \frac{\hat{\alpha}_{T_{ts}}}{\gamma F}\right) \tag{8}$$

$$\dot{m}(o) = 4\pi\rho D r_\ell(o) C_v ln\left(1 + \frac{\hat{\alpha}_{F_{ts}} - \hat{\alpha}_{F\ell}}{\hat{\alpha}_{F\ell} - \varepsilon_F}\right) \tag{9}$$

where C_v is the vaporization correction factor calculated as follows

$$C_v = \left[1 - \varsigma_{ts} + G_s \int_{\eta_{co}}^{\eta_{ts}} \int_{\eta_{co}}^{\eta} K(\eta|\xi)\mu(\xi) d\xi d\eta \right]^{-1} \tag{10}$$

in which

$$K(\eta|\xi) = \eta^{-2}\xi^2 g(\xi) \tag{11}$$

$$\mu = \frac{\dot{m}(\eta)}{\dot{m}(o)} = \frac{4\pi\rho D r_\ell \ell n\left[1 + \frac{\hat{\alpha}_{T_c}(\eta)}{\gamma_F}\right]}{\dot{m}(o)[1 - \varsigma_c(\eta)]} \tag{12}$$

where $\alpha_{F\ell}$ is given by

$$\alpha_{F\ell} = \frac{\hat{\alpha}_{F_{ts}} + \hat{\alpha}_{T_{ts}} + \varepsilon_F}{\hat{\alpha}_{T_{ts}} + \gamma_F} \tag{13}$$

The vaporization laws (8) and (9) remain incomplete until the distribution of μ is provided.

The analytical procedure for the determination of μ is described in the following. To begin with, the mean gas phase temperature of a field droplet is estimated from the volumetric averaged temperature of the gas confined in the bubble of a field droplet. By using this mean gas temperature one can subsequently calculate the radial distribution of μ from the energy equation, Eq.(2) with i=T. Finally, the resulting expression of $\mu(\eta)$ is inserted into Eq.(10) to obtain the following scaling factor C_v

$$C_v = \left\{1 - \varsigma_{ts} + G_s \int_{\eta_{co}}^{\eta_{ts}} \int_{\eta_{co}}^{\eta} K(\eta|\xi)\theta(\xi)\left[W_1(\xi) + BW_2(\xi)\right] \left(\frac{\eta_{co} - 1}{\eta - 1}\right)\left(\frac{\eta}{\eta_{co}}\right) d\xi d\eta\right\}^{-1} \tag{14}$$

where

$$B = \left[\frac{1}{\Lambda} - W_1(\eta_{ts})\right] / W_2(\eta_{ts}) \tag{15}$$

and

$$\Lambda = \frac{1 - \varsigma_{ts} - G_s \int_{\eta_{co}}^{\eta_{ts}} \int_{\eta_{co}}^{\eta} K(\eta|\xi)\theta(\xi)\frac{1-\varsigma_{co}}{1-\varsigma(\xi)}\frac{W_2(\xi)}{W_2(\eta_{ts})} d\xi d\eta}{1 - \varsigma_{ts} + G_s \int_{\eta_{co}}^{\eta_{ts}} \int_{\eta_{co}}^{\eta} K(\eta|\xi)\theta(\xi)\frac{1-\varsigma_{co}}{1-\varsigma(\xi)}\left[W_1(\xi) - \frac{W_1(\eta_{ts})}{W_2(\eta_{ts})}W_2(\xi)\right] d\xi d\eta} \tag{16}$$

Two functions, W_1 and W_2, which are the homogeneous solutions of the energy equation (2), satisfy the following canonical boundary conditions: $W_1(\eta_{co}) = 1$, $\frac{dW_1}{d\eta}(\eta_{co}) = 0$, $W_2(\eta_{co}) = 0$, $\frac{dW_2}{d\eta}(\eta_{co}) = 1$.

NUMERICAL RESULTS AND DISCUSSION

The structure and scaling laws of RND and their dependence on the principal collective interaction parameters of a stationary cloud of n-octane droplets are determined for the droplets with the following fuel properties; $\rho_l = 707\ kg/m^3$, $T_b = 398.7\ K$, $L = 71.7 kcal/kg$, and $W_F = 114.14\ kg/kg - mole$.

(1) Pair-Distribution Function and Canonical Bubble

In the absence of experimental data, a pair-distribution function is constructed on the basis of the geometrical distribution of molecules in a dense liquid. The following two parameter function is adopted for the numerical calculation.

$$g(\eta) = 1 + a\exp(-b\eta)\cos(2\pi\eta_{co}), \qquad 2\eta_{co} < \eta \leq \eta_{ts} \tag{17}$$

where a and b are constants to be determined from the experimental data. In the present analysis a=1.8, b=0.65 are chosen. The resulting distribution patterns are shown in Fig.2. The corresponding signatures of the inverse of the radius of canonical bubbles with η_{co}=5, 10, and 15 are shown in Fig. 3.

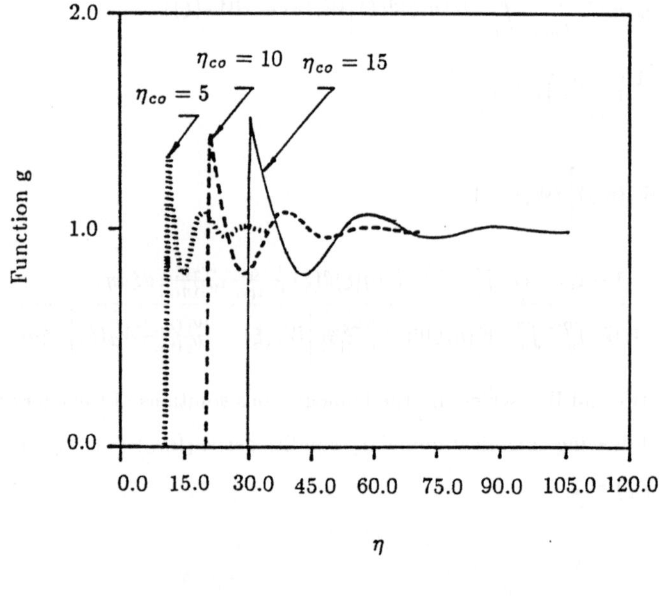

Fig.2 Pair-distribution function of quasi-droplets in transition sphere for three selected test droplet bubble sizes.

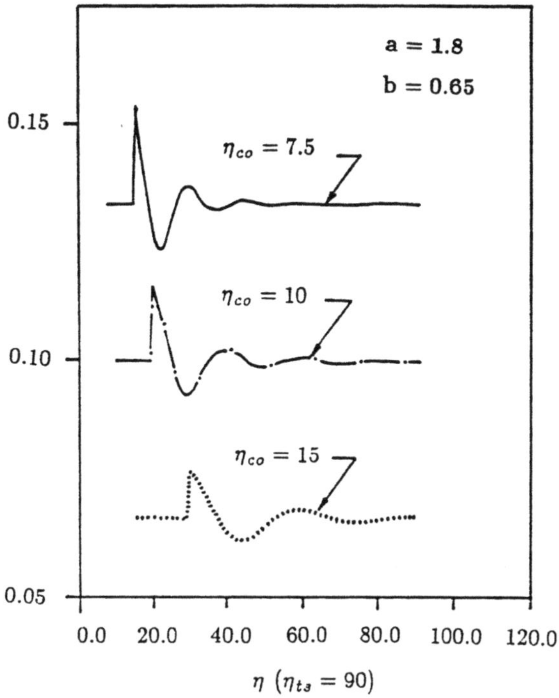

Fig.3 Signature of the inverse of cannonical bubble radius for three selected test droplet bubble sizes.

(2) Vaporization Shape Factor

A pronounced increase of μ in the radial direction is observed for a smaller value of droplet spacing, i.e. η_{co}=7.5 in Fig.4. The ratio of the vaporization rate of the droplet located at the first coordination shell, for the case of η_{co}=7.5, is only 5% of the corresponding value of RND with η_{co}=15. This trend of a higher increasing rate of μ at smaller canonical bubble radii; μ-stratification, is a common feature for a small droplet spacing in non-dilute sprays, as confirmed for a case study of various RND clouds featured with different values of a and b that appear in the pair distribution function.

(3) Temperature Distribution

High μ-stratification at a smaller droplet spacing, as illustrated in Fig. 4 is associsted with steep radial temperature gradients in the transition sphere, shown in Fig.5. Indeed, the comparison of "μ-T stratifications" suggests that (1) the rapid vaporization of droplets in the outer layer of the cloud collectively quench as the environment and thereby reduces the vaporization of the test droplet and (2) the increase in η_{ts}, at a fixed droplet spacing, tends to reduce the inward heat transfer rate and thus suppresses the vaporization of the test droplet.

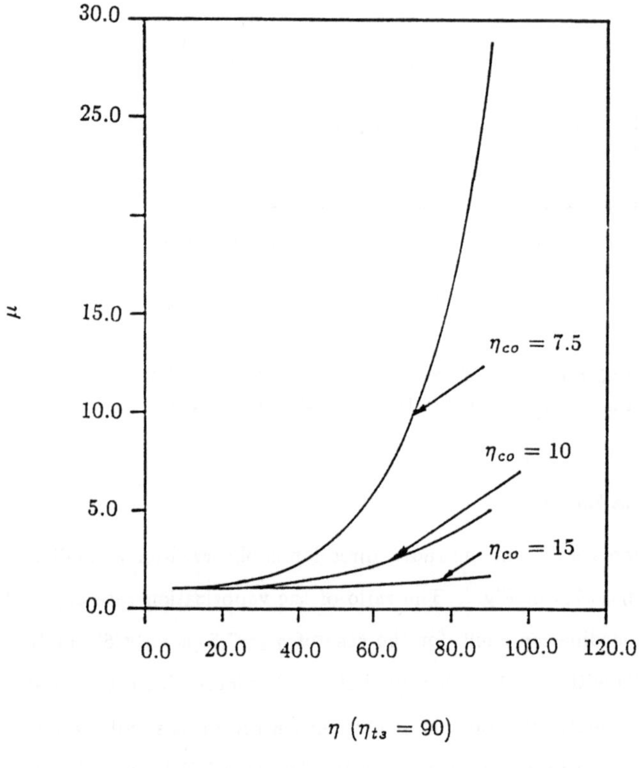

Fig.4 μ-stratification; Vaporization shape factor distribution in transition sphere for three selected test droplet bubble sizes.

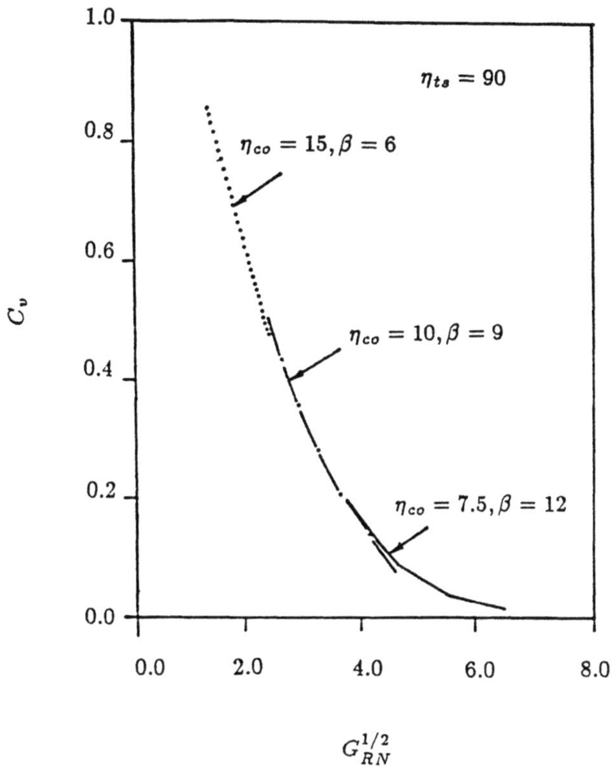

Fig.5 T-stratification = gas temperature distribution in transition sphere for three selected test droplet bubble sizes.

(4) Vaporization Rate-Correction Factor

The correction factors of RNDs for three selected values of η_{co}, Fig. 6, are found to decrease monotonically as the group combustion number of RND increases. Saturation is projected to occur when $G_{RN} = 30 \sim 40$ with η_{co}=7.5. While the group combustion number is a primary factor controlling the magnitude of C_v, Fig. 6 shows a small variation in the correction factor for two RND's which have the same group combustion number but a different value of renormalization number, $\beta = \eta_{ts}/\eta_{co}$.

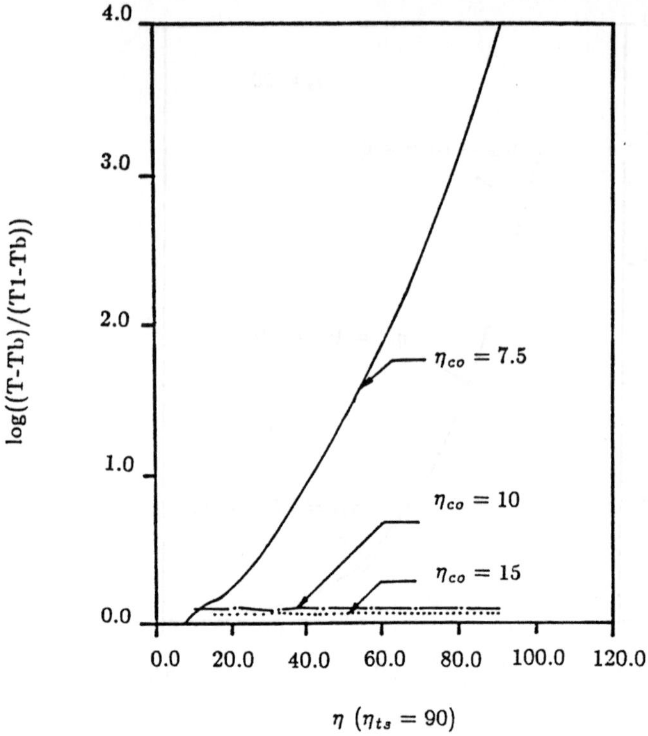

Fig.6 Correction factor for vaporization rate vs group combustion number for three selected test droplet sizes: Exact Solution.

(5) Scaling Laws with Linear Stratification Model

In order to facilitate the practical application of the scaling law, the correction factor is integrated by adopting the following functional form of $g\mu$

$$g\mu = \overline{g\mu} K_1 \eta$$

where K_1 is a stratification coefficient, and $\overline{g\mu}$ is the mean value of the weighted vaporization shape factor. The correction factor predicted by the linear stratification model described above, is given by

$$C_v = \left\{1 - \varsigma_{ts} + G_{RN}\left[(1-\frac{1}{\beta})(1+\frac{1}{\beta}-\frac{2}{\beta^2}) + \frac{1}{2}K_1\eta_{ts}(1-\frac{1}{\beta})(1+\frac{1}{\beta}+\frac{1}{\beta^2}-\frac{1}{\beta^3})\right]\right\} \quad (18)$$

where

$$G_{RN} = \frac{1}{6} G_s \eta_{ts}^2 \overline{g\mu}$$

(6) Spray Classification

Numerical assessment of the scaling law suggests the following structural classification of non-dilute sprays.

1. Diffusively Dense Cloud

In a moderately dense cloud, say, η_{co} equals to 10 or 15, shown in Fig.4, RND exhibits little or no μ-stratification. The reduction in the vaporization is attributed to a moderate thermal quenching. However, the thermal diffusion is large such that the temperature remains nearly constant throughout the transition sphere, see Fig.5, η_{co} =10 and 15. The corresponding correction factor is a function of the group combustion and a minor correction term of the order of $\frac{1}{\beta^2}$.

$$C_v \cong (1 + G_{RN})^{-1} + O(\frac{1}{\beta^2}) \tag{18 - b}$$

2. Densely Stratified Cloud

This dense cloud, with a smaller droplet spacing, as shown in Fig.4 with $\eta_{co} = 7.5$, is featured with strong μ and T stratifications in a transition sphere casued by an intense collective quenching that exceeds the rate of thermal transfer in a transition sphere. The renormalization number η_{ts}/η_{co} is larger than unity so that the correction factor is given by

$$C_v \cong [1 + G_{RN}(1 + \frac{1}{2}k_1\eta_{ts})]^{-1} + O(\frac{1}{\beta^2}) \tag{18 - c}$$

3. Sharply Dense Cloud - Fine Structure

When the coordination shells contain a large number of droplets and the renormalization number is not excessively large compared with unity, the correction factor depends on all the collective parameters: $G_{RN}, \eta_{ts}/\eta_{co}$ and η_{ts}. Two sharply dense clouds with the same group combustion number will exhibit structural variation when the renormalization number is different, for example, at $G_{RN} = 4.2$ the RND with $\beta = 12$ has a larger value of C_v than that of $\beta = 9$, see Fig.6.

CONCLUSION

A droplet law that accounts for the collective interaction of the neigbouring droplets on the vaporization of a droplet in a collesionless, non-dilute cloud is obtained. The law is given by the product of a classical single droplet law multiplied by the correction factor that is a function of the local spray structural parameters; a local group combustion number and a renormalization number.

The vaporization scaling law reveals that the droplet vaporization is retarded primarily by the collective thermal quenching and the formation of high vapor concentration in the transition sphere. The retardation is scaled by the combined topological-thermochemical parameters: group combustion number, renormalization number and droplet spacing. The scaling laws suggest that non-dilute clouds or sprays can be classified into: (1) "Diffusively Dense" clouds in which the vaporization reduction is attributed to a moderate thermal quenching that can be scaled by the group combustion number alone; (2) "Densely Stratified" clouds, with a marked radial stratification and a large temperature gradient in the transition sphere; and (3) "Sharply Dense" clouds, featured with a relatively small bubble, and radial stratifications so that the scaling law depends on G_{RN}, η_{ts} and β. Selected case studies show that the correction factor falls monotonically as the group combustion number increases and the saturation is expected to occur when the G_{RN} is of the order of 30-40, with the droplet spacing of approximately 7.5 times of the droplet size and environment temperature of 500 K.

REFERENCES

1. Chiu, H.H. and Liu, T.M.: COMB. SCI. AND TECH., 17, 127, 1(1977).
2. Labowsky, M. and Rosner, D.E., Advances in Chemistry Series No 166 Amer. Chem. Soc., (1978).
3. Twardus, E.M., and Brzustowski, T.A.: Comb. Sci. Tech 17, 215, (1978).
4. Tishkoff, J.M.: Int. J. Heat and Mass Transfer 22, 1407, (1979).
5. Labowsky, M.: Combust. Sci. Tech. 22, 127, (1980).
6. Umemura, A.: Eighteenth Symposium (International) on Combustion, 1335, The Combustion Institute, 1981.
7. Chiu, H.H., Kim, H.Y., and Croke, E.J.: Nineteenth Symposium (International) on Combustion, 971, The Combustion Institute, 1982.
8. Bellan, J. and Cuffel, R.: Combust. Flame 51, 55, (1983).
9. Sirignano, W.A.: Prog. Energy Combust. Sci. 9, 291, (1983).
10. Sichel, M. and Panlaniswamy, S.: Twentieth Symposium (Intenational) on Combustion, 1789, The Combustion Institute, 1984.
11. Chiu, H.H.: "Theory of Bipropellant Combustion, Part I-Conjugate, Normal and Composite Combustion Phenomena:" AIAA-86-0220 Presented at the AIAA 24th Aerospace Sciences Meeting, Reno, Nevada, 1986.
12. Chiu, H.H.: "Theory of Renormalized Droplets I Law of Droplet Vaporization Rate Under Short-Range Droplet Interaction" Presented at First Annual Conference on Liquid Atomization and Spray Systems ILASS Americas, Madison, Wisconsin, 1987.
13. Jang, S.D. and Chiu, H.H.:"Theory of Renormalized Droplets, II, Nonsteady Vaporization Droplet in Non-Dilute Sprays" AIAA-88-0639, presented at the AIAA 26th Aerospace Sciences Meeting, Reno Nevada, 1988.
14. McCreath, C.G. and Chigier, N.A.: Fourteenth Symposium (International) on Combustion, 1355, —The Combustion Institute, 1973.
15. Miyasaka, K. and Law, C.K.: Eighteenth Symposium (International) on Combsution, 283, The Combustion Institute, 1981.
16. Yule, A.J. and Bolado: Combust Flame 55, 1, (1984).
17. Xiong, T.Y., Law, C.K. and Miyasaka, K.: Twentieth Symposium (International) on Combustion, 1781, The Combustion Institute, 1984.
18. Roshland, C.P. and Bowman, C.T.: Twentieth Symposium (International) on Combsution, 1979, The Combustion Institute, 1984.
19. Chiu, H.H.: "Droplet laws in Spray Combustion Processes" under Preparation, 1990.

Experimental Investigation of the Reduction of Burning Rate Due to Finite Spacing Between Droplets

N. Roth, K. Anders, A. Frohn
ITLR Institut für Thermodynamik der Luft- und Raumfahrt
Universität Stuttgart
Pfaffenwaldring 31, 7000 Stuttgart 80

INTRODUCTION

Many combustors in technical applications are supplied with liquid fuel. The fuel is atomized in order to improve the evaporation process. In such fuel sprays the droplet diameter ranges typically from $10\,\mu m$ up to $500\,\mu m$. If the distances between the droplets are small, the mutual interactions of the droplets have to be taken into account for the description of the physical and chemical processes in the combustor. Usually it is very difficult to investigate these processes quantitatively in real sprays in a detailed manner. Often only mean or global quantities can be examined. To overcome these difficulties monodisperse equally spaced droplets may be applied as a useful tool for studying basic droplet phenomena as for instance the mutual interaction in spray combustion.

DROPLET SYSTEMS

Monodisperse droplet arrays consisting of one or more parallel droplet streams with defined droplet spacings have been examined. These monodisperse droplet streams have been produced by vibrating orifice droplet generators [1,2]. The initial droplet size d_{To}, velocity v_{To}, and distance between the droplets s_{To} can be chosen in a wide range. A single droplet stream is called a linear droplet array. To obtain planar monodisperse arrays several parallel droplet streams arranged in one plane are used. The distance s_A between neighbouring droplet streams of one particular array configuration was constant. In Fig. 1 the main parameters characterizing a planar droplet array are explained. All droplets are moving with the velocity v_T. By measuring the droplet size d_T and the droplet velocity v_T at different distances x from the point of ignition the temporal evolution of size and velocity along each droplet path can be determined. This is possible because the initial conditions of each droplet are the same and each droplet undergoes the same history along a particular stream. To study droplet combustion the droplet streams are ignited by a heated coil of filament. After ignition the droplet streams are surrounded by a stationary laminar flame. In Fig. 2 a flame surrounding two droplet streams is shown. The traces

R. S. L. Lee · J. H. Whitelaw · T. S. Wung (Eds.)
Aerothermodynamics in Combustors
IUTAM Symposium Taipei, Taiwan, 1991

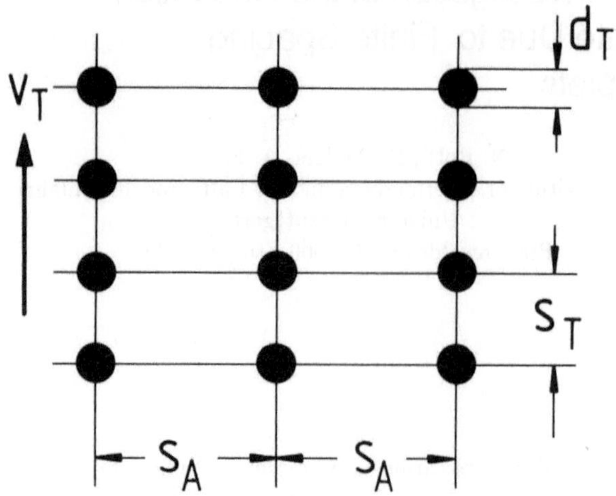

Fig. 1: Sketch of a planar array with configuration parameters.

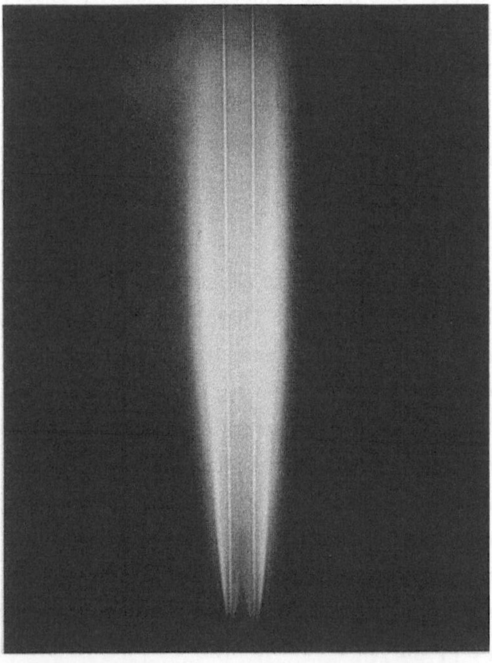

Fig. 2: Flame surrounding two droplet streams with distance $s_A = 400\,\mu m$. The traces of the droplets can be seen due to a continous external illumination.

of the droplets along the streams can be seen due to a continuous external illumination. It depends on the spacing between the droplet streams, whether they are surrounded by a common flame front from the beginning as shown in Fig. 1 or whether each stream is surrounded by an individual flame front. In the latter case, which can be observed for larger spacings, the individual flame fronts merge after a certain distance into one common front for all streams.

MEASURING TECHNIQUES

To investigate the interaction between the droplets a very precise sizing method is necessary. At ITLR a laser interference method was developed for determining droplet sizes with an accuracy of about 2 %. This instrument is independent of the intensity of the incident light. The light scattered by the droplets in the forward region is recorded by a linear CCD camera over a small angle. This light intensity distribution is digitized by an A/D-converter and evaluated by a computer. For scattering angles from approximately 30° to 80° the scattered light shows regular fringes. The distance between the fringes is a measure for the droplet size. Detailed information about this sizing method has been published elsewhere [2,3]. The droplet velocity is determined simultaneously using a conventional LDV-technique [4]. Here the Doppler signals are detected by a photodiode. The signals are received by a transient recorder and then evaluated by a computer. They can be displayed on the computer screen, to control the regularity of the droplet stream and to detect coagulations between droplets.

RESULTS

Linear arrays consisting of one single droplet stream and planar arrays consisting of two, three, or five parallel droplet streams have been studied. In all experiments ethanol was used as fuel. In the experiments with the single droplet stream the measurements were performed for different initial spacings s_T. An example is shown in Fig. 3. There the droplet diameter d_T is plotted as a function of the time t, where t is the time a droplet needs from the ignition point to the measuring point. As the droplets are decelerated by the ambient atmosphere the velocity along the droplet path has to be measured in order to obtain the temporal evolution of the droplet size from the measurements performed at different locations x along a droplet stream. As can be seen from Fig. 3 the burning rate decreases with decreasing initial droplet spacing s_{T_o} as a result of mutual interaction.

To study the mutual interaction in planar droplet arrays the temporal evolution of droplet size has been determined along each droplet stream within an array. As an example the results of a planar droplet array with three parallel streams, the center stream and the

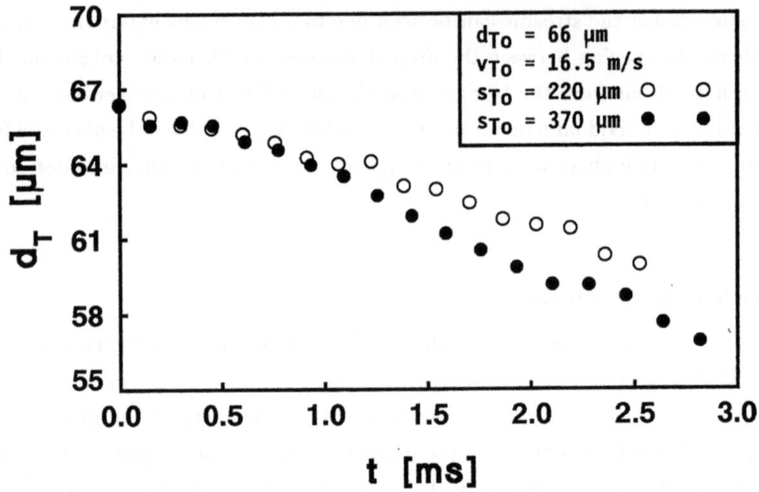

Fig. 3: Droplet diameter d_T as a function of time t for linear droplet arrays with different initial spacings s_{T_o} between neighbouring droplets. In both cases the initial droplet diameter $d_{T_o} = 55\,\mu m$ and the initial droplet velocity $v_{T_o} = 16.5\,m/s$ where the same.

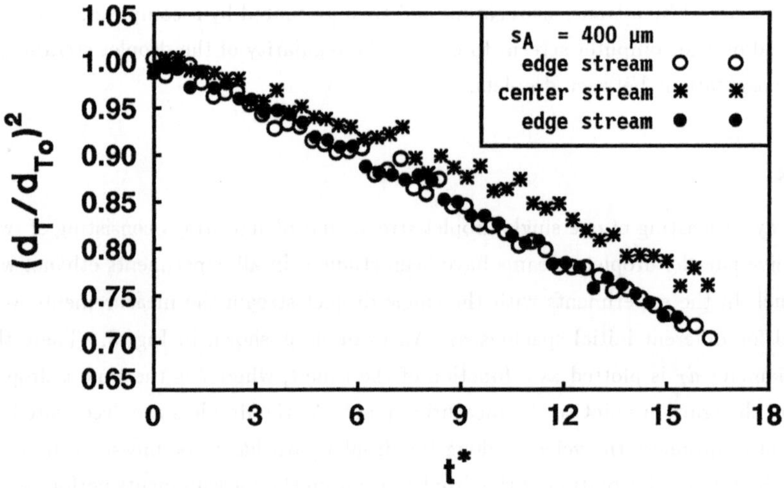

Fig. 4: Square of the dimensionless droplet diameter versus dimensionless burning time t^*. The diagram shows results of planar droplet arrays of three droplet streams with spacing $s_A = 400\,\mu m$ between neighbouring droplet streams. The behaviour of the droplets in the center stream differs significantly from the behaviour in the edge streams.

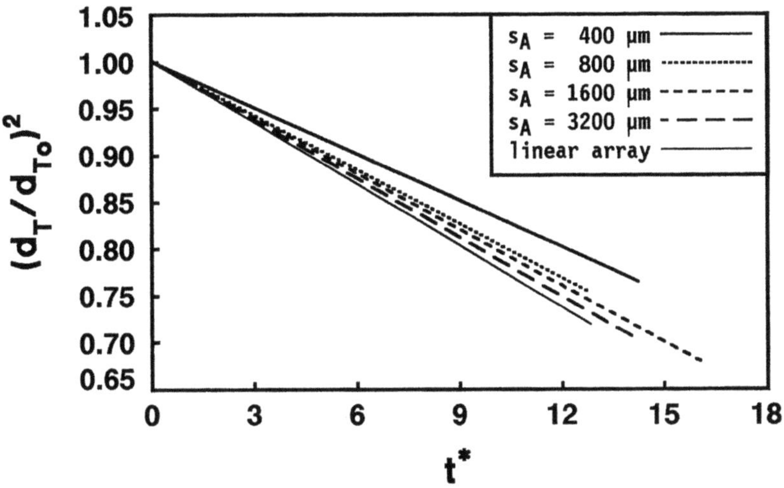

Fig. 5: Square of the dimensionless droplet diameter versus dimensionless burning time t^*. The diagram shows results of planar droplet arrays of two droplet streams with different spacing s_A. Both streams of a particular array configuration are represented by the mean of the individual streams. Each line represents best fits to the measured data. For comparison the results of the linear array are shown in the same manner.

two edge streams, are shown in Fig. 4. In this figure the square of the dimensionless droplet diameter is represented as a function of the dimensionless time $t^* = tD/d_{T_o}^2$. Here is d_{T_o} the initial droplet diameter. The quantity D stands for the coefficient of diffusion for the fuel, here ethanol, against air. The value of D was $13.37 \cdot 10^{-6}\, m^2/s$ [5]. This representation of the measured data can be derived using similarity rules and corresponds to the well-known d^2-law [6]. The d^2-law is valid under the assumption of constant droplet temperature. In the experiments shown here this condition is not fullfilled along the whole path of the droplets. During a instationary initial phase the droplets are heated up to their equilibrium temperature close to their boiling temperature. One can see that a best fit through the data of Fig. 4 should have a slight curvature for small values of t^* which is due to this instationary initial phase. Nevertheless a linear approximation has been used to get an average burning rate. It should be emphasized, that this effect is small. The dimensionless representation of the measured data in Fig. 4 eliminates the influence of different initial droplet sizes. It can be seen from Fig. 4, that due to symmetry the edge streams of this configuration show the same behaviour. This indicates that the results of measurements on droplet streams, which have a symmetric position within the droplet configuration, can be combined and treated as the results obtained from measurements

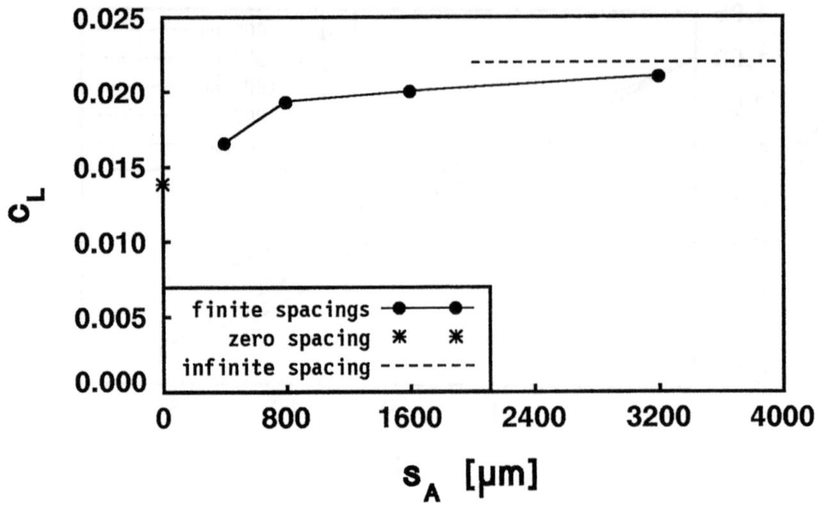

Fig. 6: Slope c_L as a function of distance s_A between the two droplet streams of the array. The dotted line marks the level of c_L if the spacing is infinite. This level is represented by the results of the linear array. The star on the c_L-axis marks the limit for zero spacing between the streams. This limit is derived from the results of the linear array. The ininitial droplet diameter was 61 μm.

on one stream. In Fig. 5 results for arrays consisting of two parallel streams are shown for different spacings s_A. In this diagram the results obtained from measurements of both streams in a particular array configuration are combined in one line. The lines are best fits to the measured data. The slopes c_L of the different lines characterize the burning rate or vaporating rate. It should be mentioned that this representation does not account for the non linear initial phase of droplet burning as mentioned above. For further comparison the result of a linear array having the same initial spacing s_{T_o}, the same initial velocity v_{T_o} and the same initial droplet diameter d_{T_o} is plotted in Fig. 5. The values c_L are shown in Fig. 6 as a function of the spacing s_A between the streams. The dotted line represents the value of c_L for the linear droplet array, which can be considered as a two stream configuration with infinite spacing s_A. Furthermore the result for an array configuration of two droplet streams with zero spacing can be derived from the result of the linear array due to symmetry. This value of c_L is marked on the c_L-axis with a star.

Results of planar droplet arrays consisting of three droplet streams are shown in Fig. 7. Plotted is again the slope c_L as a function of the droplet spacing s_A between neighbouring streams. The dotted line represents results for an infinite spacing s_A like in Fig. 6. In Fig. 7 the results of the edge streams of the configurations are marked by closed symbols.

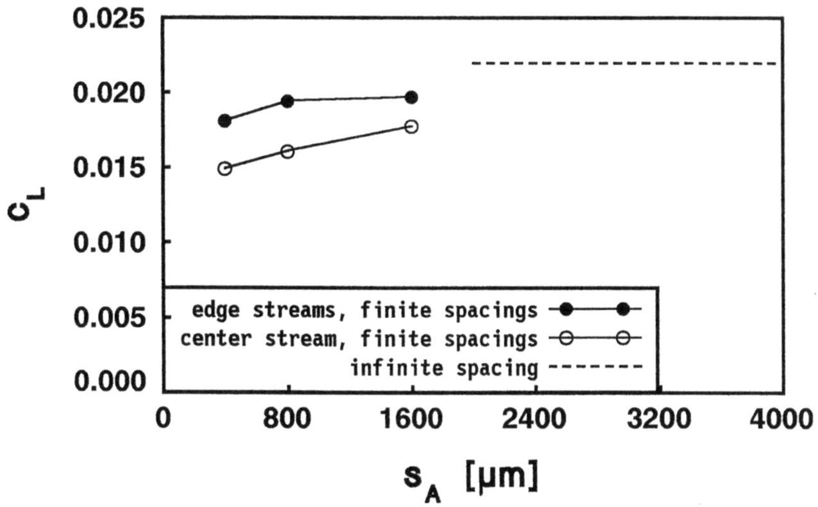

Fig. 7: Slope c_L as a function of distance s_A between neighbouring droplet streams of planar array of three streams. Results are shown for droplets of the edge streams marked by closed symbols and of the center stream marked by open symbols. The dotted line marks the level of c_L if the spacing is infinite. This level is represented by the results of the linear array. The initial droplet diameter was 61 μm.

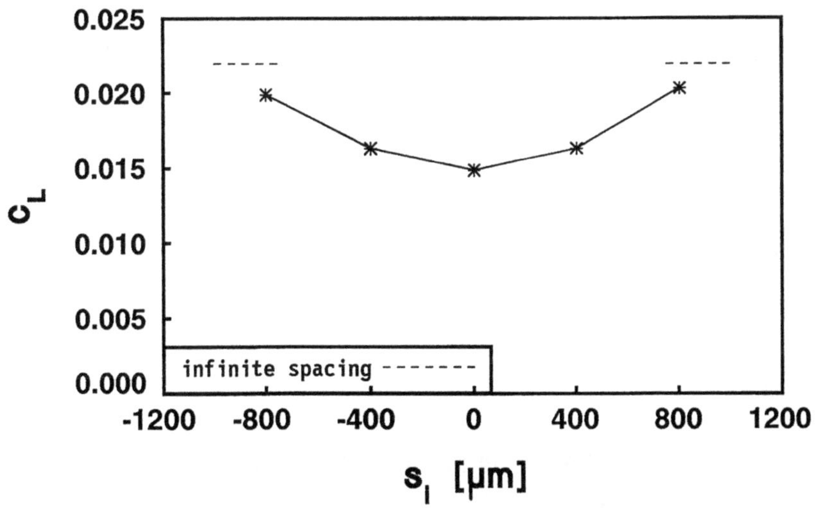

Fig. 8: Slope c_L as a function of the distance s_i from the center stream of planar droplet array consisting of five droplet streams. The distance s_A between neighbouring droplet streams was 400 μm. The dotted line shows the value of c_L for linear droplet array.

For each configuration the result of the left and the right edge stream are presented by one common symbol. The corresponding results of the center streams are marked by open symbols. This diagram shows, that droplets within the center streams, which have neighbours on both sides, show a smaller burning rate than droplets in the edge streams. In all cases of planar arrays discussed above the initial values of droplet diameter d_{To}, velocity v_{To}, and droplet spacing s_{To} were chosen the same in order to allow comparisons between the results.

Finally a droplet configuration consisting of five droplet streams was examined. The spacing between the droplet streams was $s_A = 400\,\mu m$. In Fig. 8 the slope c_L for each of the five streams is plotted against the distance s_i from the center stream.

CONCLUSIONS

The aim of the present investigation was to study the mutual interaction in droplet combustion using well defined droplet configurations. Monodisperse droplet arrays have been examined as linear and planar arrays. The diagrams of Figs. 6 and 7 show clearly that for wider spacings s_A between parallel droplet streams the results approximate the results for a linear array consisting of a single droplet stream. Furthermore the diagrams of Figs. 4, 7 and 8 show, that for the droplet streams in or near the center of a planar configuration, the burning rate is lower than for all other streams. These results show that the combination of the above mentioned measuring techniques with well defined droplet configurations is a very suitable instrument for studying of basic combustion phenomena. The results obtained may be considered as a experimental basis for improvement of numerical simulations.

ACKNOLEDGEMENT

This work has been sponsored by "Stiftung Volkswagenwerk".

REFERENCES

[1] Berglund, R.N., Liu, B.Y.H., (1973), Generation of Monodisperse Aerosol Standards, ENV. SCI. TECH. 7, pp. 147-153.

[2] König, G., Anders, K., Frohn, A., (1986), A New Light Scattering Technique to Measure the Size of Periodically Generated Moving Particles, J. AEROSOL SCI. 17, pp. 157-167.

[3] Roth, N., Anders, K., Frohn, A., (1989), Simultaneous Measurement of Temperature and Size of Droplets in the Micrometer Range, JOURNAL OF LASER APPLICATIONS 2, No. 1, pp. 37-42

[4] Durst, F., Ernst, Zaré, M., (1975), Laser-Doppler-Measurements in Two-Phase-Flows, Proc. LDA Symp. Copenhagen, P. O. Box 70, DK-2740 Skovlunde, Denmark 1976.

[5] Weast, R.C.(Ed.), (1978), *CRC Handbook of Chemestry and Physics, 59th edition,* CRC Press Inc., Boca Raton.

[6] Godsave, G.E.A., (1953), Studies of the Combustion of Drops in a Fuel Spray - The Burning of Single Drops of Fuel, 4th Int. Symp. on Combustion, Williams and Wilkins, Baltimore, pp. 818-830.

Blue Flame Combustion in a Jet-Mixing-Type Spray Combustor

Hiroyuki HIROYASU, Masataka ARAI, Kaoru NAKAMORI and Shinji NAKASO

Department of Mechanical Engineering,
University of Hiroshima, Japan.

Summary

A new combustion system called a jet-mixing-type combustor was designed to obtain a blue flame from a kerosene spray. A spray was injected by a conventional-type swirl atomizer into the combustor and combustion air was introduced through a baffle plate with 16 inlet holes. The basic mechanism of this combustion method was revealed as a rapid mixing of the air and the spray, which was achieved by a high-speed air jet. Combustion characteristics including combustion stability, temperature distribution and exhaust emissions were compared with a conventional swirl-type combustor. NO and THC emission levels exhausted from this blue flame combustor were lower than those exhausted from the swirl-type combustor.

Introduction

Turbulent diffusion combustion systems in which sprays of liquid fuels are burned have been widely utilized in many industries as heating equipment. However, owing to a characteristic of the diffusion flame [1], these conventional combustion systems have been exposed to the danger of an exhausted soot emission which is one of the most undesirable of combustion products [2,3] for users.

To be free from this soot emission problem, a blue flame burner system has been desired, and a few small burner systems which produce a blue flame have been successfully developed [4-6]. Also there has been some minimal experimental work [7-9] concerning the relation of the soot emission and the visual appearance of the flame, such as yellow and blue flames or luminous and nonluminous flames. While a soot-free combustion system has been strongly demanded, there is no blue flame system which can allow the use of the conventional spray system.

A new combustion system called a jet-mixing-type combustor has been developed by the authors according to the demand for a nonluminous blue flame combustion system to attain a soot-free combustion. In this report, the principle mechanism of the combustor and exhaust emission characteristics are discussed.

R. S. L.Lee · J. H. Whitelaw · T. S. Wung (Eds.)
Aerothermodynamics in Combustors
IUTAM Symposium Taipei, Taiwan, 1991

Combustor and Experimental Set-up

Experimental set-up and outline of the combustor are shown in Fig. 1. The combustor was made of stainless steel pipe with an inner diameter of 134mm and a length of 560mm. The combustion air and fuel spray were supplied from the one end of the combustor. The combustion air supplied from a blower was introduced into the combustion chamber as air jets through a baffle plate with 16 air holes. Kerosene fuel was supplied from an accumulator. The fuel nozzle used in this study was a swirl atomizer of a hollow cone type. Its cone angle was 60 deg. and the Sauter mean diameter of its spray was about 25 μm.

In this study, combustion characteristics of nonluminous blue flame, flow field of air jets and fuel spray, temperature distribution in the combustor and exhaust gas emissions were investigated to understand the mechanism and combustion characteristics of the blue flame combustor. In order to make visible the flow field of air jets and fuel spray in the cold state, MgO powder was introduced into the air jets and a quartz glass combustor were used to take a photograph. Temperature distribution in the combustor was measured by a Pt-PtRh13% thermocouple with no compensation for radiation. The gas was sampled through a water cooled sampling probe, and total hydrocarbon (THC) and NO concentrations in the gas were analyzed by FID and chemiluminescence analyzers, respectively.

The baffle plate and relative position of fuel spray and air jets are shown in Fig. 1 (a). This baffle plate was one of the most important elements in this blue flame combustion system. The air holes of the baffle plate were specified by a hole position D, hole diameter d, number of holes n and a relative impinging angle θ between the air jet and the spray sheath. After many trials, these specifications of the air holes were finalized as D = 50mm, d = 5mm and n = 16 to obtain a blue

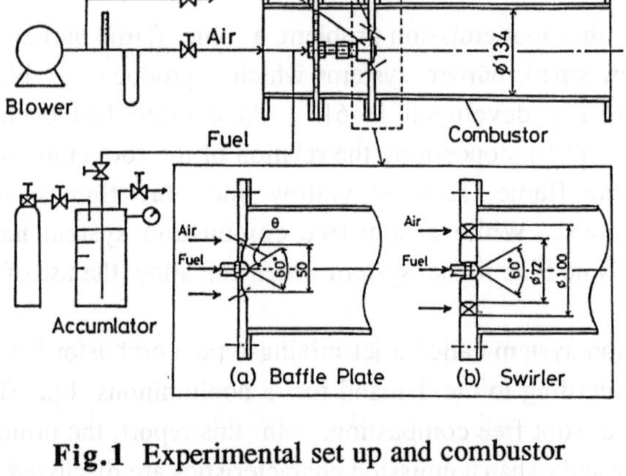

Fig.1 Experimental set up and combustor

flame under wide flow rates of fuel and air. In order to investigate the effect of a rapid mixing process of fuel and air on the combustion process, the relative impingement angle θ was varied from 0 deg. to 90 deg., because the mixing process was strongly affected by the location of the impingement point and the angle between the air jets and the fuel spray. Two kinds of swirlers, of which swirl number S were 0.5 and 0.87, were also used to obtain reference data for this jet-mixing-type combustor. The size of the swirler and its relative position to the nozzle are shown in Fig. 1 (b).

Results and Discussion

Combustion characteristics

The combustion characteristics of this jet-mixing type combustor were thoroughly investigated by changing the relative impinging angle. It was found that the blue flame combustion state appeared under an impinging angle in the range of 30 deg. to 90 deg. The combustion characteristics obtained by the combustor using the baffle plate with a relative impinging angle of 60 deg. is shown in Fig. 2 (a), where Mf is the fuel flow rate and Ma is the air flow rate. The combustion state in the diagram was divided into three zones according to the visual appearance of the flame. In this figure, zone (1) shows the partially blue and partially yellow flame. Both zones (2) and (3) show a kind of blue flame.

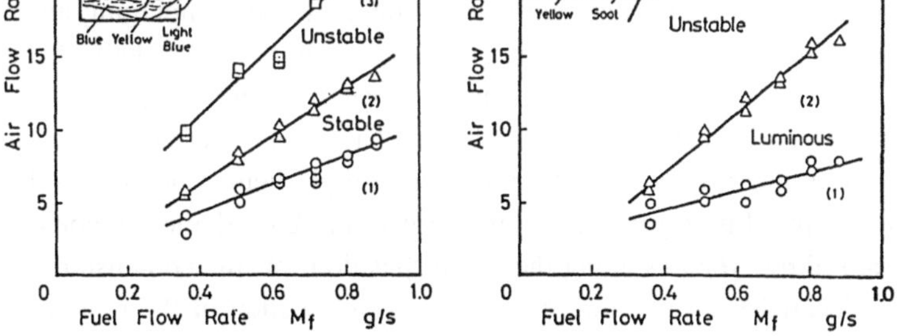

(a) Flame appearance in the blue flame combustor with the 60 deg. baffle plate

(b) Flame appearance in the yellow flame combustor with the 0 deg. baffle plate

Fig.2 Combustion characteristics of the blue and yellow combustors

The blue flame shown in zone (2) was a stable nonluminous flame that was followed by a violet-blue flame at the exhaust end of the combustor, and zone (3) shows a relatively unstable blue flame owing to a lean mixture ratio of fuel and air. With a further increase of the air fuel ratio, a blow off limit was suddenly reached.

Figure 2 (b) shows the combustion characteristics obtained by the combustor with a baffle plate of which the relative impinging angle was 0 deg. The condition of this relative angle means that the sheath of the hollow cone spray and the air jets were flowing in a parallel way and they did not impinge together. No blue flame zone was obtained in this combustor, because a slow and weak mixing process of fuel and air was attained, producing a yellow flame.

The direct photographs of blue and yellow flames are shown in Fig. 3. The combustion photograph taken in the blue flame combustor with a 60 deg. baffle plate shows three nonluminous flame zones. The first flame appeared just alongside the air jets as a violet-blue colored flame. The second flame appeared as the main blue flame in the middle of the combustor. The last one was violet-blue colored and located at the end of the combustor. This last flame only appeared under an equivalence ratio higher than the stoichiometric mixture ratio. The spray supplied in the combustor with 0 deg. baffle plate, was burned as a yellow diffusion flame owing to the slow mixing process of fuel and air. As a reference, the typical flame obtained with the conventional swirl type combustor is also shown in this figure. The swirler had good potential for stabilizing a flame because it caused a strong recirculating flow inside the hollow cone of the spray. However this recirculating flow usually created a yellow diffusion flame owing to the low oxygen concentration and high temperature gas recirculating from the main diffusion flame.

The upper half of Figure 4 shows the isothermal lines of the blue flame combustion state obtained in the combustor with the 60deg. relative impinging angle. The flow rate condition here was the same as Fig. 3 In the lower half of the figure, visual flame zones are shown. The high temperature zone and the main flame location corresponded closely to each other. The flame that appeared alongside the nozzle, was located in the high temperature gradient zone around the air jet. According to the this figure, the combustor in the blue flame combustion state divided into three regions along the axial position of the combustor. The first region was located from the nozzle to the mixing section of the fuel and air. The second was the mixing section itself and the last was the main flame.

Mechanism of the blue flame combustion

The range of the equivalence ratio that allows the blue flame state was obtained as a function of the relative impinging angle to investigate the role of a flow field.

(a) Baffle Plate , θ = 60 deg

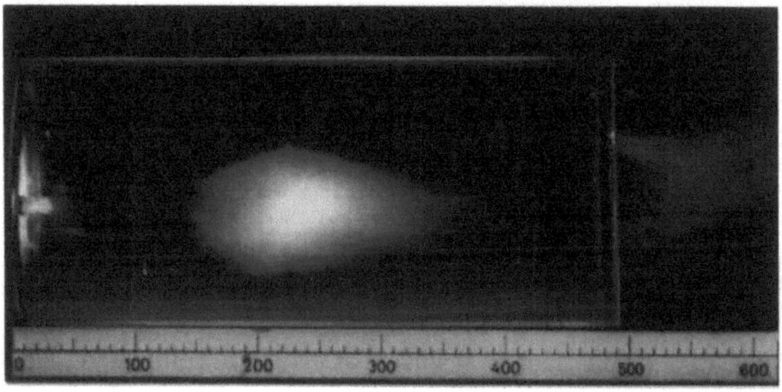

(b) Baffle Plate , θ = 0 deg

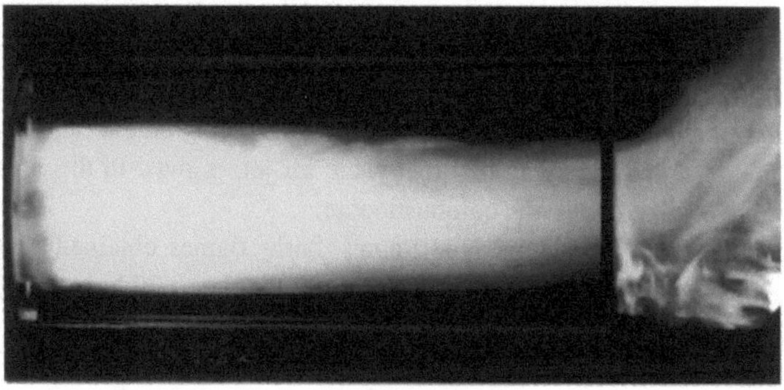

(c) Swiler , S = 0.5

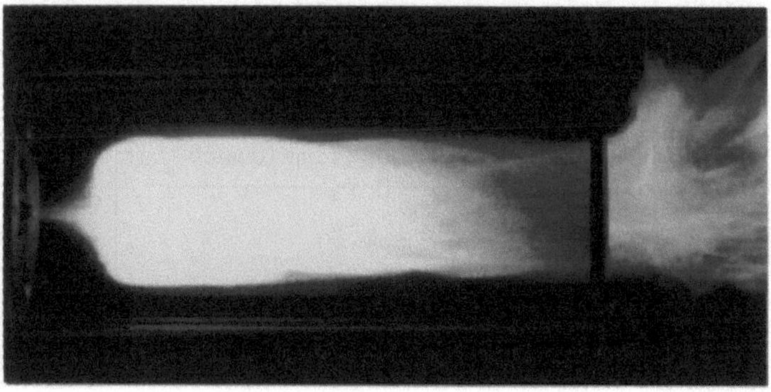

Fig.3 Combustion photographs; Air flow rate M_a = 10 g/s, Fuel flow rate M_f = 0.74 g/s and Equivalence ratio ϕ = 1.1,

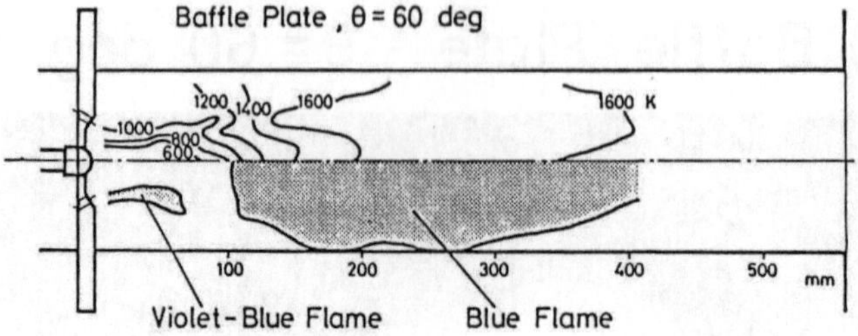

Fig.4 The isothermal lines and flame locations;
M_a = 10 g/s, M_f = 0.74 g/s, ϕ = 1.1

The range of the stable blue flame was widest at the relative impinging angle of 60 deg., as shown in Fig. 5. The flow field made by the air jets with the impinging angle 60 deg. was considered to be suitable in promoting the better mixing that caused the blue flame. To confirm the role of the flow field, the air jets and spray motions in a cold state were made visible using MgO powder as a tracer. From an image analysis of the direct photograph, it was revealed that the air jets flowing through the 16 air holes impinged to the spray sheath and were concentrated together around the combustor axis. It became a single convergent and divergent flow around the axis, and the spray that met each air jet, moved to the center axis and was thoroughly mixed with the combustion air.

In order to compare the yellow and blue flames, both flames obtained from the combustors which had the relative impinging angles 20deg. and 60deg.

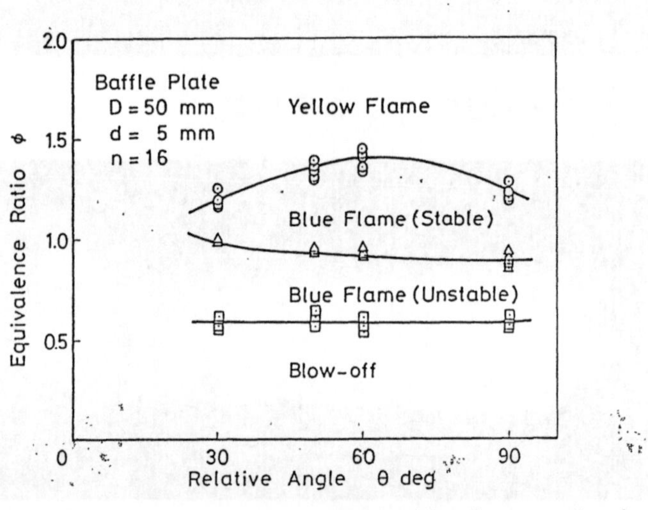

Fig.5 Effect of relative angle of a baffle plate on a blue flame combustion state;
M_a = 10 g/s

respectively were looked at under the same flow condition. The fuel flow rate of 0.74 g/s and equivalence ratio of 1.4 were set as the flow condition where the stable blue and yellow flames appeared in each combustor.

The axial temperature distributions along the center axis of the combustors are shown in Fig. 6. The temperature rise in the blue flame combustor (60 deg.) was retarded compared with that obtained in the yellow flame combustor (20 deg.). This retarded temperature rise was considered to be caused by the convergent and divergent flow field in the blue flame combustor, and, moreover, this was regarded as a characteristic feature of this blue flame mechanism.

The axial distributions of THC are also shown in Fig. 6. The THC level in the yellow combustor shows a very high concentration near the nozzle but it decreased quickly. However, in the blue flame combustor, the THC concentration was not so high near the nozzle. While the THC increased slightly in the main flame zone, it finally decreased to a lower level than that in the yellow combustor at the combustor end.

From the temperature and the THC distributions, it was found that in the yellow flame combustor, a vaporized over-rich mixture was created and it burned as a turbulent diffusion flame. On the other hand, there was no vaporized over-rich mixture in the blue flame combustor; the evaporation of the fuel and thorough mixing with sufficient combustion air occurred rapidly at the impinging area of the fuel and air. Furthermore, the ignition was retarded owing to the high velocity of the convergent and divergent flow of the mixture.

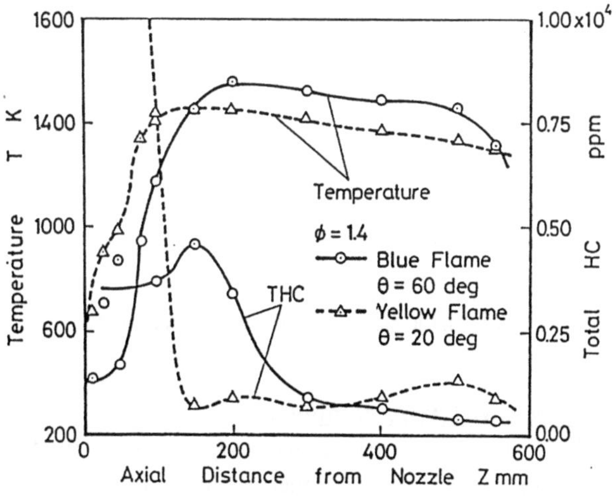

Fig.6 Effects of the relative angle of the baffle plate on the Temperature and THC concentration along the combustor axis; $M_a = 12.7$ g/s, $M_f = 0.74$ g/s, $\phi = 1.4$

According to these investigations, the mechanism of this blue flame combustor is summarized in Fig. 7 as follows.

(1) From the nozzle to the convergent flow section of the air jets: The impingement of the air jets and the fuel spray induced their rapid mixing. In the same instance, a hot lean burned gas supplied from the blue flame around the air jets was entrained into the mixture and promoted the evaporation of the spray.

(2) From the divergent flow section of the air jets to the flame front: Mixing and evaporating processes occurred in the highly turbulent divergent flow. However there was no flame propagation from the main flame, because the axial flow velocity of the mixture in this zone was higher than the flame propagation velocity.

(3) The main flame zone : The main combustion took place here as a blue nonluminous flame.

(4) Out side of the combustor : When the mixture of higher equivalence ratio than the stoichiometry was burned inside the combustor, the supplemental violet-blue flame appeared outside of the combustor.

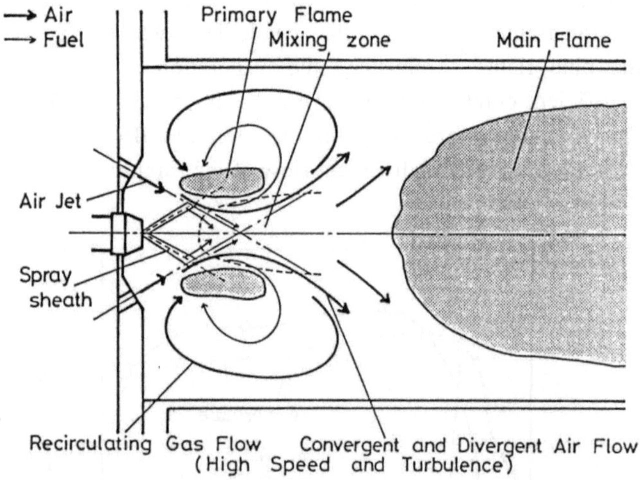

Fig.7 Blue flame combustion mechanism

Comparison with the swirl type combustor

The temperature, THC and NO distributions along the axis in the blue flame combustor are shown in Fig. 8. The isothermal lines in the combustor are also shown in the upper part of this figure. The THC concentration took the maximum value at around z = 100 mm where the main combustion was started. After that, it decreased gradually in the main flame zone. The NO concentration

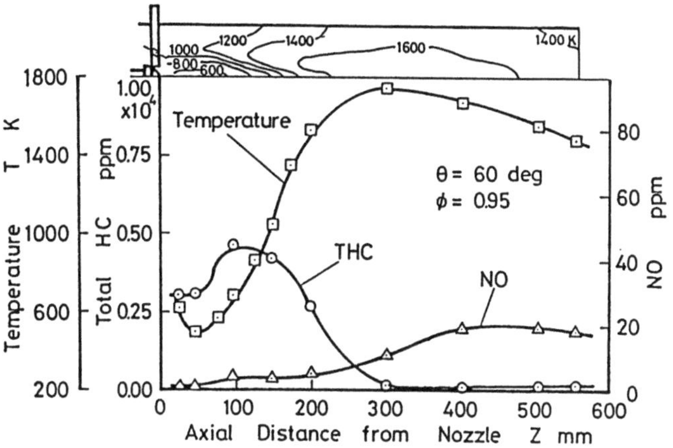

Fig.8 Isothermal lines and axial distributions of THC and NO in the blue flame combustor; $M_a = 10$ g/s, $M_f = 0.64$ g/s, $\phi = 0.95$

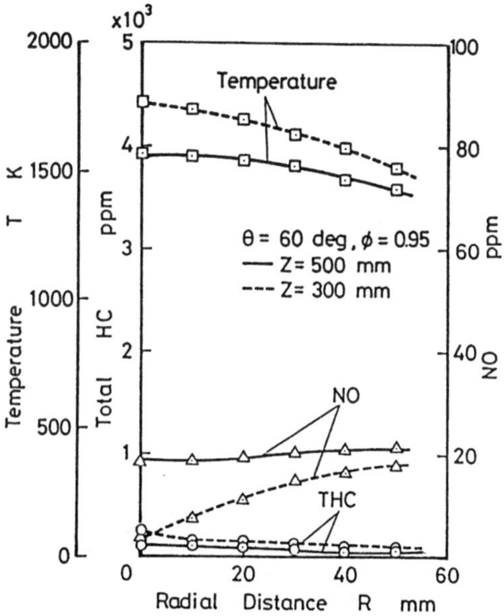

Fig.9 Radial distributions of THC, NO and temperature in the blue flame combustor; $M_a = 10$ g/s, $M_f = 0.64$ g/s, $\phi = 0.95$

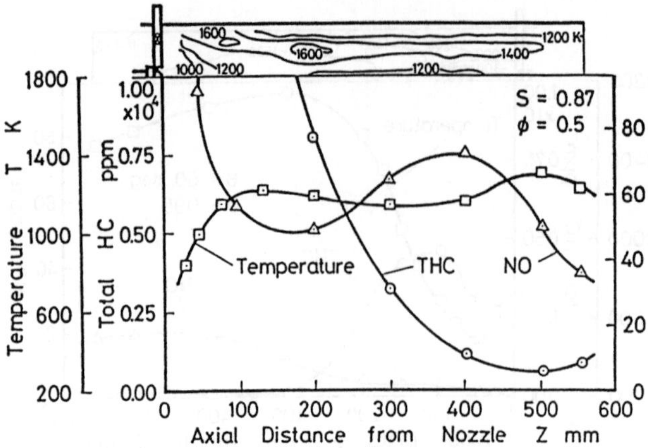

Fig.10 Isothermal lines and axial distributions of THC and NO in the yellow flame combustor with swirler (S = 0.87); M_a = 19 g/s, M_f = 0.64 g/s, ϕ = 0.5

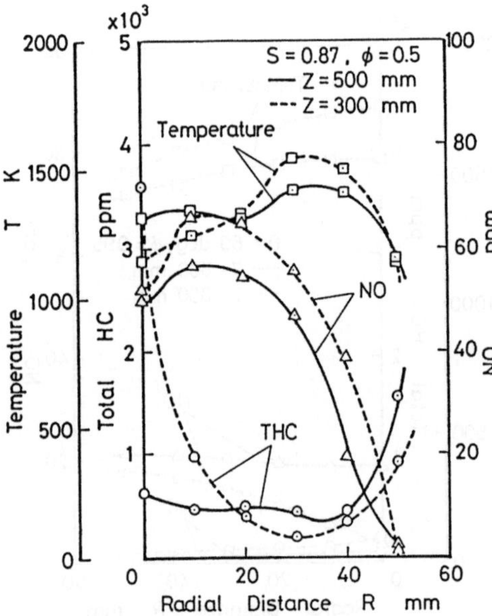

Fig.11 Radial distributions of THC, NO and temperature in the yellow flame combustor with swirler (S = 0.87); M_a = 19 g/s, M_f = 0.64 g/s, ϕ = 0.5

increased gradually in the main flame zone but even its highest value was not so considerable. Figure 9 shows the radial distributions of the THC, NO and temperature at the axial positions of Z = 300 mm and 500 mm. Both the THC and NO levels in the blue flame combustor show uniform distributions at z = 500 mm. The isothermal lines and THC and NO concentrations along the axis which were measured in the yellow flame obtained in the swirl type combustor are shown in Fig. 10. The fuel flow rates of the conditions in Figs. 8 and 10 were the same. However the air flow rates were different because they required different flow conditions to obtain a stable flame. The high temperature zone in the swirl combustor was not located on the combustor axis. Both the THC and NO levels in the blue flame combustor (Fig. 8) show lower levels than those in the swirl combustor (Fig. 10).

Figure 11 shows the radial distributions of the THC and NO at the axial positions of Z = 300mm and 500mm. Both the THC and NO levels in the blue flame combustor (Fig. 9) show lower values than these in the swirl type combustor (Fig. 11). While the equivalence ratio of the swirl combustor was 0.5, almost half of the blue flame combustor, and fuel flow rates of them were the same, the NO and THC exhausting levels of the blue flame combustor at Z = 500 mm were lower than those of the swirl combustor. It suggested that the blue flame combustion system had a greater potential in reducing the NO and THC emissions as well as soot emission.

Conclusion

A new combustion system called a jet-mixing-type combustor was developed to obtain a blue flame combustion system. The combustion characteristics such as combustion stability, temperature distribution, and emission characteristics were investigated in this new combustion system. It was revealed that the combustion mechanism of rapid mixing and retarded ignition was caused by the convergent and divergent air flow in this combustor. This combustion system showed a high potential for attaining soot-free high efficiency combustion with low NO and THC emissions.

References

1. GAYDON, A. G. : The Spectroscopy of Flames 2nd Ed. p.41, Chapman and Hall, 1974.
2. LEFEBVRE, A. H.: Gas Turbine Combustion, p.473, Hemisphere Pub. Co.,1983.
3. LEVY, A., Nineteenth Symposium (International) on Combustion, p.1223, The Combustion Institute, 1982.
4. TANASAWA,Y., Eighth Symposium(International) on Combustion, p.1003, The Combustion Institute, 1960.
5. NAKAYAMA, M., Trans. of JSME, 39-318, 704 (1973).
6. NAKAYAMA, M. and SUZUKI, T.,Trans. of JSME (B), 48-434, 2116(1982).

7. HIROYASU, H.,KADOTA, T. and ARAI, M., Proceeding of the 3rd World Hydrogen Energy Conference, Tokyo Japan, p.1199 (1980).
8. RICHARD, G. A.,SOJKA, P. E. and LEFEBVRE, A. H., J. of Engineering for Gas Turbines and Power, (Trans. of ASME), 111, 84 (1989).
9. SIVATHANU, Y. R., KOUNALAKIS, M. E., GORE, J. P. and FAETH, G. M.: Radiation from Turbulent Nonluminous and Luminous Diffusion Flames. NIST-GCR-88-553 (1988).

The Influence of Upstream Prevaporization on Flame Extinction of One-Dimensional Dilute Sprays

Chi-Chang Liu and Ta-Hui Lin

Department of Mechanical Engineering
National Cheng-Kung University
Tainan, Taiwan 70101 R.O.C.

Abstract

The influence of upstream prevaporization on flame extinction of one-dimensional dilute sprays is analyzed by using large activation-energy asymptotics. The heat loss mechanism is composed of the external heat loss represented by the bulk heat conduction form the spray to a tube wall, and the internal heat loss associated with the droplet gasification process. The enhancement of upstream prevaporization is achieved by varying the wall temperature, and thereby the spray experiences simultaneously the external heat gain and loss in upstream. Results show that the enhancement of upstream prevaporization weakens the partially prevaporized burning of both lean and rich sprays, but has no influence on the completely prevaporized sprays. On the basis of a critical droplet radius, it is found that the extinction curve of a spray having smaller droplets is composed of a upper branch representing the partially prevaporized burning, and a lower branch showing the completely prevaporized burning; and that the flame extinction occurs at the completely prevaporized burning which is originally identified to be unstable in the absence of the enhancement of upstream prevaporization. Conversely, for a spray having larger droplets, the flame extinction is totally controlled by the partially prevaporized burning, and the S-shaped curve shows the influence of the internal heat loss from liquid droplets. The critical droplet radius and the minimum wall temperature are parametrically analyzed and discussed for both lean and rich sprays.

Introduction

The flame extinction of a dilute spray burning in a steady, one-dimensional, low-speed, sufficiently off-stoichiometric, two-phase flow was initially reported by Huang et al. [1]. The mathematical technique was based on the large activation-energy asymptotics introduced to spray studies by Lin et al. [2] and Lin and Sheu [3]. Results showed that a continuous increase of the liquid fuel loading, or decrease of the initial droplet size, results in flame extinction for a rich spray, while no flame extinction occurs for a lean spray. The S-shaped extinction curve (a triple-valued function) for a rich spray was in agreement with Mitani's analysis [4] on the heterogeneous inhibition of premixed flames by chemical inert dust or spray. By introducing the external heat loss represented by the buck heat conduction from the system to the surrounding, to compete with the internal heat loss associated with the droplet gasification process, Liu and Lin [5] found that there only exists C-shaped extinction curves (double-valued functions) for both lean and rich sprays. Furthermore, the flame flux at extinction for a lean spray was first increased, then decreased, and finally approached to exp(−0.5) with increasing

R. S. L. Lee · J. H. Whitelaw · T. S. Wung (Eds.)
Aerothermodynamics in Combustors
IUTAM Symposium Taipei, Taiwan, 1991

the droplet radius; but the correspondingly asymptotic value of the external heat loss was slightly larger than exp(−1.0). For a rich spray, the flame flux at extinction was monotonically increased with increasing the initial droplet radius or decreasing the liquid fuel loading.

From the previous study [5], it is clear that the influence of the internal heat loss on flame extinction is weakened by the introduction of the external heat loss to the system. To understand the whole spectrum on the upstream interaction between external and internal heat losses, we now try to enhance the upstream prevaporization of liquid fuel by varying the surrounding temperature in this study. The detailed discussion will be focussed on the flame extinction curve, the flame flux at extinction, and the critical condition of liquid fuel at extinction.

Configuration and Assumptions

We consider a one-dimensional system in which a plane flame sits at the origin ($x = 0$), the two-phase combustible mixture comes from $x = -\infty$ and equilibrium reaction products move away toward $x = +\infty$, as shown in Fig. 1. The combustible mixture

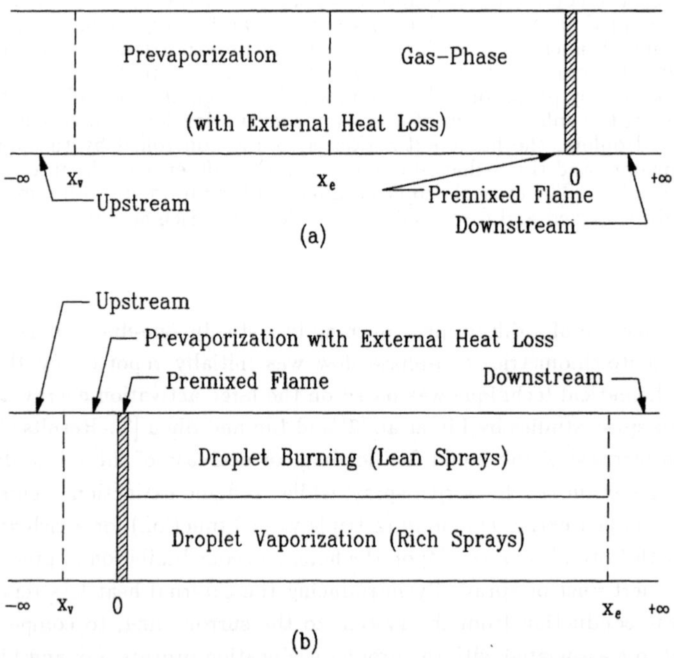

Fig. 1 Schematic diagram of (a) completely prevaporized and (b) partially prevaporized burning sprays.

consists of various concentrations of oxidizer, inert, fuel vapor, and fuel droplets of a certain radius. To achieve the asymptotic analysis, we assumed that the spray is monodisperse and dilute, with the amount of liquid loading being $O(\varepsilon)$ of the total spray mass. The small parameter of expansion ε is the ratio of thermal energy to activation energy in the combustion process.

In the problem, the critical initial droplet radius (r'_c) for the droplet to achieve complete vaporization at the premixed flame front can be identified. Thus for the cases of $r'_{-\infty} \leq r'_c$ and $r'_{-\infty} > r'_c$, we have respectively the completely prevaporized burning and the partially prevaporized burning, as shown in Fig. 1a and 1b, respectively. Depending on the availability of the oxidizer in the downstream region, the downstream mixture has either droplet burning for lean sprays or droplet vaporization for rich sprays in the partially prevaporized burning model (Fig. 1b). We further assume that the droplet temperature is constant, and its motion is in phase with that of the gas. Therefore, the droplets will start to vaporize, at $x = x_v$, only when the gas temperature has reached the boiling point of the liquid. Droplets then ignite upon crossing the flame, and will extinguish at $x = x_e$ upon complete depletion of the oxygen in the gas mixture. In the case of rich sprays only droplet vaporization is possible.

We shall assume the external heat loss is proportional to $(T - T_u)$, representing loss through heat conduction from the spray to a tube wall maintained at the constant temperature, T_u. For simplicity, we shall also assume that external heat loss being of $O(\varepsilon)$ occurs only in the upstream region of x_v to 0. In the previous study, Liu and Lin [5], the wall temperature (T_u) is assigned to be equal to the upstream temperature ($T_{-\infty}$). However, the selected values of T_u in this analysis will be higher than $T_{-\infty}$ to allow for the simultaneous existence of the external heat gain and loss in upstream. It obviously results in the enhancement of upstream prevaporization for the spray.

Finally, we assume that the fuel and oxidizer reaction for the bulk premixed flame is one-step overall, that the fuel droplets burn in the flame-sheet limit, and that constant property simplification apply. The comments on the assumptions were generally discussed in Lin et al. [2].

Formulation and Results

The dimensional conservation equations for heat and mass are referred to Lin et al. [2] following Williams [6]. We designated the extent of gas-phase heterogeneity by the parameter $Z = \rho'_G/\rho'$ such that $Z = 1$ represents the completely vaporized state. Then the non-dimensional equations for gas-phase continuity, and conservations of fuel, oxidizer and energy are respectively given by

$$Z\dot{m}\frac{dZ}{dx} = \frac{A}{T}(1 - Z_{-\infty})^{\frac{2}{3}}(1 - Z)^{\frac{1}{3}}F(T, Y_O) \qquad (1)$$

$$\frac{d}{dx}(Z\dot{m}Y_F - \frac{d}{dx}Y_F) = \dot{\xi} + f_F\dot{m}\frac{dZ}{dx} \tag{2}$$

$$\frac{d}{dx}(Z\dot{m}Y_O - \frac{d}{dx}Y_O) = \dot{\xi} + f_O\dot{m}\frac{dZ}{dx} \tag{3}$$

$$\frac{d}{dx}(Z\dot{m}T - \frac{d}{dx}T) = -\dot{\xi} + f_T\dot{m}\frac{dZ}{dx} - \varepsilon K(T - T_u)H(0) \tag{4}$$

where

$$A = 3\left(\frac{\lambda'}{r'_{-\infty}\dot{m}'_p}\right)^2\left(\frac{P'\tilde{\tilde{M}}'}{\tilde{R}c'_{PG}Q'\rho'_L}\right) \tag{5}$$

$$\dot{\xi} = -\left(\frac{B'\sigma}{\tilde{M}'_O}\right)\left(\frac{P'\tilde{\tilde{M}}'}{\tilde{R}}\right)^2\left(\frac{\lambda'}{c'_{PG}\dot{m}'^2_p}\right)Y_OY_F\exp\left(-\frac{T_a}{T}\right) \tag{6}$$

and the function

$$H(0) = \begin{cases} 1 & \text{for } x_v \le x \le 0 \\ 0 & \text{for } x > 0 \end{cases} \tag{7}$$

while x is the nondimensional distance expressed in units of the preheat zone thickness. In equations (1) – (4), the function $F(T, Y_O)$ and the constant parameters f_F, f_O and f_T are respectively $\ln[1 + (T - T_b)/h_{LG}], 1, 0, -h_{LG}$ for the vaporizing droplet and $\ln[1 + (T - T_b - Y_O)/h_{LG}], 0, -1$ and $(1 - h_{LG})$ for the burning droplet. K and $\dot{m}$ denote the heat transfer coefficient for the external heat loss, and the flame propagation flux normalized by the premixed value, $\dot{m}'_p$.

For the present problem it is more expedient to use an alternate density parameter,

$$\varsigma = \frac{1 - Z}{1 - Z_{-\infty}}$$

instead of Z. Thus in a dilute spray we can expand $Z_{-\infty} = 1 - \varepsilon\gamma$ such that $Z = 1 - \varepsilon\gamma\varsigma$ where $\varepsilon = T_\infty/T_a$.

Performing the inner and outer expansions, and folllowing the detailed matching procedure of ref. [5] to match the inner and outer solutions, we have final results as follows

$$\dot{m}^2 = \exp[T_1^+(0)] \tag{8}$$

indicating that the flame propagation flux is exponentially affected by the first order temperature downstream near the flame, $T_1^+(0)$. The amount of $T_1^+(0)$,

$$T_1^+(0) = \left(\frac{1 - e^{-\dot{m}x_v}}{e^{-\dot{m}x_e} - e^{-\dot{m}x_v}}\right) \quad \left[T_{1\infty}e^{-\dot{m}x_e} + \frac{\gamma}{T_\infty}(\alpha - Y_{i-\infty})(1 - e^{-\dot{m}x_e})\right.$$

$$+\frac{\dot m\gamma}{T_\infty}(T_\infty + h_{LG} - \alpha)\int_0^{x_e} \varsigma_0^+ e^{-\dot m x}dx\Big] + \left(\frac{1-e^{-\dot m x_e}}{e^{-\dot m x_e} - e^{-\dot m x_v}}\right)$$

$$\left\{\frac{\dot m\gamma}{T_\infty}\int_{x_v}^{0}[(T_\infty + h_{LG})e^{-\dot m x} + Y_{i-\infty}]\varsigma_0^- dx - \frac{KY_{i-\infty}}{\dot m^2 T_\infty}(1 - \dot m x_v - e^{-\dot m x_v})\right.$$

$$\left.+\frac{K(T_{-\infty} - T_u)}{\dot m^2 T_\infty}(1 - \dot m x_v e^{-\dot m x_v} - e^{-\dot m x_v}\right\} \tag{9}$$

is governed by the combined influence of internal and external heat losses. Setting $Z = 1$ to approach the condition of a homogeneous premixture, we obtain

$$\dot m^2 \ln(\dot m^2) = -Q_L \tag{10}$$

where the parameter,

$$Q_L = K\left[\frac{(T_\infty - T_b)}{T_\infty} - \frac{(T_{-\infty} - T_u)}{T_\infty}\ln\left(\frac{T_b - T_{-\infty}}{T_\infty - T_{-\infty}}\right)\right] \tag{11}$$

shows the combined effect of the external heat loss and gain for the case of $T_u > T_{-\infty}$. It is noted that the upstream spray experiences the external heat gain and loss when $T_u > T_0^-$, and $T_u < T_0^-$ respectively. Therefore, Q_L represents the net influence of external heat transfer on flame extinction, and will be discussed with the other two parameters of the internal heat loss such as the initial droplet radius ($r'_{-\infty}$) and the liquid fuel loading (γ) in the following calculations.

Discussions

Sample calculations for n-octane (C_8H_{18}) burning in air are now considered to illustrate the interaction between external and internal heat losses on the flame extinction of dilute sprays by using equations (8) and (10). In addition to three parameters of $\gamma, r'_{-\infty}$ and Q_L used before, Liu and Lin [5], we vary the wall temperature T_u in the analysis. As mentioned earlier, choosing a higher value of T_u will result in a region near x_v and a region ahead of the plane flame having external heat gain and loss, respectively. However, the higher value of T_u will accompany with a larger value of K to provide enough Q_L for flame extinction. To avoid the participant of the learning effect in the gas-phase mixture, discussed by Lin et al. [2], we perform calculations in a non-conserved manner which maintains the initial gas-phase composition the same but varies the liquid fuel loading systematically.

Lean Sprays

The responses of flame flux ($\dot m$) on the external heat loss (Q_L) are presented in Figs. 2 and 3 with various values of the initial droplet size ($r'_{-\infty}$) and the wall temperature (T_u) for a lean spray of $\phi_G = 0.8$ and $\gamma = 0.02$. The C-shaped extinction curves of the homogeneous premixtures and the completely prevaporized sprays are shown by dot-dash lines and solid lines, respectively. For the completely prevaporized sprays, the critical value of $\dot{m}$ at extinction ($\dot{m}_E$) and the corresponding asymptotic value of the external heat loss are higher than exp(−0.5) and exp(−1.0), respectively. This is caused by the additional heat gain through burning the secondary gasified fuel from the droplet gasification process for a lean spray. It is further found that extinction curves of the homogeneous premixture and the completely prevaporized spray are independent of the wall temperature. A spray having a given burning strength requires a fixed amount of external heat loss to achieve flame extinction. Therefore, the additional heat gain associated with a higher value of T_u will be suppressed by the following heat loss enhanced by a larger value of K, resulting in the same amount of Q_L for extinction.

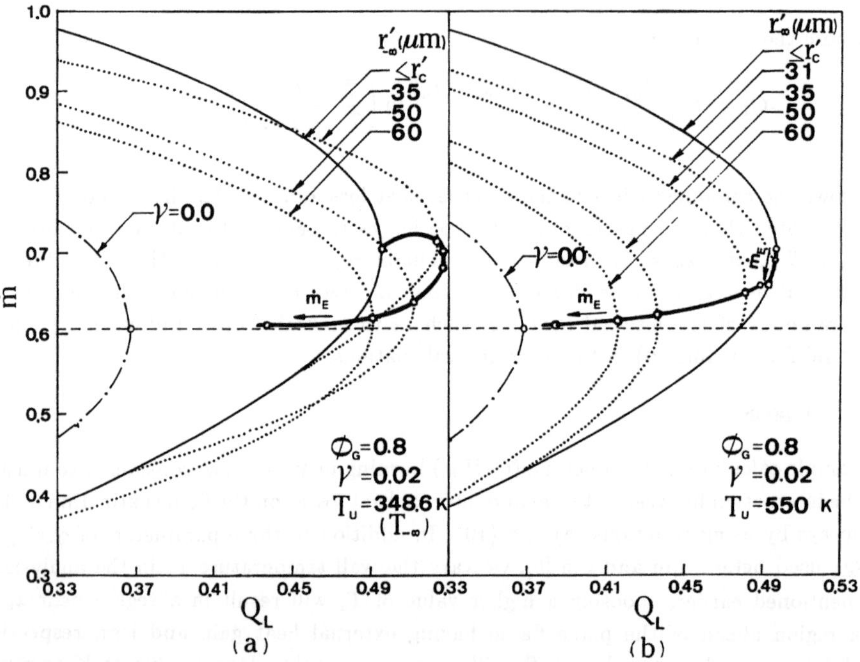

Fig. 2 The extinction curves of the partially prevaporized spray with external heat loss Q_L for various values of $r'_{-\infty}$ and T_u, (a) $T_u = 348.6K$ (b) $T_u = 550K$, for a lean spray.

Fig. 2a is basically adopted from the previous study, Liu and Lin [5], for comparison. Results show that the upper branch of $\dot{m}$ curve corresponding to the partially prevaporized spray, shown by dotted lines in Fig. 2a, is initially lower than that of the completely prevaporized burning, and has a cross-over value of Q_L, across which the partially prevaporized burning becomes stronger than the completely prevaporized one.

Considering the same value of $r'_{-\infty}$ in Fig. 2, the heat transfer coefficient K for the case of T_u = 550K is larger than that for the case of T_u = 348.6K at the same Q_L. It is noted that a larger value of K promotes the upstream prevaporization when the spray experiences the external heat gain, enhances the following heat loss near the plane flame, and suppresses the possibility of droplet burning in downstream. Therefore, the extinction curve of the partially prevaporized spray with a certain droplet radius shifts inside that of the completely prevaporized spray. The flame extinction then occurs at the condition of the completely prevaporized burning accompanying with a relatively small Q_L, e.g. $r'_{-\infty} = 31\mu m$ in Fig. 2b.

In Fig. 2b, the flame flux at extinction moves along the lower branch of the extinction curve for the completely prevaporized spray, as we increase the initial droplet size. A spray having large droplets has a small propagation flux, i.e. a large preheating thickness and a shallow temperature gradient at the plane flame. Thus the spray has weak burning intensity, and can be extinguished under a small value of Q_L. Increasing $r'_{-\infty}$ up to $33\mu m$ in Fig. 2b, the flame extinction occurs at the condition of the partially prevaporized burning instead of that of the completely prevaporized burning. By assigning

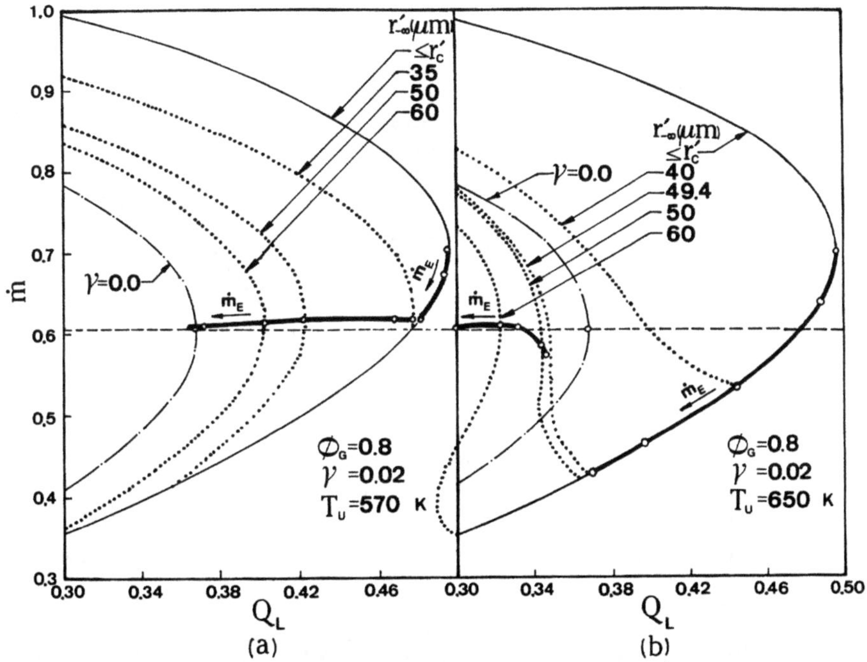

Fig. 3 The extinction curves of the partially prevaporized spray with external heat loss Q_L for various values of $r'_{-\infty}$ and T_u, (a) $T_u = 570K$ and (b) $T_u = 650K$, for a lean spray.

$r'^*_{-\infty} = 33\mu m$, it is suggested that flame extinction of a lean spray is controlled by the completely and partially prevaporized spray for the case of $r'_{-\infty} < r'^*_{-\infty}$ and $r'_{-\infty} \geq r'^*_{-\infty}$, respectively. It is worth to emphasize that as $r'_{-\infty} < r'^*_{-\infty}$, the spray may have partially prevaporized burning under a certain amount of Q_L, but must be identified by completely prevaporized burning on extinction. As we increase $r'_{-\infty}$ further in Fig. 2b, the extinction curve of the lean spray moves toward that of the homogeneous premixture because the liquid fuel in the spray loses its identify.

The above discussion is further explored in Fig. 3. For a very large value of T_u, e.g. $650K$ in Fig. 3b, the $\dot{m}_E$ line has a large jump from the completely prevaporized burning to the partially prevaporized burning at the condition of $r'^*_{-\infty}(49.4\mu m)$. The S-shaped extinction curve is now reproduced and inserted into the C-shaped extinction curve as $r'_{-\infty} = 60\mu m$ in Fig. 3b. It is therefore understood that under the influence of the external heat loss, the characteristics of flame extinction by the internal heat loss (or gain) can be emphasized by the enhancement of upstream prevaporization. In Fig. 3a, it is of interest to note that the flame flux on extinction ($\dot{m}_E$) is almost independent of the initial droplet size and the external heat loss, and closely equal to $\exp(-0.5)$ as $T_u = 570K$.

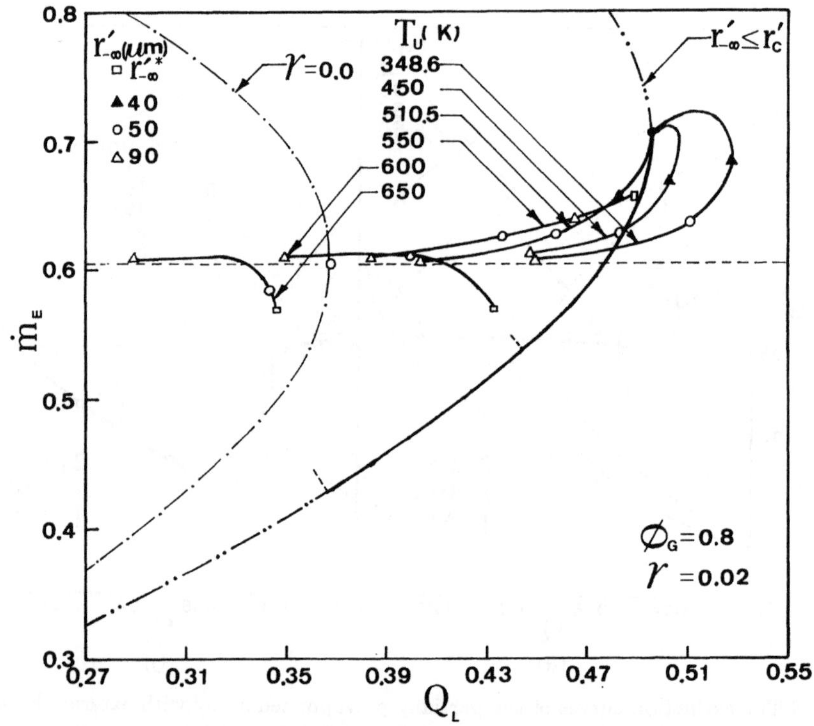

Fig. 4 The flame flux of extinction $\dot{m}_E$ as functions of the external heat loss Q_L and the initial droplet radius $r'_{-\infty}$ and the wall temperature T_u, for a lean spray.

The flame flux on extinction as a function of Q_L is shown in Fig. 4 for various values of $r'_{-\infty}$ and T_u in a lean spray. It is shown that with increasing the initial droplet size, the $\dot{m}_E$ and its associated Q_L on flame extinction are first increased, then decreased, and finally approaches to exp(−0.5) and exp(−1.0) respectively for small values of T_u. Increasing the wall temperature T_u, the $\dot{m}_E$ line will shift inside the envelope of the extinction curve for the completely prevaporized spray. At the critical condition of $T_u = 510.5K$, the $\dot{m}_E$ is monotonically decreased with increasing the initial droplet size. As $T_u > 510.5K$, the $\dot{m}_E$ curve is broken by a jump from the completely prevaporized burning to the partially prevaporized burning at $r'^{*}_{-\infty}$. In Fig. 4, it is generally concluded that the unstable branch of the C-shaped extinction curve for the completely prevaporized spray may be realistically approached by considering a partially prevaporized spray having a higher value of T_u; and that lean sprays having larger droplet size or experiencing larger T_u will be extinguished under a smaller Q_L.

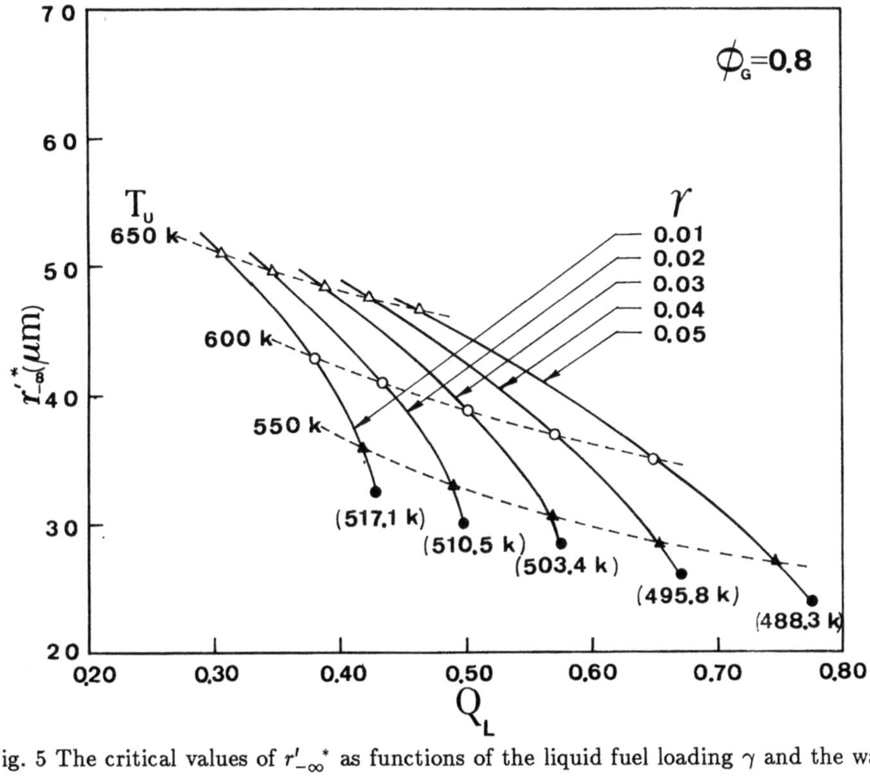

Fig. 5 The critical values of $r'^{*}_{-\infty}$ as functions of the liquid fuel loading γ and the wall temperature T_u, for a lean spray.

The variation of $r'^{*}_{-\infty}$ for a lean spray $\phi_G = 0.8$ is presented in Fig. 5 with various values of γ, T_u and Q_L. According to Fig. 5, the value of $r'^{*}_{-\infty}$ is increased with increasing T_u for a given amount of γ. This points out that the external heat gain enhancing the upstream prevaporization, allows the spray having larger droplets to undergo completely prevaporized burning. Increasing the liquid fuel loading, the $r'_{-\infty}{}^*$ curve moves to experience higher Q_L because of the additional burning of the secondary gasified fuel for a lean spray. It is further found that the minimum value of T_u and the corresponding $r'_{-\infty}{}^*$, allowing a jump from the completely prevaporized burning to the partially prevaporized burning on extinction, are decreased with increasing the liquid fuel loading. Finally, it seems that all curves of $r'_{-\infty}{}^*$ have a tendency to converge together at a very high T_u, as shown in Fig. 5.

Rich Sprays

Variations of the extinction curves with various of $r'_{-\infty}$ and T_u are shown in Figs. 6 and 7 for a rich spray of $\phi_G = 2.0$ and $\gamma = 0.3$. In contrast to the lean spray, the $\dot{m}_E$ and the corresponding Q_L for the completely prevaporized burning are smaller than exp(−0.5) and exp(−1.0), respectively, as shown in Figs. 6 and 7. The extinction curve of the completely prevaporized spray is located inside the envelope of the extinction curve for the homogeneous premixture. As mentioned earlier, the liquid fuel of a rich spray

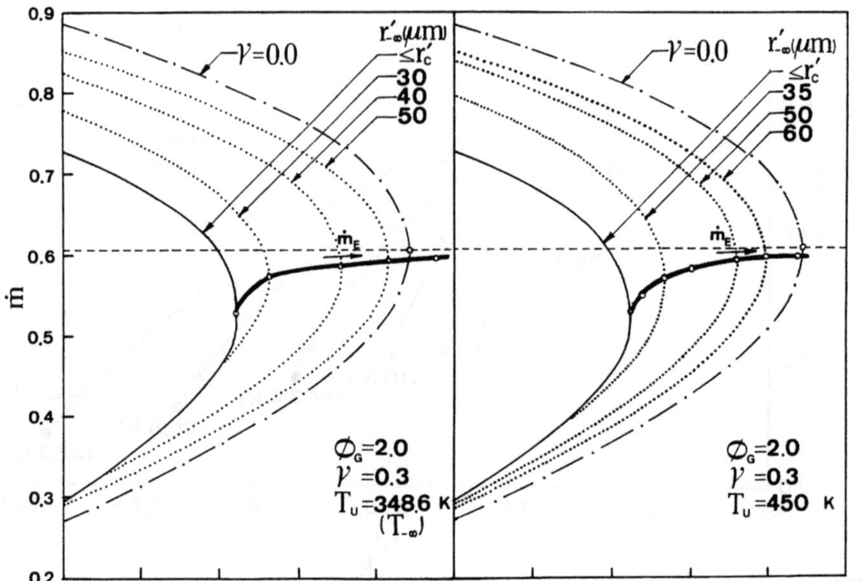

Fig. 6 The extinction curves of the partially prevaporized spray with external heat loss Q_L for various values of $r'_{-\infty}$ and T_u, (a) $T_u = 348.6K$ and (b) $T_u = 450K$, for a rich spray.

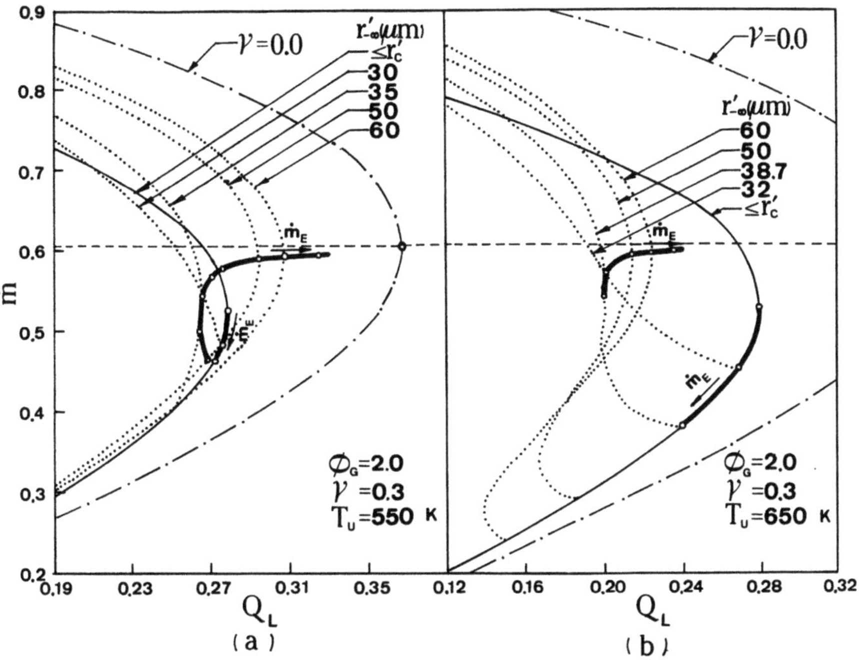

Fig. 7 The extinction curves of the partially prevaporized spray with external heat loss Q_L for various values of $r'_{-\infty}$ and T_u, (a) $T_u = 550K$ and (b) $T_u = 650K$, for a rich spray.

absorbs heat for vaporization without any contribution on burning. For the case of $T_u = T_{-\infty}$ implying no external heat gain, Fig. 6(a) shows that increasing the initial droplet radius accompanied with less droplet vaporization and smaller internal heat loss, the $\dot{m}$ curve for the partially prevaporized spray shifts upward and its turning point related to flame extinction occurs with larger amount of external heat loss. However, as we increase the wall temperature to strengthen the upstream prevaporization, the $\dot{m}$ curve will move closer to the extinction curve for the completely prevaporized spray, e.g. comparing with Figs. 6(a) and 6(b). Increasing the value of T_u, the $r'^{*}_{-\infty}$ is identified to be $32\mu m$ and $38.7\mu m$ for the cases of $T_u = 550K$ and $T_u = 650K$ respectively, as shown in Fig. 7. As $r'_{-\infty} < r'^{*}_{-\infty}$ the extinction curve is composed of a upper branch which is stable and corresponding to the partially prevaporized burning, and a lower branch which is unstable and associated with the completely prevaporized burning. Furthermore, the condition of flame extinction, which is originally identified to be unstable in the absence of the enhancement of upstream prevaporization, can be properly approached by a partially prevaporized spray with a certain amount of T_u. Conversely for the case of $r'_{-\infty} \geq r'^{*}_{-\infty}$, the flame extinction is totally controlled by the partially prevaporized

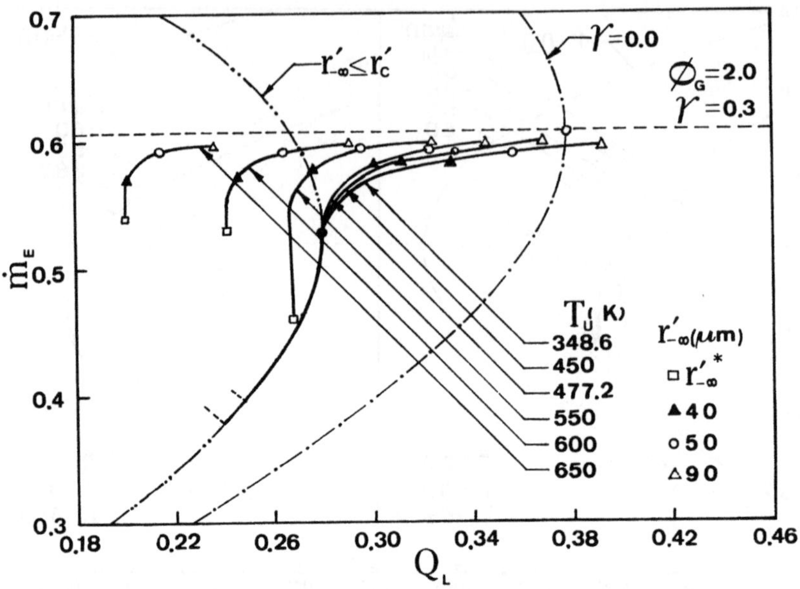

Fig. 8 The flame flux at extinction $\dot{m}_E$ as functions of the external heat loss Q_L and the initial droplet radius $r'_{-\infty}$ and the wall temperature T_u, for a rich spray.

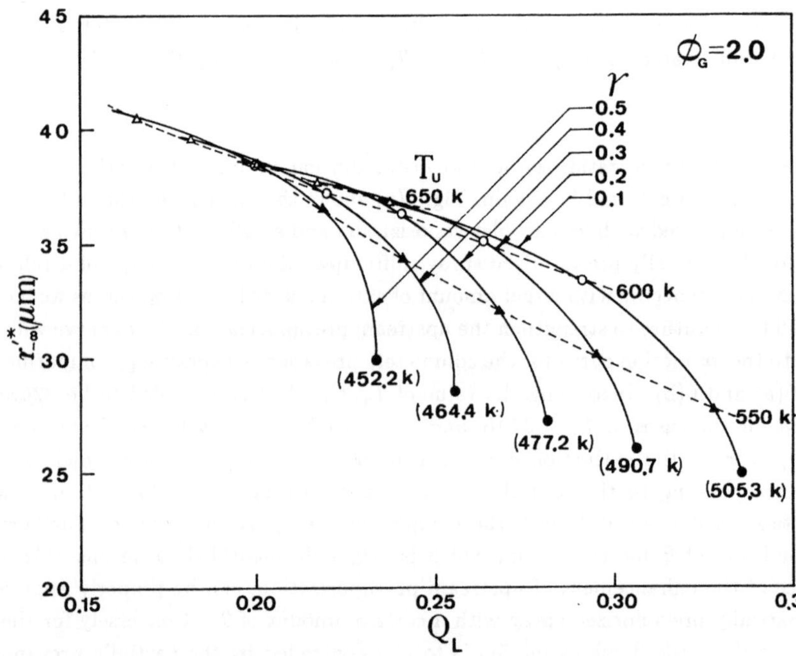

Fig. 9 The critical values of $r'^{*}_{-\infty}$ as functions of the liquid fuel loading γ and the wall temperature T_u, for a rich spray.

burning, and the S-shaped extinction curve is floated to show the influence of the internal heat loss from liquid droplets.

Fig. 8 shows the flame flux on extinction as a function of Q_L for various values of $r'_{-\infty}$ and T_u in a rich spray. It is clearly shown that there is a minimum wall temperature, $T_u = 477.2K$ in Fig. 8, to produce a jump from the completely prevaporized burning to the partially prevaporized burning on extinction. Additionally, by increasing the initial droplet size or decreasing the wall temperature, the $\dot{m}_E$ of the spray will approach to that of the homogeneous premixture because of the spray having less internal heat loss. The jump condition of $r'_{-\infty}{}^*$ as functions of γ, T_u and Q_L for a rich spray is plotted in Fig. 9. For a given amount of liquid fuel loading, a higher value of T_u promotes the external heat gain, enhances the upstream prevaporization, and thereby the spray has a larger $r'_{-\infty}{}^*$, as shown in Fig. 9. Finally, Fig. 9 further shows that the $r'_{-\infty}{}^*$ curve moves toward regions having smaller Q_L with increasing the liquid fuel loading, and asymptotically converges for very high value of T_u.

Conclusion

We have further investigated the interaction between the external and internal heat losses on the flame extinction of dilute sprays by considering the enhancement of upstream prevaporization. The spray flame is burning in a steady, one-dimensional, low-speed, sufficiently off-stoichiometric, two-phase flow, and enduring heat loss through heat conduction from the spray to a tube wall in upstream. Varying the wall temperature promotes the upstream prevaporization, and allows the spray to simultaneously experience the external heat gain and loss in this analysis.

For the completely prevaporized burning of a lean spray, the flame flux at extinction and the corresponding asymptotic value of the external heat loss are respectively higher than exp(−0.5) and exp(−1.0) which are referred to flame extinction of a homogeneous premixture, while the opposite holds for a rich spray. The extinction curve of the completely prevaporized spray is independent of the enhancement of the upstream prevaporization. However, the enhancement of upstream prevaporization influenced by a higher value of the wall temperature, weakens the partially burning of both lean and rich sprays. A critical value of the initial droplet radius ($r'_{-\infty}{}^*$) is then determined to show that the extinction curve of a spray having $r'_{-\infty} < r'_{-\infty}{}^*$ is composed of a upper branch representing the partially prevaporized burning, and a lower branch showing the completely prevaporized burning; and that the flame extinction is controlled by the completely prevaporized burning which is originally identified to be unstable in the absence of the enhancement of upstream prevaporization. For a spray having $r'_{-\infty} \geq r'_{-\infty}{}^*$, the flame extinction is totally controlled by the partially prevaporized burning, and the S-shaped curve shows the influence of the internal heat loss from liquid droplets.

Finally, the minimum wall temperature required to identify the $r'_{-\infty}{}^*$ and $r'_{-\infty}{}^*$ itself are parametrically analyzed and discussed in terms of the liquid fuel loading and the external heat loss for both lean and rich sprays.

Nomenclatures

A equation (5)

B' frequency factor

c'_{PG} specific heat

h_{LG} latent heat of vaporation

K heat transfer coefficient

$\bar{\bar{M}}'$ average molar mass

$\dot{m}$ $\dot{m}/m'_p$, normalized mass flux

Q' heat of combustion per unit mass of fuel

Q_L external heat loss parameter

$\tilde{R}$ universal gas constant

T $T'c'_{PG}/Q'$, temperature

T_a activation temperature

x transformed coordinate

Y $Y_F = Y'_F$ and $Y_O = Y'_O/\sigma$

Z ρ'_G/ρ'

α $\alpha = 1$ and 0 for lean and rich sprays

γ $(1 - Z_{-\infty})/\epsilon$

ϵ small expansion parameter, T_∞/T_a

ς $(1 - z)/(1 - z_{-\infty})$

σ stoichimetric ratio

ϕ equivalence ratio

$\dot{\xi}$ equation (6)

Superscripts and Subscripts:

$*$ critical conditions

$'$ dimensional quantities

b boiling state

e state at which droplet is completely gasified

E state at extinction

f flame front

F,O Fuel and oxygen

$_{G,L}$ gas and liquid phases

$_i$ $i = F$ and O in lean and rich sprays

$_j$ $j = O$ and F in lean and rich sprays

$_v$ state at which vaporization initiates

$_{0,1}$ zeroth and first orders

$_{-\infty,\infty}$ initial and final states

References

1. Huang, C. L.; Chiu, C. P.; Lin, T. H.: The influence of liquid fuel on the flame propagations of dilute sprays in non-conserved systems. J. of CSME 10 (1989) 333–343.

2. Lin, T. H.; Law, C. K.; Chung, S. H.: Theory of laminar flame propagation in off-stoichiometric dilute sprays. Int. J. Heat and Mass Transfer 31 (1988) 1023–1034.

3. Lin, T. H.; Sheu, Y. Y.: Theory of laminar flame propagation in near-stoichiometric dilute sprays. Combust. Flame 84 (1991) 333–342.

4. Mitani, T.: A flame inhibition theory by inert dust and spray. Combust Flame 43 (1981) 243-253.

5. Liu, C. C.; Lin, T. H.: The interaction between external and internal heat losses on the flame extinction of dilute sprays. Combust. Flame 85 (1991) 468–478.

6. Williams, F. A.: Combustion theory. 2nd Ed. Menlo Park, CA: Benjamin/Cummings 1985.

Acknowledgement

This work was supported by the National Science Council, Tainan, R.O.C. under the contract NSC79-0401-E006-20.

Part IV

Unsteady Flows

Control of Oscillations in Ducted Premixed Flames

E.W. Hendricks[1], S. Sivasegaram[2] and J. H. Whitelaw[2]

[1]Naval Ocean Systems Center
San Diego, Ca 92152-5000, USA

[2]Department of Mechanical Engineering
Imperial College, London SW7 2BX, UK

Abstract

The paper presents the results of a study of active control of oscillations with a single dominant frequency in premixed flames stabilised behind a disk and a rearward-facing step in ducts without and with an exit nozzle, by the oscillation of the pressure field and by the oscillatory injection of secondary fuel upstream of the flame holding plane, and quantifies the influence of heat release rate on attenuation. Control was not sensitive to power input greater than the minimum necessary to achieve control and to phase difference between the fluctuating wall static pressure and the oscillatory input for phase within 20° of the optimum and within 10° for oscillation of fuel. Acoustic frequencies in the range between 100 and 200 Hz with rms pressures of up to 3 kPa at the pressure antinode and heat release rates of around 50 kW were suppressed by the oscillation of the pressure field and rms pressures of up to 6 kPa with heat release around 100 kW by the oscillation of fuel. Control of bulk mode oscillations in ducts with nozzles was less effective because the oscillations were associated with poor flame stabilisation.

Knowledge-based control was shown to be more effective than the use of a constant amplitude of oscillatory input. The influence of the gain parameter was found to be small so that the precise determination of transfer functions is not likely to be of critical importance to adaptive control of a single dominant frequency. The use of an adaptive filter with a loudspeaker resulted in the attenuation of rms pressures of up to 2.5 kPa by 20 dB. Adaptive control was less effective with control by oscillation of fuel due to the sensitivity of the filter to the "time of flight" of the cancellation input to the feedback signal.

1. Introduction

Oscillations in ducted premixed flames are usually associated with the acoustic and bulk-mode frequencies of the duct, and rough combustion, associated with high amplitudes of oscillation, is dominated by a single frequency. The dominant acoustic frequency can be attenuated by passive methods such as the attachment of a Helmholz resonator or a quarter-wave tube close to the pressure antinode of the frquency [1, 2], the use of sound absorbent material [3, 4], or the strategic location of a nozzle or an orifice [4, 5] within the duct. Bulk-mode frequencies tend to dominate in the presence of a nozzle and may be averted by improving flame stabilisation by the addition of swirl [6]. Modifications to the fuelling arrangement [7, 8] and to the geometry of the flame holder [9, 10] can, over a limited

R. S. L. Lee · J. H. Whitelaw · T. S. Wung (Eds.)
Aerothermodynamics in Combustors
IUTAM Symposium Taipei, Taiwan, 1991

range of flow conditions, reduce susceptibility to rough combustion. Passive attenuation of oscillations invariably involves changes to the combustor geometry and the need for active control arises when changes to the geometry are inadmissible or large variations in flow conditions are considered.

Active control of combustion oscillations is achieved by imposing oscillations on the pressure field, the flow or the heat release in a way which counteracts the thermoacoustic source. The imposed oscillations are much smaller in amplitude than the oscillations they are intended to suppress and the energy input necessary to impose and maintain control is a negligible fraction of the combustor heat release. Active control by the oscillation of the pressure field using a loudspeaker [11, 12] was demonstrated in combustors with heat release rates of up to 25 kW and by the oscillation of the mean flow [13] for heat release rates of up to 250 kW. The oscillation of the mean flow is hard to accommodate in practical combustors and the use of loudspeakers implies undesirable design features as well as uncertainties in the acoustic boundary conditions. Relatively high levels of oscillatory input may be needed at the large heat release rates and are more readily met by the oscillation of all or part of the fuel [14, 15, 16].

Attenuation of oscillations can be further enhanced by improving the response of the input to the feedback and by adaptive control. The existence of a universal transfer function for all regimes of combustion and different rates of heat release is doubtful and the problem is even more complicated for oscillatory injection of fuel which is subject to variable damping and time delay between the points of injection and burning. It is likely that each control system and combustor will have a unique transfer function to suit a limited set of flow conditions, and a knowledge-based or adaptive controller is desirable with the oscillatory input regulated by a microcomputer on the basis of a periodically monitored feedback signal.

The present work concerns active control of combustion oscillations in premixed flames stabilised behind a disk of area blockage ratio 0.25 in ducts of diameter between 40 and 80 mm and behind a rearward-facing step in ducts of 51 and 80 mm diameter for heat release rates of up to 150 kW and dominant frequencies in the range between 100 and 200 Hz. Control was achieved by the oscillation of a secondary fuel supply constituting up to 18 % of the total fuel by a needle valve driven by a vibrator at the dominant frequency of the combustion instability with the phase and amplitude of the oscillatory input and a gain parameter of the control system as variables. Tests were carried out with the secondary fuel injected radially into the duct a short distance upstream of the flame-holder and with the fuel added to secondary air which was injected either radially or axially into the duct. To quantify phase and amplitude of oscillation with precision, active control by the oscillation of the pressure field using a rigid diaphragm attached to the duct wall and using a loudspeaker was used in the development of the knowledge-based system and the results served as a basis for comparing the performance of control by the oscillation of fuel.

Digital adaptive filters are suitable for the control of several discrete frequencies and have been used succesfully in the active cancellation of noise in liquid filled pipes [18]. This approach was examined in some detail in arrangements where attenuation was not highly sensitive to phase.

The following section describes the experimental method and the equipment. The results are presented the third section, and discussed in the final section.

2. The Flow and Control Arrangements and Instrumentation

The combustor arrangements and the important dimensions are shown in figure 1. Figure 2a shows the control circuit for the knowledge-based control system and figure 2b that for the adaptive filter. In all the arrangements of figure 1, excluding 1b, the main air and fuel supplies were premixed in a swirl register which acted as an acoustically closed end and flowed past a honeycomb flow straightener and a flame-arresting wire-mesh screen before combustion downstream of the flame-holding disk or rearward-facing step. In the arrangement of figure 1b, radial jets of air and fuel mixed by impingement close to the acoustically open upstream end and flowed past wire-mesh screens before combustion downstream of a flame-holding disk.

Active control was achieved by the oscillation of a rigid 35 mm diameter metal diaphragm located at the wall of the combustor duct and attached to a 200 W vibrator (Ling, model 403) driven by a power amplifier (Derritron TA 120) as shown in figure 1a and also by a 100 W, wide-range loudspeaker of cone diameter 300 mm, as shown in figure 1b. Control by the oscillation of heat release involved the oscillation of the flow rate of secondary fuel, comprising up to 18% of the total fuel, by a needle valve attached to the vibrator. The fuel was injected radially into the combustor duct in the arrangement of figure 1c and into a stream of secondary air upstream of the combustor duct as in figures 1d and 1e. The total air and fuel flow rates varied from 1.0 and 0.05 kg/min, respectively, in 40 mm ducts to 4.0 and 0.2 kg/min in 80 mm ducts, and the flow rates of main and secondary air and fuel were metered separately using calibrated flow meters.

Figure 2a shows the circuit for knowledge-based control. The feedback signal from a pressure transducer (Kistler, type 6121 with charge amplifier type 5007) with a constant gain over the range of frequencies of interest was band-pass filtered (Krohn-Hite model 3202) to eliminate aliasing, and phase shifted and preamplified by a custom-built phase shifter before input to the power amplifier driving the actuator. The phase shifter monitored the filtered feedback signal to give a synthesised sinusoid locked to the feedback with a switch-selectable phase shift at 256 equal intervals between 0 and 360° and was identical to that described in greater detail in ref. [17], and the software-controlled output voltage could be incremented in the ratio $2^{n/4}$, where n was any integer between 0 and 63. The characteristics of the vibrator and the loudspeaker determined that their peak-to-

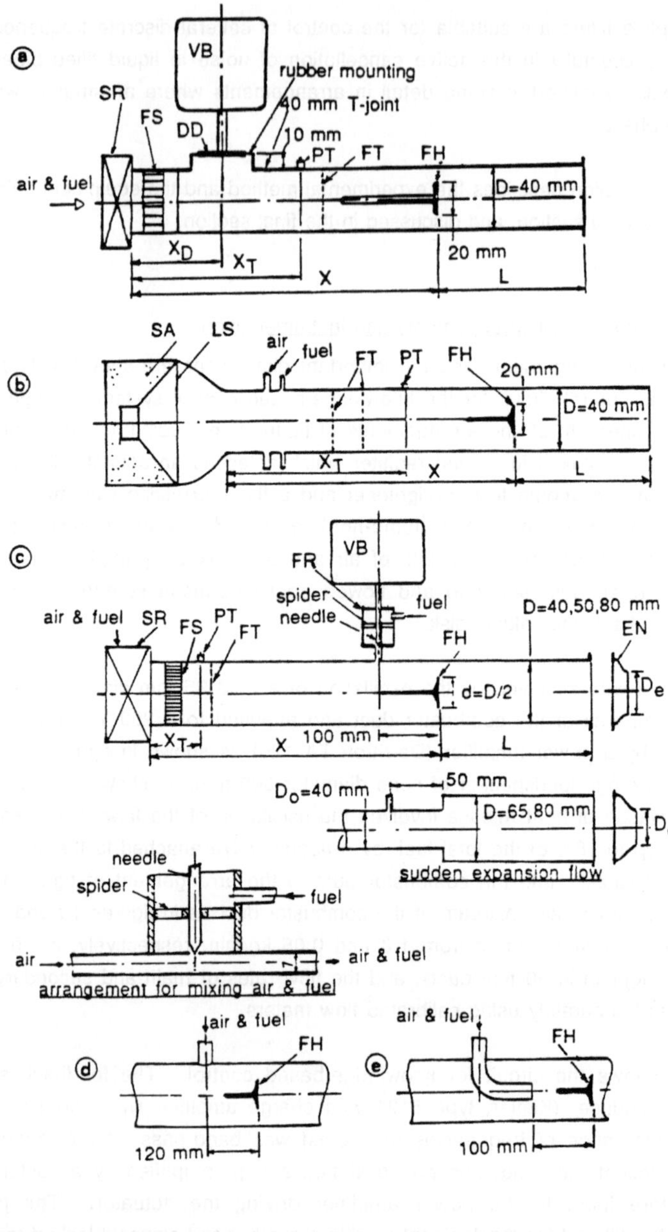

Figure 1: Burner Geometries. (a) arrangement with oscillating diaphragm; (b) with loudspeaker; (c) with direct injection of fuel; (d) modification of 1c for radial injection of secondary air and fuel; (e) modification of 1c for axial injection. DD disk diaphragm, EN exit nozzle, FH flame holder, FT flame trap, FR fuel receiver, FS flow straightener, LS loudspeaker, PT pressure transducer, SA sound absorbent packing, SR swirl register, VB vibrator

peak displacements were proportional to the input voltage so that the acoustic power inputs in the flow arrangements of figures 1a and 1b were incremented by $2^{n/2}$. The oscillation of secondary fuel flow rate in the arrangements of figures 1c, 1d and 1e was incremented by $2^{n/4}$ at small amplitudes of oscillation of the needle and by a smaller ratio at higher amplitudes. The variable n determined the voltage increment and was a gain parameter of the control system with $n=0$ implying an output of constant amplitude and n>>1 an on/off condition.

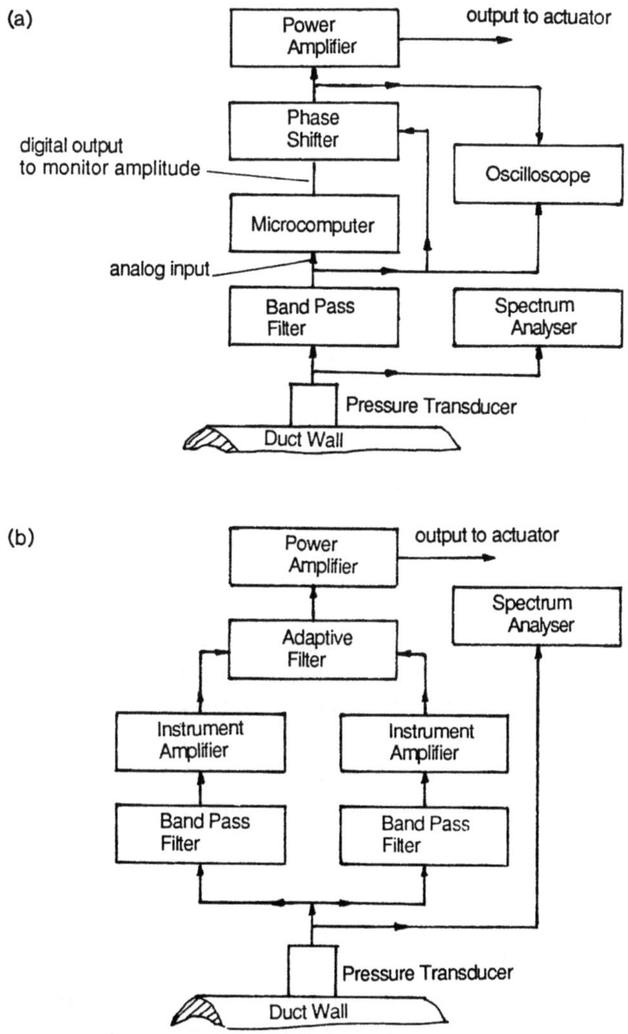

Figure 2: Control circuits. (a) knowledge-based control; (b) control using adaptive filter.

The pressure signal from the transducer was digitised (Data Translation, A/D board DT2824-PGL) and sampled over a time interval just exceeding half the period of the dominant frequency (5 ms for a frequency of 120 Hz) by a microcomputer (Tandon, model TM7104) and the amplitude of a peak or a crest calculated and compared with that for the preceding sample. The computing time was estimated at 4 ms, so that the delay due to the processing of the signal was of the order of one period. The voltage output from the phase-shifter was incremented through an interface board (Amplicon PC-14A) when the feedback amplitude increased and decremented when the amplitude decreased. The output voltage was maintained constant at its maximum value for feedback amplitudes greater than a preset threshold value which, like other control parameters such as the maximum and minimum voltage output levels and the voltage increment *n*, could be reset while the control system was in operation.

Figure 2b shows the control circuit for adaptive control. The adaptive controller was a modified adaptive filter (Adaptive Digital Systems, Inc. Model MASP416) which utilised a digital implementation of the Wiener-Hopf least-mean-squares algorithm for adaptation [19]. The filter was modified to allow differencing of the performance and input signals externally. The bandpass filtered pressure transducer signal was amplified by instrument amplifiers (Fylde, model 351UA) to produce the detector and error signals with voltage gains of around 10 (20 dB) over the filtered signal. In the course of the experiments it was observed that the filter performance was sensitive to a time delay setting which accounts for the "time of flight" of the cancellation input to the feedback from the pressure transducer, and detailed measurements were restricted to control by loudspeaker because of the variation of optimum phase with attenuation and the sensitivity of attenuation to phase in control by oscillation of fuel. This is not an inherent limitation of the control technique since Warnaka et al. [20, 21] have shown that the sensitivity of filter performance to the time delay setting could be rectified by adjusting this delay by a second adaptive filter.

The amplitude and frequency of the feedback were also measured using an FFT spectrum analyser (Spectral Dynamics, model 340) and the phase shift monitored with a dual-beam oscilloscope (Tektronix, model 2225). The input and feedback signals were recorded using an eight-channel data recorder (Hewlett-Packard, model 3968A) and radiated sound levels 20 duct diameters away from the combustor axis and in the exit plane of the combustor measured with a condenser microphone (Bruel and Kjaer, type 4134 with meter type 2215).

3. Results

Control by Oscillation of the Pressure Field

In the arrangement of Figure 1a with a 40 mm duct and no exit nozzle, the acoustic power input to the combustor from the oscillating diaphragm was proportional to the power input to the vibrator, but the limiting driving force on the moving coil of the vibrator determined that the maximum acoustic power was inversely proportional to the square of the frequency for values greater than 100 Hz. The upstream duct length was selected to give a dominant frequency between 100 and 125 Hz so as to maximise the acoustic input which was about 2 mW for a power input of 120 W to the vibrator.

Figures 3 and 4 show the influences of phase and power input on the attenuation of rms pressures of around 2 kPa close to the pressure antinode of a 120 Hz quarter-wave frequency and for a heat release rate of 45 kW. The attenuation was not sensitive to phase within 30° of the optimum and to power input greater than the minimum necessary to achieve control and was around 10 dB with an acoustic input of constant amplitude close to the pressure antinode where it was more effective than at locations close to the node. The attenuation in sound level, as in other measurements reported in this paper, was nearly the same as that in the rms pressure. Smaller levels of attenuation were achieved with rms pressures greater than 2 kPa in longer ducts and larger heat release rates owing to the small amplitude of the acoustic input.

The use of the 100 W, wide-range loudspeaker in the arrangement of figure 1b resulted in an increase in acoustic input to 1 W but the 300 mm cone of the loudspeaker implied that the location of the acoustic input corresponded to a pressure node. Figure 5a shows that half-wave oscillations at 120 Hz with rms pressures of up to 3 kPa close to the pressure antinode in the 40 mm duct with a heat release rate of 45 kW were attenuated by 10 dB with a constant amplitude of acoustic input. The attenuation was not sensitive to phase within 20° of the optimum and tolerance to error in phase was extended to within 40° of the optimum for rms pressures of around 1.6 kPa with input power about twice the minimum necessary for control. Figure 5b shows that a periodic variation in feedback signal amplitude resulted in a temporary loss of lock between the input and the feedback when the amplitude decreased below 500 Pa so that the oscillations were augmented by the input. Once the amplitude of the feedback recovered, lock was swiftly restored and control was effective. This behaviour hindered the possibility of further improvement in attenuation with an input of constant amplitude. The use of knowledge-based control resulted in a 5 dB improvement in attenuation to 15 dB.

Attempts to avert oscillations of high amplitude by imposing control at equivalence ratios close to transition and associated with small amplitudes of oscillation were not successful with a constant and high amplitude of input which tended to induce oscillations, and small amplitudes of input were inadequate to attenuate oscillations at higher equivalence ratios. Knowledge-based control was free from this problem and could be imposed even before transition by presetting the phase at its optimum so that a gradual increase in equivalence ratio ensured that high amplitudes of oscillation were averted.

Control using the adaptive filter resulted in an attenuation of rms pressures of up to 2.5 kPa by 20 dB compared with 15 dB using knowledge-based control. The adaptive controller did not depend on a phase-lock loop to track the frequency of the oscillations and as such was not limited by the threshold level of that circuitry as was the knowledge-based controller. Higher amplitudes of oscillation were associated with greater sensitivity to phase, and the sensitivity of filter performance to the time delay setting resulted in levels of attenuation similar to those with knowledge-based control.

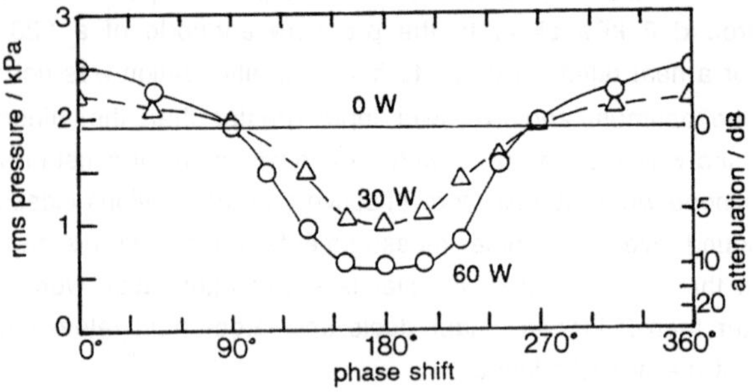

Figure 3: Influence of phase on oscillations. Geometry 1a, D=40 mm, X/D=17.5, L/D=7, $X_T/D=5$, $X_D/D=2$, U=mean upstream velocity=16 m/s, Re=UD/ν=42 000, ϕ=equivalence ratio=0.85, heat release rate=45 kW.

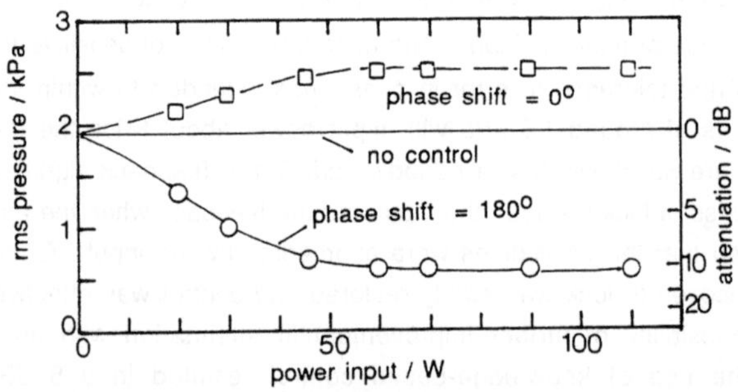

Figure 4: Influence of power input. Flow conditions as for figure 3; optimum phase.

Control by Oscillation of Secondary Fuel

Of the three fuelling arrangements for control by the oscillation of secondary fuel shown in figures 1c to 1e, axial injection of the secondary fuel-air mixture (figure 1e) was the most effective since the damping of the oscillation of fuel due to effects of convection and diffusion were less than in the arrangements with radial injection. All fuelling arrangements altered the stability limits since the primary and secondary flows were not of the same mean composition and the direct injection of fuel sometimes led to early blow-off. Extensive measurements were therefore carried out with the arangement of figure 1e.

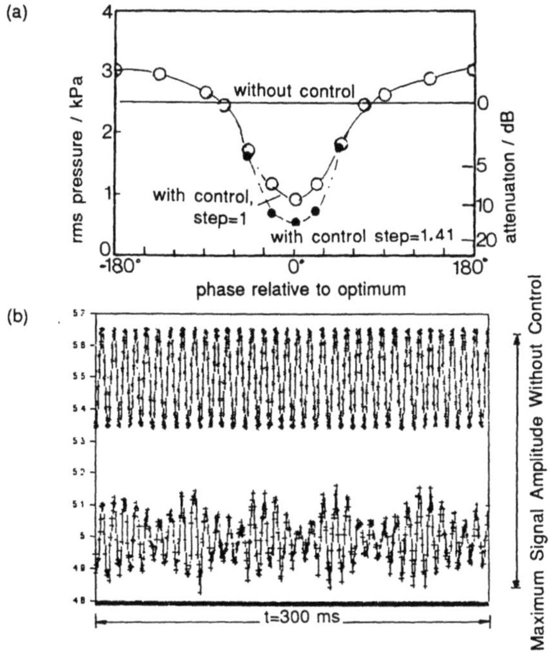

Figure 5: Control using loudspeaker. Geometry 1b, D=40 mm, X/D=35, X_T/D=17, L/D=8, U=16 m/s, Re=42 000, ϕ=0.83, heat release=45 kW. (a) influence of phase on rms pressure; (b) input (upper) and feedback (lower) signal traces for control with constant amplitude input.

The oscillation of the supply rate of the secondary fuel was regulated by varying the amplitude of oscillation of the needle valve. Combustion oscillations corresponding to rms pressures of up to 3 kPa at the pressure antinode of a 120 Hz quarter-wave in the 40 mm duct and heat release rates of up to 50 kW were attenuated by the oscillation of secondary fuel constituting around 12 % of the total. Higher values of rms pressure required higher amplitudes of oscillation of fuel which could not be achieved by increasing the travel of the needle and an increase in the proportion of secondary fuel to 18 % was necessary to control an rms pressure of 4 kPa with heat release rates of around 60 kW.

The phase relationship between the output of the phase shifter and the oscillated heat release was marred by the time delay of the fuel flow between the needle valve and the burning zone and uncertainties in the combustion process so that the optimum phase shift was individually determined for each flow condition. Sensitivity to phase was greater than in control using a loudspeaker, possibly due to uncertainties relating to the additional time delay, estimated to be of the order of 5 ms, due to the flow between the needle valve and the burning zone, the mixing between the primary and secondary streams and to the chemical reaction.

Figure 6 shows that tolerance to error in phase was within 10° of the optimum for amplitudes of oscillation greater than 3 kPa but, as in control by oscillation of the pressure field, extended to within 30° of the optimum for an amplitude of 1.6 kPa. The figure also shows that attenuation with knowledge-based control was around 15 dB compared with 10 dB for control with an input of constant amplitude. The power input to the actuator to sustain control was of the order of 10 W compared with 30 W to impose control for this flow condition, and this implied that it was possible to impose control before the onset of oscillations of high amplitude. But the sensitivity of attenuation to phase and the dependence of the optimum phase on the amplitude of the feedback, although small, made this more difficult to achieve than with the loudspeaker. The uncertainties relating to optimum phase also made it difficult to impose control using the adaptive filter at rms pressures greater than 2.5 kPa at the antinode.

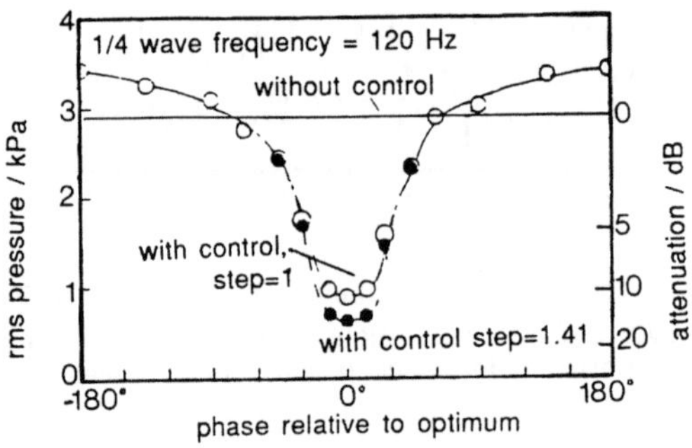

Figure 6: Control by oscillation of fuel. Geometry 1e, disk-stabilised flame without exit nozzle, D=40 mm, X/D=16.5, L/D=8, $X_T/D=2$, U=16 m/s, Re=42 000, $\phi=0.9$ (12% of fuel injected as secondary fuel at $\phi=1.3$), heat release= 50 kW.

Figure 7 shows the feedback traces for three values of the gain parameter *n* and figure 8 the influence of the gain parameter on attenuation for the flow conditions of figure 6 and optimum phase. The use of a gain parameter *n* of around 2, corresponding to a voltage step ratio of 1.4, resulted in the smallest variation in signal amplitude and in maximum attenuation. Figure 8 also shows that the dependence of the attenuation on the gain parameter *n* is small so that the precise determination of transfer functions may not be critical to the active control of a single dominant frequency.

Knowledge-based control by oscillation of secondary fuel attenuated bulk mode oscillations associated with poor flame stabilisation in the 40 mm duct with an acoustically closed end and an exit nozzle with rms pressures of up to 2 kPa by 8 dB. Figure 9 shows that the attenuation was poorer at higher amplitudes.

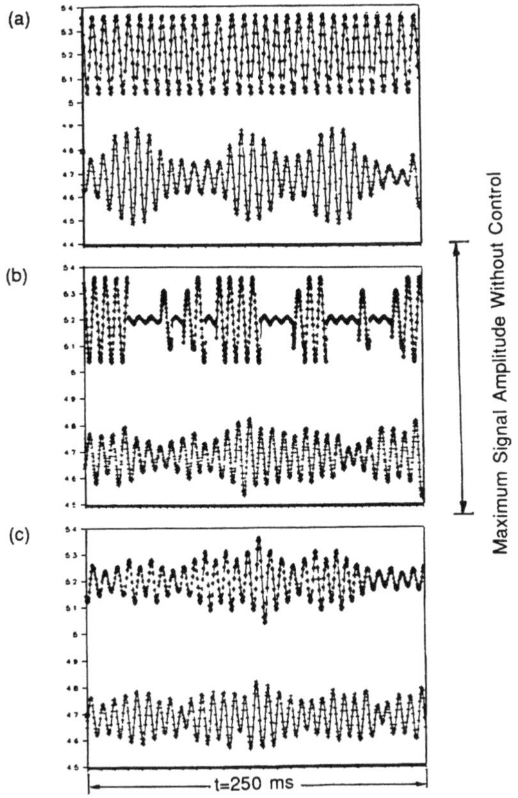

Figure 7: Influence of gain parameter on feedback signal. Flow conditions of figure 6. The upper trace is input and the lower feedback. (a) constant input amplitude, n=0; (b) on/off control, n>>0; (c) input voltage step ratio 1.41, n=2.

Figure 10 shows the influence of heat release rate on the attenuation of oscillations in ducts of 40, 51 and 80 mm diameter and for a range of mass flow rates. The amplitude of oscillations depends on the rate and distribution of heat release which are influenced by the equivalence ratio, flame confinement, flow rate and duct dimensions, so that comparison of results for different flow conditions and duct dimensions needs to be made with caution. The results of figure 10 were obtained with the equivalence ratios of the main and secondary air-fuel mixtures maintained constant for each duct geometry. The amplitude of oscillations increased with duct diameter, and rms pressures close to the antinode in the 80 mm duct for an equivalence ratio of 0.85 were around 10 kPa and could not be effectively attenuated. The results shown for that duct diameter are therefore for an equivalence ratio of 0.75 close to transition to rough combustion with rms pressures of around 6 kPa. The figure clearly shows that an increase in heat release results in an increase in the amplitude of oscillations and a decrease in attenuation and that effective control is not possible with the present arrangement for the oscillation of secondary fuel for heat release rates greater than around 100 kW.

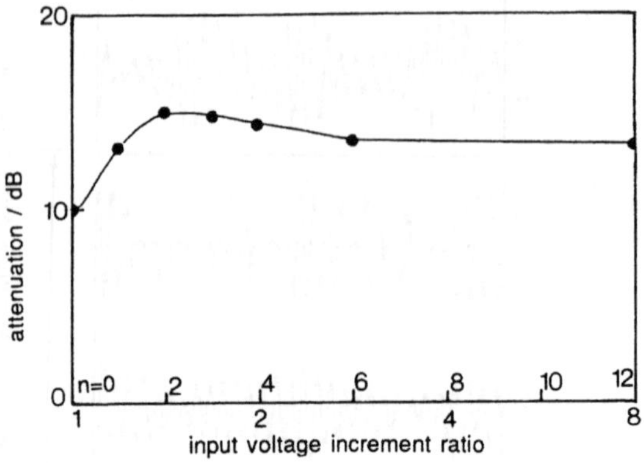

Figure 8: Influence of step size on attenuation. Flow conditions of figure 6; optimum phase.

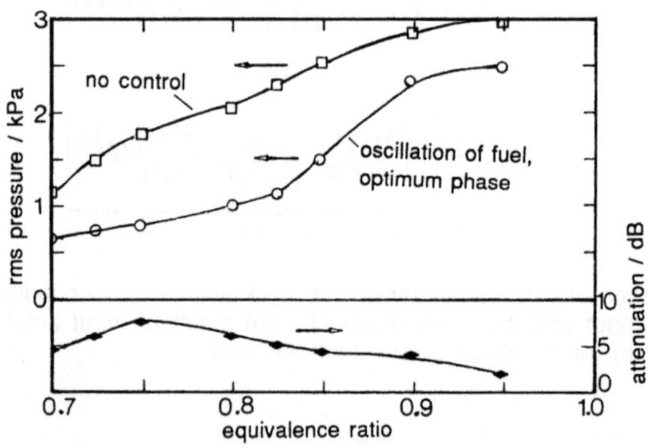

Figure 9: Control of bulk mode oscillations. Geometry 1e, disk-stabilised flame with exit nozzle, D=40 mm, X/D=10, L/D=6, $D_e/D=0.7$, $X_T/D=2$, U=16 m/s, Re=42 000, phase optimum. Approximately 10% of fuel injected as secondary fuel at $\phi=1.0$.

Control of quarter-wave oscillations at a frequency of 200 Hz and a heat release rate of 80 kW in disk-stabilised flame resulted in attenuation by about 10 dB at rms pressures of up to 2.5 kPa. The smaller attenuation was possibly due to reduced oscillation of fuel at the higher frequency and the distribution of heat release over a longer distance downstream of the flameholder at the higher flow rate.

Quarter-wave oscillations at 200 Hz with rms pressures of around 4 kPa near an antinode in sudden expansion flows with an area ratio of 2.5 and heat release rates of around 100 kW were attenuated by between 10 and 15 dB. Effective attenuation at these large heat release rates is possibly due to the release of the secondary air-fuel mixture at a location close to the flame holding plane. Control was less effective, however, for an area ratio of 4.0 due to higher amplitudes of oscillations with rms pressures of the order of 10 kPa at the pressure antinode.

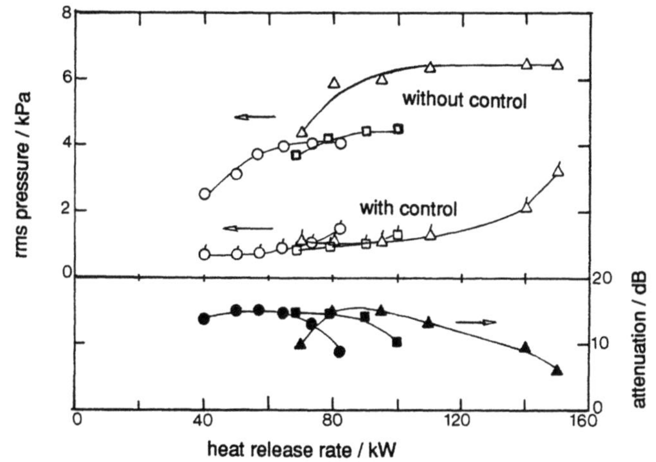

Figure 10: Influence of heat release rate on attenuation. Geometry 1e, disk-stabilised flame without exit nozzle; 18 % of fuel injected as secondary fuel at ϕ=1.2. ○ - D=40 mm, X/D=16.5, L/D=8, X_T/D=2, ϕ=0.9 ; □ - D=50 mm, X/D=13.0, L/D=8, X_T/D=1.5, ϕ=0.9; △ - D=80 mm, X/D=8.5, L/D=8, X_T/D=1, ϕ=0.75. Open symbols without flags indicate rms pressures without control, with flags the attenuated values, and solid symbols the attenuation.

4. Discussion

The results demonstrate that oscillatory injection of fuel is appropriate for the attenuation of high amplitudes of oscillation and that rms pressures of up to 6 kPa with heat release rates of up to 100 kW can be attenuated by 15 dB using knowledge-based control compared with 10 dB using a control system with a constant amplitude of input. The need for a feedback signal to sustain control implies residual oscillations and it is likely that control will result in amplitudes similar to those at equivalence ratios close to transition to rough combustion. The amplitude of oscillation increases with heat release rate in smooth and in rough combustion and that of residual oscillations at the larger heat release rates in practical combustors would be greater than in laboratory combustors. But the ratio of the amplitude in rough combustion to that close to transition also increases with heat release, for example from around 6 (15 dB) in 40 mm ducts and heat release around 50 kW to 10 (20 dB) in 80 mm ducts and 150 kW heat release, so that higher levels of attenuation may be expected at larger heat release rates.

Table 1 compares the maximum attenuation in rms pressure close to a pressure antinode in the present work with those reported in earlier work. Levels of attenuation greater than 20 dB have been reported by Poinsot et al. [11] and Gulati and Mani [12] for heat release rates less than 25 kW with rms pressures less than those associated with rough combustion. At higher heat release rates the maximum attenuation is limited to around 15 dB for rms pressures of up to 4 kPa owing to the small amplitude of the oscillatory input. The smaller level of attenuation reported by Langhorne et al. [14] is possibly due to difficulties in modulating the oscillation of fuel in response to variations in the feedback amplitude. The 20 dB attenuation of an rms pressure of 2.5 kPa in the 40 mm duct using the adaptive filter and the 15 dB attenuation of an rms pressure of 6 kPa in the 80 mm duct by knowledge-based control were achieved at equivalence ratios close to transition. Larger amplitudes of oscillation associated with higher equivalence ratios and heat release rates will probably require higher levels of input and a smaller delay betwen the feedback and the input. Control by oscillatory injection of part or whole of the fuel close to the burning zone using an adaptive filter with reduced sensitivity to time delay settings would enable attenuation of oscillations with higher amplitudes and heat release rates and should also reduce damping of the oscillatory input so that less power will be necessary to drive the actuator. The heat release could be oscillated by oscillating the fuel flow rate or by discrete injection of part of the fuel with phase and proportion of fuel as variables. Care is needed, however, to ensure that the method of injection does not adversely affect flame stabilisation and combustion efficiency.

Large pressure fluctuations close to the location of fuel injection, for example in dump combustors where a pressure antinode exists close to the flame holding plane, could hamper the imposition of control, and time delays and uncertainties relating to the combustion may also limit the effectiveness of the control system. Bulk-mode frequencies relate to poor flame stabilisation and are less amenable to the present type of active control and may require a combination of active and passive methods for their attenuation. The use of active control in conjunction with passive methods which could partially ameliorate the oscillations and the augmentation of the performance of passive devices such as Helmholz resonators by active excitation of the devices could be cost-effective and deserve further investigation.

Acknowledgments

This work was carried out with funds made available under ONR Grant N-00014-89-J-1721. Assistance in developing software from Dr R. Hockey and technical assistance from Mr J. Laker are acknowledged with thanks.

TABLE 1: Attenuation of Oscillations in Ducted Flames

	Technique	Heat Release, kW	RMS Pressure, kPa	Attenuation, kPa	(dB)
Poinsot et al. [11]	Loudspeaker	25	1.7	1.6	(22)
Gulati & Mani [12]	Loudspeaker	1	0.2	0.19	(33)
Bloxsidge et al. [13]	Oscillation of flow	250	4.0	3.2	(15)
Langhorne et al. [14]	Oscillation of fuel	250	4.0	2.5	(8)
Wilson et al. [15]	Oscillation of fuel	1000	small	-	(6)
Knowledge-based control Disk-stabilised, 40 mm duct	Oscillation of diaphragm	45	2.0	1.4	(10)
	Loudspeaker	50	3.0	2.5	(15)
	Oscillation of fuel	60	4.0	3.3	(15)
Disk-stabilised, 80 mm duct	Oscillation of fuel	90	6.0	5.0	(15)
Sudden-expansion, 80 mm duct	Oscillation of fuel	100	4.0	3.3	(15)
Control using adaptive filter Disk-stabilised, 40 mm duct	Loudspeaker	45	2.5	2.2	(20)

References

1. Putnam, A.A.; Murphy, M.J.; Locklin, D.W.: Burner Technology Bulletin: Control of Combustion Noise for Small Gas Burners. Topical Report, GR1-85/210, Gas Research Institute, Chicago 1985.

2. Sivasegaram, S.; Whitelaw, J.H.: Oscillations in Axisymmetric Dump Combustors. Comb. Sci. Tech. 52 (1987) 413-426.

3. Heitor, M.V.; Taylor, A.M.K.P.; Whitelaw, J.H.: Influence of Confinement in Combustion Instabilities of Premixed Flames Stabilised on Axisymmetric Baffles. Comb. Flame 57 (1984)109-121.

4. Sivasegaram, S.; Whitelaw, J.H.: Oscillations in Confined Disk-Stabilised Flames. Comb. Flame 68 (1987) 121-129.

5. Sivasegaram, S.; Whitelaw, J.H.: Suppression of Oscillations in Disk-Stabilised Flames. J. Prop. Power 3 (1987) 291-295.

6. Sivasegaram, S.; Whitelaw, J.H.: The Influence of Swirl on Oscillations in Ducted Premixed Flames. Comb. Flame (to appear) 1991.

7. Schadow, K.C.; Gutmark, E.; Parr, T.P.; Parr, D.M.; Wilson, K.J.: Passive Shear Flow Control to Minimise Ramjet Instabilities. 23rd JANNAF Combustion Meeting 1986.

8. Whitelaw, J.H.; Sivasegaram, S.; Schadow, K.C.; Gutmark, E.: Oscillations in Non-Axisymmetric Dump Combustors. AGARD CP-450, paper 15 (1989).

9. Gutmark, E.; Schadow, K.C.; Nina, M.N.R.; Pita, G.P.A.: Suppression of Buzz Instability by Geometrical Design of the Flame Holder. AIAA Paper 90-1966 (1990).

10. Sivasegaram, S.; Whitelaw, J.H.: Combustion Oscillations and Their Control. ONR Review Meeting, Penn. State University, April 1991.

11. Poinsot, T.; Bourienne, F.; Candel, S.; Esposito, E.; Lang, W.: Suppression of Combustion Instabilities by Active Control. J. Prop. Power 5 (1989), 14-20.

12. Gulati, A.; Mani, R.: Active Control of Unsteady Combustion-Induced Oscillations. AIAA Paper 90-0270 (1990).

13. Bloxsidge, G.J.; Dowling, A.P.; Hooper, N.; Langhorne, P.J.: Active Control of Reheat Buzz. AIIA J. 26 (1988) 783-790.

14. Langhorne, P.J.; Dowling, A.P.; Hooper, N.: Practical Active Control System for Combustion Oscillations. J. Prop. Power 6 (1990) 324-330.

15. Wilson, K.J.; Gutmark, E.; Schadow, K.C.; Smith, R.A.: Active Control of a Dump Combustor with Fuel Modulation. AIAA Paper 91-0368 (1991).

16. Hockey, R.; Sivasegaram, S.: Knowledge-Based Active Control of Oscillations in a Premixed Disk-Stabilised Flame. Report FS/90/17, Mech. Eng. Dept, Imperial College, London 1990.

17. Sivasegaram, S.: Experiments in Active Control of Oscillations in a Premixed Disk-Stabilised Flame. Report FS/89/30, Mech. Eng. Dept, Imperial College, London 1989.

18. Culbreth, W.G.; Hendricks, E.W.; Hansen, R.J.: Active Cancellation of Noise in a Liquid Filled Pipe Using an Adaptive Filter. J. Acou. Sound 83 (1988) 1306-1310.

19. Widrow, B.; Glover Jr., J.R.: Adaptive Noise Cancelling: Principles and Applications. Proc. IEEE, 63 (1975) 1692-1716.

20. Tichy, J.; Warnaka, G.E.; Poole, L.A.: Active Noise Reduction Systems in Ducts. ASMA 84-WA/NCA-75 (1984).

21. Warnaka, G.E.: Active Acoustic Attenuator. US Patent No. 4, 473,906 (1984).

Catalytic Ignition Model in Monolithic Reactor with In-Depth Reaction

TA-CHING TIEN AND JAMES S. T'IEN

Department of Mechanical and Aerospace Engineering
Case Western Reserve University
Cleveland, Ohio 44106

Summary

A transient model has been developed to study ignition in a monolithic catalytic reactor. The special features are the inclusion of thermal and species structures in the porous catalytic layer and the resolution of small-time-scale events. Catalyst loading is used as a parameter to study these structure variations before and after ignition. Regions where the catalytic reaction is kinetically controlled and regions where the reaction is diffusionally controlled are identified in both steady and transient states.

I. Introduction

Catalytic igniter is a passive ignition device which requires no external energy source. The granular bed catalytic igniter has been in use for many years in small hydrazine rocket thrusters for the altitude control of satellites [1]. The use of bipropellant (e.g., H_2/O_2) in granular bed has been tested [2] and modelled [3]. On the other hand, monolithic bed catalytic combustors have been favored in gas turbine application [4, 5] and recently there has been interest in adopting them for the hydrogen and oxygen system for rocket and space power applications [6].

Previous modeling work on monolithic catalytic combustors all assume that the catalytic reaction is a surface reaction [7-9]. Mathematically, the reaction rate is given by the rate of reactant consumed per unit area. This type of data is difficult to obtain experimentally. In reality, a catalytically active porous layer of finite thickness exists in which temperature and species profiles exist and the reaction is generally distributed. One of the main objectives in this modeling work is to be able to investigate the thermal and concentration structures in this layer during ignition transient and in the steady state. The second objective is to resolve small-time-scale events. Previous transient catalytic models often employ quasi-steady approximation except for the solid heat-up process (see, for example, Ref. 7) for the sake of computational economy. Since catalytic layer profiles can be

R. S. L. Lee · J. H. Whitelaw · T. S. Wung (Eds.)
Aerothermodynamics in Combustors
IUTAM Symposium Taipei, Taiwan, 1991

affected by fast transient, we have used a full transient model to investigate this effect. The full transient model can also be used to establish the limitation for quasi-steady model when in-depth reaction is included.

II. Theoretical Model

A monolithic catalytic reactor consists of a solid substrate block with parallel flow channels (e.g., honeycomb). It is assumed that all the channels are identical and the modeling of the reactor can be reduced to the analysis of one single channel unit (i.e., a tubular reactor). On the surface of the channel, a layer of porous material (i.e., alumina) is coated. Within the pores, catalysts are deposited (see Fig. 1). When the gaseous reactants flow through the channel, they diffuse toward the porous layer and into the pores. There they get absorbed and reacted. The heat released conducts through this catalytic layer to both the substrate and the gas in the channel. When the gas temperature in the channel becomes sufficiently high, gas-phase reaction is triggered.

II.1 Chemical Kinetic Rate Expressions

In this paper, the reactants are rich H_2/O_2 mixtures and the catalyst is platinum. The catalytic reaction rate is given by [10]

$$W_c^* = \beta^* 1.772\ (\rho_o^* T^*)^{0.8} \exp\ (-5230/R^* T^*) \tag{1}$$

where β^* is the catalyst loading [kg-pt/m^3].

The gas-phase reaction rate is also represented by an overall reaction rate expression which is deduced from shock tube data [11] using spontaneous ignition theory [12]. It has the following form:

$$W_g^* = K^* P^{*a} \rho_{og}^{*b} \exp(-E_g^*/R^* T_g^*) \tag{2}$$

where $K^* = 4 \times 10^{18}$, a = 1.5, b = 0.1, E_g = 79480, when $T^* \le T_{cr}^*$ and

$K^* = 9 \times 10^{18}$, a = 1, b = 1, E_g = 19780 when $T^* > T_{cr}^*$. The ignition data suggest that a two-activation-energy representation fits the data better. The breaking temperature (T_{cr}^*) is where the two rates intercept each other. In this study T_{cr}^* = 1200 K.

II.2 Assumptions

(a) Diffusion of oxygen (the deficient species) in the porous catalytic layer is governed by Fick's law with an effective diffusion coefficient D_e^*. D_e^* is assumed to be proportional to the absolute temperature so that ρD_e^* = constant.

(b) Similarly, heat conduction through the catalytic layer is governed by Fourier's law with an effective heat conduction coefficient λ_c^*.

(c) Other assumptions are the same as in [7] which include plug flow in gas channel, constant and uniform pressure, negligible heat and mass diffusion in streamwise direction in the gas and the catalytic phases, negligible radiative effect and constant specific heat.

II. 3 Governing Equations

Nondimensional equations are presented below, definitions can be found in the nomenclature section. The values of some of the parameters used in the calculation are listed at the end of the definitions. The detail of the equation derivation can be found in Ref. 13.

Gas-phase

continuity:
$$\delta_g \frac{\partial \rho}{\partial t} + \frac{\partial(\rho u)}{\partial x} = 0 \tag{3}$$

species: (oxygen)
$$\rho\delta_g \frac{\partial y_g}{\partial t} + \rho u \frac{\partial y_g}{\partial x} + J_D(y_g - y_{c,s}) = B_1^*(\rho y_g)^b \exp\left(\frac{-E_g}{T_g}\right) \tag{4}$$

energy:
$$\rho\delta_g \frac{\partial T_g}{\partial t} + \rho u \frac{\partial T_g}{\partial x} + J_H(T_g - T_{c,s}) = q_g B_2^*(\rho y_g)^b \exp\left(\frac{-E_g}{T_g}\right) \tag{5}$$

equation of state:
$$P = \rho T_g \tag{6}$$

Initial conditions:
$$y_g(x=0,t=0) = 1,\ T_g(x,t=0) = 1,\ \rho(x,t=0) = 1,\ u(x,t=0) = 1 \tag{7}$$

Solid phase

Solid phase consists of two layers, a catalytic layer on top of a substrate layer.

Oxygen species in catalytic layer:

$$\delta_m \frac{1}{T_g} \frac{\partial y_c}{\partial t} = \frac{\partial}{r\partial r}\left(r\frac{\partial y_c}{\partial r}\right) - D'_{am} y_c^{0,8} \exp\left[E_c\left(1 - \frac{1}{T_c}\right)\right] \tag{8}$$

Boundary conditions

$$r = r_d,\ \frac{\partial y_c}{\partial r} = S_n(y_c - y_g);\ r = r_c,\ \frac{\partial y_c}{\partial r} = 0 \tag{9}$$

Energy in catalytic layer

$$\delta_c\ \frac{\partial T_c}{\partial t} = \frac{\partial}{r\partial r}(r\ \frac{\partial T_c}{\partial r}) + D'_{at}\ y_c^{0,8}\exp[E_c\ (1 - \frac{1}{T_c})] \tag{10}$$

Boundary conditions

$$r = r\ ,\ \frac{\partial T_c}{\partial r} = B_c(T_c - T_g);\ r = r_c,\ \frac{\partial T_c}{\partial r} = \lambda\ \frac{\partial T_s}{\partial r} \tag{11}$$

Energy in substrate layer

$$\delta_s\ (\frac{\partial T_s}{\partial t}) = \frac{\partial}{r\partial r}(r\ \frac{\partial T_s}{\partial r}) \tag{12}$$

Boundary conditions

$$r = r_c,\ \frac{\partial T_s}{\partial r} = \lambda\ \frac{\partial T_c}{\partial r};\ r = r_s,\ \frac{\partial T_s}{\partial r} = 0 \tag{13}$$

Initial conditions

$$T_c(x,r,\ t = 0) = 1,\ Y_c(x,r,\ t = 0) = 0,\ T_s(x,r,\ t = 0) = 1 \tag{14}$$

It should be noted that only the oxygen species equation is needed. This is because oxygen is the deficient species (the mixture is hydrogen-rich) and the reaction rates are independent on the concentration of hydrogen. The initial conditions are deduced based on a start-up transient in which oxygen is injected into the entrance of the reactor channel in a stepwise manner at time t = 0.

II.4 Time Scales

There are many time scales in this problem [13]. One sees that the shortest time scale is the mass diffusion time in the porous catalytic layer (the heat conduction time in catalytic layer is taken to be the same magnitude), and the longest time is that of the solid (substrate plus catalytic layer) heat-up time. In order to resolve the fast transient event in the reactor, we use the mass diffusion time in catalytic layer, τ^*_m, as the reference time scale. Thus the time ratios (δ's) and the Damkohler numbers (D'_{am} and D'_{at}) in the above system of equations are all ratios of respective time scales in comparison to τ^*_m. One should notice that in realistic catalytic reactors, the solid heat-up time can be much greater than the value listed in the Nomenclature. The reason to choose a smaller value is to accelerate numerical computation because the time steps are limited by the mass diffusion time. Although the solid heat-up time is smaller than that of the realistic device, it is still one order of magnitude greater than the next longest time scale in the reactor. Therefore, the numerical prediction should be correct as long as the proper nondimensional time is used for interpretation.

III. Numerical Scheme

For the gas-phase part which includes four equations (3)-(6) and four unknowns - y_g, T_g, u, and ρ. The Euler explicit method is applied to the species equation and the energy equation for y_g, and T_g and the Euler backward implicit method is applied to solve the continuity equation for u. Finally, ρ is solved from the equation of state. For the solid phase part, Eqs. (8) to (14) are solved by the implicit Crank-Nicolson method. A quasi-linearization technique [8] is implemented on nonlinear heterogeneous chemical reaction rate. The resulting tridiagonal system of linear algebraic equations are solved by the LU-decompositions method. The computational procedures are:

(1) Gas-phase equations (3-6) are solved using initial conditions from Eqs. (7) and (14).

(2) Substituting the gas-phase solution into Eqs. (9) and (11), Eqs. (8) and (10) are integrated to advance y_c and T_c to the next time step.

(3) The new values for $y_{c,s}$ and $T_{c,s}$ are substituted into Eqs. (4) and (5) and the computational procedure is repeated for the next time cycle. Normally the computation is stopped when steady state is reached.

IV. Results and Discussions

In the steady-state operation, catalytic reactor exhibits spontaneous ignition characteristics in spatial direction. During start up, temperol ignition characteristics are also present. In this section, we will first look at the in-depth structure from steady-state calculations and then transient behaviour will be presented.

IV.1 Steady State

In this series, the catalyst loading factor β^* will be varied to illustrate the ignition development. When β^* is changed, the catalytic reactivity and the Damkohler numbers are varied accordingly.

For $\beta^* = 1.8$ kg/m^3, Fig. 2 gives the in-depth profiles in the solid. Fig. 2(a) shows that the solid temperature is uniform in r-direction in the

substrate ($1 < r \leq 3$). In the catalytic layer ($0 \leq r \leq 1$), the r-direction temperature is monotonic in the upstream region (small x) with the lowest temperature near the surface. In the downstream region the temperature is almost uniform in r-direction. Streamwise the solid temperature increases monotonically in the downstream direction with the largest gradient in the entrance region ($x < 0.02$) before ignition point. Fig. 2(b) shows that in the flow entrance region, oxygen penetrates deeply into the catalytic layer,in the downstream the penetration is shallower and at $x = 0.55$ only 5% of the oxygen is left on the surface. The oxygen profile is monotonic in r-direction and is always decreasing when r increases, as expected.
Fig. 2(c) gives the catalytic reaction rate contours in the catalytic layer. Maximum reaction rate occurs near the surface slightly downstream of the tube entrance ($x = 0.02$). The contours show that in the upstream the reaction is well distributed inside the catalytic layer and tends to be kinetically controlled. The position of maximum reaction rate can be regarded as the ignition location of catalytic reaction. Downstream of the ignition point the solid temperature rises and the reaction layer becomes shallower and is confined near the surface which indicates the tendency toward a diffusionally controlled regime.
If we look at a similar plot for a much lower catalyst loading factor, $\beta^* = 0.18$ kg/m^3 (ten times smaller than the previous case), Fig. 3 gives us a different picture. First, Fig. 3(c) shows that the ignition location (maximum reaction rate) occurs much farther downstream at $x = 2$ (instead of 0.2 for $\beta^* = 1.8$ kg/m^3). Secondly, both temperature and oxygen profiles show almost uniform distribution in r-direction. The reaction rate is in-depth everywhere and the whole catalytic layer participates in reaction.
A map of ignition boundary is shown in Fig. 4. The ordinate is the ignition length and the abscissa is the catalyst loading. When the combustor length and catalyst loading are too small, no ignition occurs. When they become large enough both catalytic and gas-phase ignitions can be obtained. In between, catalytic ignition occurs but gas-phase ignition does not. Notice that when β^* is greater than 1.08 kg/m^3, catalytic ignition occurs close to the entrance of the tube but gas-phase ignition position stays at 4 cm downstream. For steady operation it has no advantage to have the catalyst loading greater than 1.08 kg/m^3 although higher loading can produce a faster response in transient.

IV. 2 Start-up Transient

In transient operation, both the temperol and spatial variations become important. In this section, a start-up transient will be used to illustrate the gas phase and the solid in-depth behaviour. In the start-up transient simulation, oxygen (the deficient species) is injected into the reactor in a stepwise fashion.

The case with $\beta^* = 1.8$ kg·pt/m^3 will be used. Fig. 5 shows the short time behaviour. At $t = 10\tau_m^*$ (10 times the mass diffusion time in catalytic layer), Fig. 5(b) shows that gas-phase oxygen only reaches one third of the channel length (gas residence time in the channel is about fifty times larger than the mass diffusion time). Part of the oxygen is absorbed into the catalytic layer. Fig. 5(e) gives the oxygen mass fraction at the surface of the catalytic layer. Fig. 5(f) indicates that there is a small amount of catalytic reaction in the entrance region of the reactor where oxygen is absorbed. Fig. 5(d) shows a small surface temperature rise in the corresponding region but Fig. 5(a) indicates negligible gas-phase temperature rise. Fig. 5(c) shows no gas-phase reaction. At $t^* = 50\ \tau_m^*$, oxygen reaches the end of the channel, but because of the transient absorption into the porous solid, there is a drop of y_g at the end of the channel. At $t^* = 100\ \tau_m^*$, this transient absorption is completed as shown in Fig. 5(b) and (e). The level of catalytic reaction increases (Fig. 5(f)) and the solid temperature is rising (Fig. 5(d)). The solid transfers heat to the gas and as shown in Fig. 5(a), the gas temperature gradually increases. At this early time, Fig. 5(c) shows negligible gas-phase reaction. Fig. 6 gives the contour plot of the solid in-depth structure at $t = 100\ \tau_m^*$. Fig. 6(a) indicates that there is a temperature gradient in r-direction in the catalytic layer and that the highest temperatures are at the surface and decrease toward the interior. This is contrary to that in the steady state where the r-direction gradient is in the opposite direction. Fig. 6(b) shows that large amounts of oxygen are present in-depth in the catalytic layer. Fig. 6(c) shows that although catalytic reaction rates are relatively low, they occur everywhere in the catalytic layer. This of course indicates a chemical-kinetic controlled ignition period where solid temperatures are still low.

At a later time when one solid heat-up time is reached, Fig. 7(a) and (6) show that the solid surface and gas temperatures have risen substantially, at $\tau^* = 2\tau_{HT}^*$, the gas temperature becomes high enough that some gas-phase reaction begins to show up downstream of the reactor as shown in Fig. 7(c).

At $t^* = 3\tau^*_{HT}$, steady state is essentially reached. Fig. 7(c) indicates peak gas-phase reaction occurs at x = 0.65. At the same position, Fig. 7(a) shows an abrupt gas temperature rise and Fig. 7(b) shows an abrupt oxygen drop. This all suggests that gas-phase ignition is accomplished. The solid-phase structure at $t^* = 3\tau^*_{HT}$ has been presented in the section on steady state.

V. Concluding Remarks

A full transient model has been developed to study ignition in a monolithic catalytic layer. The special features are the inclusion of thermal and species structures in the porous catalytic layer and the resolution of small-time-scale events. The computed results show that with low catalyst loading, the radial temperature and species profiles are nearly uniform in the catalytic layer and catalytic reaction occurs in-depth. With high catalyst loading, reaction occurs in-depth only before catalytic ignition (both streamwise and timewise). The reaction tends to be chemical-kinetically controlled because of low temperature.

After ignition, the catalytic reaction becomes confined near the surface of the catalytic layer, is diffusion controlled, and there are substantial temperature and species non-uniformity in the radial direction in the catalytic layer. The in-depth temperature profile can also be different during transient where non-monotonic profile can occur in some part of the reactor when the catalyst loading is sufficiently large.

In addition to the intrinsic interest in knowing the thermal and species structures, the introduction of a finite catalytic layer appears to be the natural way of introducing the catalyst loading into the mathematical model. Catalyst loading is an important practical parameter in designing the catalytic reactors.

The present transient model is computationally intensive because the reference scale is the mass diffusion time in the catalytic layer. If small-time-scale event is not sought, an energy integral method can be used which provides better computational economy [13]. Another possibility is to combine the two models - using the present full transient formulation when small-time-scale event needs to be resolved and followed by the energy integral formulation when the only time-controlling process is the thermal inertia of the solid. Such an example can be found in Ref. 13.

Acknowledgement

The research was supported by NASA Grant NAG 3-809. The consultation with

Mr. Robert Zurawski and Ms. Elizabeth Roncace are appreciated. The present address for Ta-Ching Tien is: Aeronautical Research Laboratory, Chung-Shan Institute of Science and Technology, Taichung, Taiwan, ROC.

References

1. Sackheim, R. L.: Survey of Space Applications of Mono-propellant Hydrazine Propulsion System, Tenth Int. Sym. in Space Technology and Science, Tokyo, 1973.
2. Zurawski, R. L. and Green, J.: Catalytic Ignition of Hydrogen/Oxygen, Presented at the AIAA/ASME/SAE/ASEE 24th Joint Propulsion Conf., pp. 11-13, July 1988.
3. Kesten, A. S. and Sangiovanni, J. J.: Transient Model of Hydrogen/Oxygen Reactor, NASA CR-120799, February, 1971.
4. Kelly, J. T., Kendall, R. M., Chu, E. and Kesselring, J. P.: Development and Application of the PROF-HET Catalytic Combustor Code, Paper Presented at the 1977 Fall Meeting Western States Section, The Combustion Institute.
5. Pfefferle, L. D. and Pfefferle, W. C.: Catalysis in Combustion, Catal. Rev. Sci. Eng., Vol. 29, pp. 219-267, 1987.
6. Zurawski, R. L.: Personal communication.
7. T'ien, J. S.: Modeling of Transient Operation of Catalytic Combustion, Combustion Science and Technology, Vol. 26, p. 65, 1981.
8. Shina, N., Bruno, C., and Bracco, F. V.: Two-Dimensional Transient Combustion of CO/Air on Platinum, J. Physicochem. Hydrodynamics, 1982.
9. Prasad, R., Kennedy, L. A., and Ruckenstein, E.: A Model for the Transient Behavior of Catalytic Combustor, Combustion Science and Technology, Vol. 30, p. 59, 1983.
10. Maymo, J. A. and Smith, J. M.: Catalytic Oxidation of Hydrogen-Intrapellet Heat and Mass Transfer, A. I. Ch. E. J., Vol. 12, p. 845, 1966.
11. Musiak, J. D.: Experimentally Measured Ignition Delay Times and Proposed Kinetic Model for Hydrogen-Oxygen, M. S. Thesis, Case Western Reserve University, 1989.
12. Kanury, A. M.P.: Introduction to Combustion Phenomena, Gordon and Beach, New York, 1982.
13. Tien, T. C.: Catalytic Ignition Model in a Monolithic Reactor with In-depth Reaction, Ph.D. Thesis, Department of Mechanical and Aerospace Engineering, Case Western Reserve University, 1991.

NOMENCLATURE

B^*_1 $L^*K^*P^{*a}\rho^{*b}(o,o)Y_g^{*b-1}(o,o)/\rho^*(o,o)u^*(o,o)$

B^*_2 $B^*_1Y_g(o,o)$

C^* Specific heat of catalytic layer or substrate layer, 180 cal/Kg/^{o}K

D^*_e Oxygen effective diffusivity in catalytic layer, m^2/sec

D'_{am} $K^*_c h_c^{*2}/\rho^*(o,o)D^*_e(o,o)Y_g(o,o)$, Damkohler number of gas

D'_{at} $q_c K^*_c h_c^{*2}/\rho^*(o,o)D^*_e(o,o)$, Damkohler number of solid

E^* Activation energy, cal/mole

J_D $J_H(Le)^{2/3}$

J_H $4Nu_x(\alpha_g^*(o,o)/d^{*2})\ (L^*/u^*(o,o))\ (\lambda_g^*/\lambda_g^*(o,o))$

h_c^* Thickness of catalytic layer, 3.3×10^{-3} m

h_s^* Thickness of substrate layer, 6.7×10^{-3} m

L^* Reference reactor bed length, 5×10^{-2} m

Le Lewis number of oxygen

Nu Nusselt number

M Molecular weight, g/mole

P^* Pressure, 9 atm

P $P^*/P^*(o,o)$

q^* Heat of combustion per unit mass of mixture, cal/Kg

T $T^*/T_g^*(o,o)$, $T_g^*(o,o) = 400$ K

t t^*/τ_m^*

u^* gas velocity, 10 m/s

u $u^*/u^*(o,o)$, nondimensional gas velocity

x x^*/L^*, nondimensional axial distance

Y Mass fraction of oxygen

y_g $Y/Y(o,o)$

α^* Thermal diffusivity, m^2/sec

β^* Catalyst loading, Kg-pt/m^3

λ^* Heat conductivity, cal/m/sec/^{o}K

ρ^* Gas density, Kg/m^3

ρ $\rho^*/\rho^*(o,o)$

ϕ hydrogen/oxygen equivalence ratio, 8

ε Porosity of Catalytic layer, 0.5

τ_m^* $\varepsilon h_c^{*2}/D_e^*(o,o)$, Mass diffusion time in catalytic layer, 4.3×10^{-5} sec

τ_{HT}^* $2\int_{r_d^*}^{r_s^*} \rho^* c^* r^* dr^*/Nu_\infty \lambda_g^*(o,o)$, Solid heat-up time, 2×10^{-2} sec

τ_t^* h_c^{*2}/α_c^*, Thermal diffusion time in catalytic layer, 4.3×10^{-5} sec

δ_c τ_t^*/τ_m^*

δ_s $h_c^{*2}/\alpha_s^*/\tau_m^*$

δ_g $L^*/u^*(o,o)/\tau_m^*$

Subscript

g Gas-phase property

c Catalytic layer property

o Oxygen

s Substrate layer property

c,s property on catalytic layer surface

Superscript

* Dimensional quantity

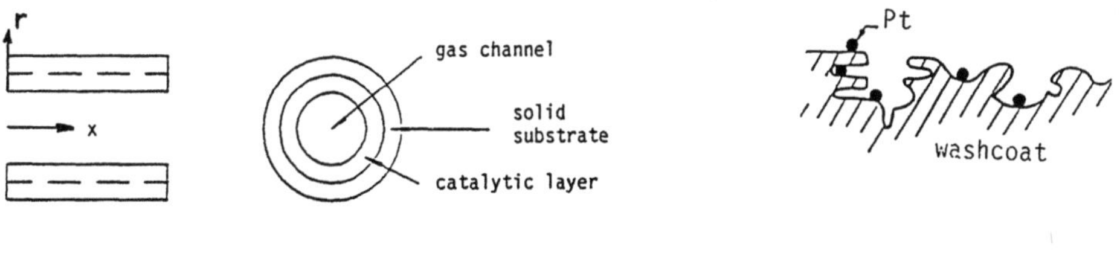

(a) Tubular catalytic reactor

(b) Illustration of catalytic layer structure

Figure 1. Monolithic reactor model description

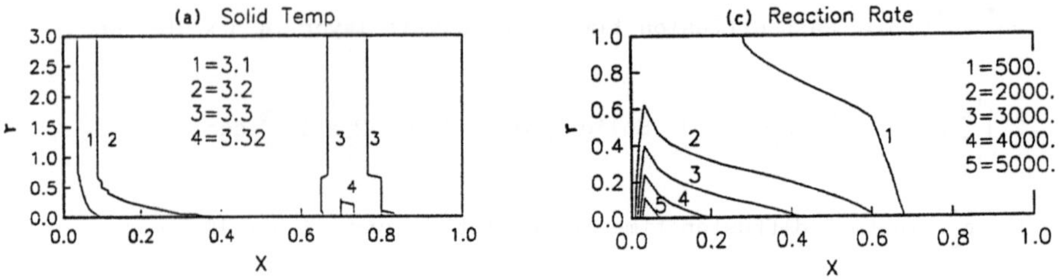

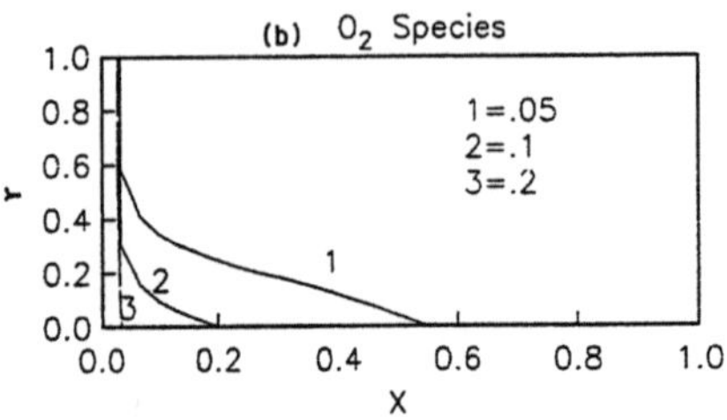

Figure 2.

Solid contours for $\beta^* = 1.8$ kg-Pt/m^3 in steady state

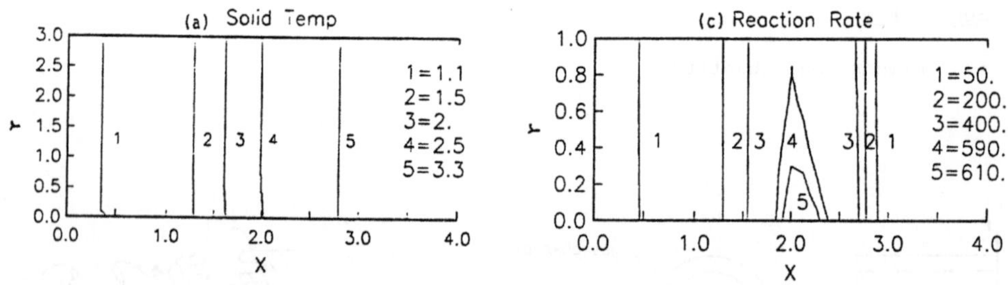

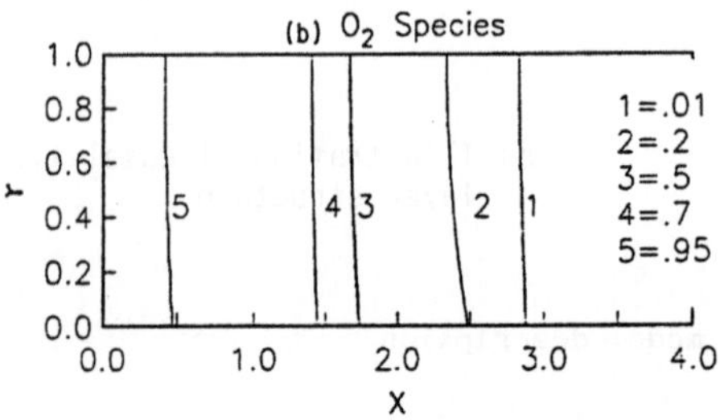

Figure 3. Solid contours for $\beta^* = 0.18$ kg-Pt/m^3 in steady state

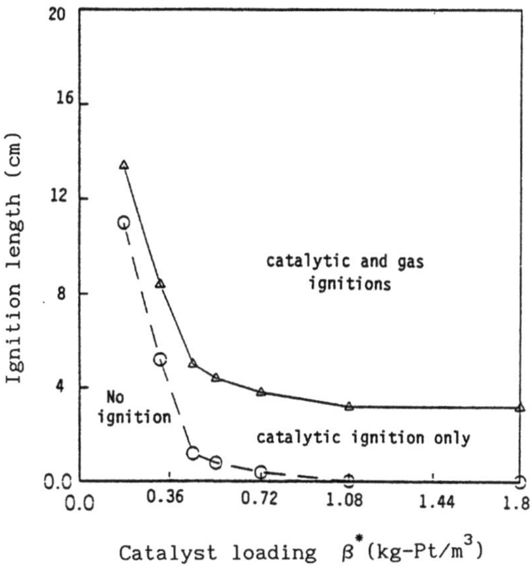

Figure 4.

Ignition boundary map (u_∞ = 10 m/sec)

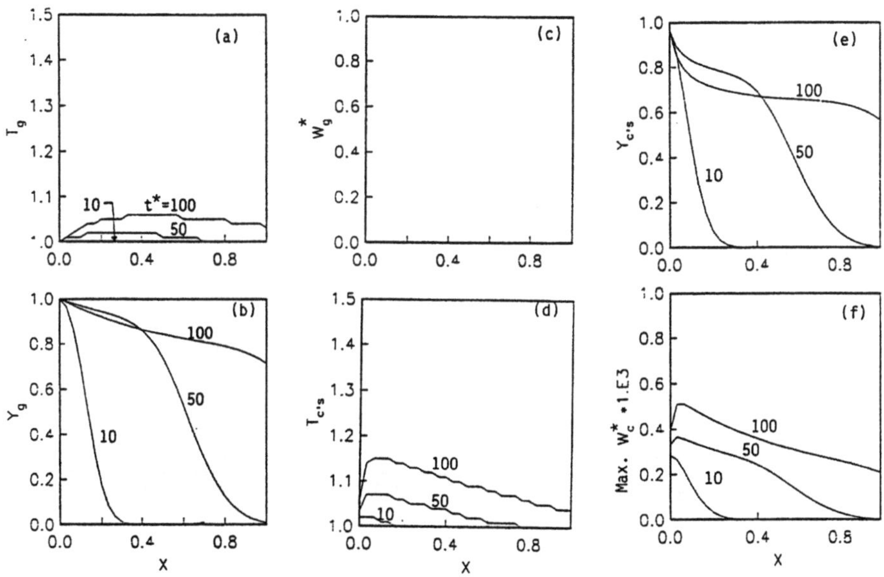

Figure 5. Transient profiles for β^* = 1.8 kg-Pt/m^3 at t^* = 10, 50, 100 τ^*_m (catalytic layer mass diffusion time)

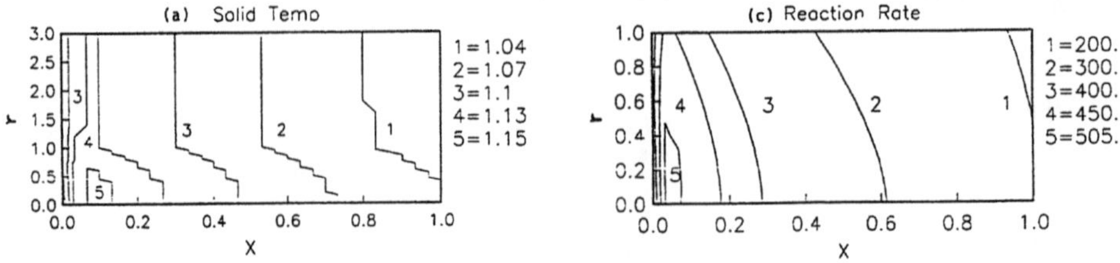

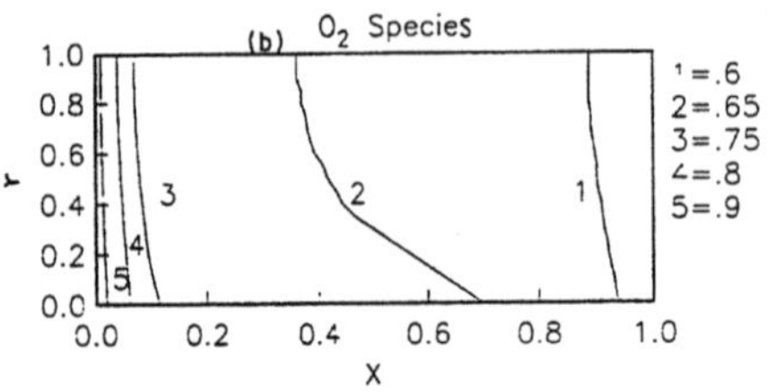

Figure 6.

Solid contours for $\beta^* = 1.8$ kg-Pt/m^3 at $t^* = 100\ \tau_m^*$

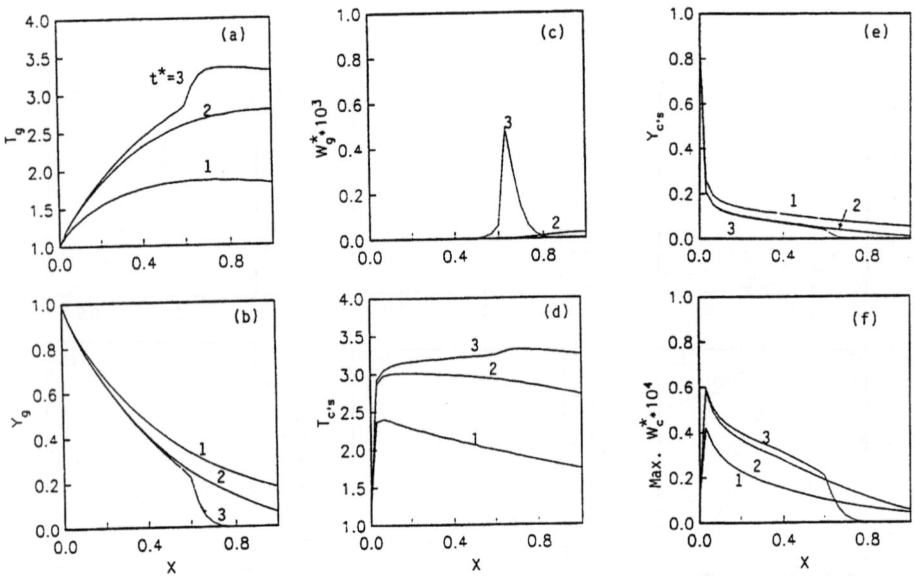

Figure 7. Transient profiles for $\beta^* = 1.8$ kg-Pt/m^3 at $t^* = 1, 2, 3\ \tau_{HT}$ (solid heat-up time)

Combustion of Heavy Fuel Oils in a Rijke Type Pulse Combustor with a Tangential Injection Stream

T. Bai, S. Shani, B. R. Daniel and B. T. Zinn

School of Aerospace Engineering
Georgia Institute of Technology
Atlanta, Georgia 30332

Abstract

This paper describes an investigation of the combustion of heavy fuel oils in a recently developed, Rijke type, pulse combustor that utilizes a tangential fuel and air injection system. This study was stimulated by the need to develop combustors that can burn low quality fuels with high combustion efficiencies and low pollutant emissions. Tests with fuel oil Nos. 5 and 6 revealed that they can be burned in the developed Rijke pulse combustor with combustion efficiencies higher than 99 percent while utilizing less than 10% excess air, which decreases exhaust flow losses and increases the thermal efficiency. Furthermore, the developed combustor could be operated over a wide range of air/fuel ratios, including fuel lean and rich operating conditions. Emissions measurements revealed that the exhaust flow NO_X concentrations are low over the whole range of investigated air/fuel ratios and that CO and particulates emissions are low for fuel lean operation. Finally, the paper describes the dependence of the amplitude of the combustor pulsations upon its operating conditions.

Introduction

This paper describes an investigation of the combustion of heavy fuel oils in a recently developed Rijke pulse combustor. Liquid fuels supply a large percentage in the total energy consumed in the world today[1,2]. Unfortunately, a large fraction of the utilized liquid fuels consists of high grade, light fuel oils that are easy to combust. However, since light liquid fuel oils are relatively scarce, there exists a need to develop efficient methods for the combustion of heavy fuel

* This research was supported by DOE Contract No. DE-AS04-85AL31881; Mr. Marvin Gunn, Contract Monitor.

R. S. L. Lee · J. H. Whitelaw · T. S. Wung (Eds.)
Aerothermodynamics in Combustors
IUTAM Symposium Taipei, Taiwan, 1991

oils whose utilization to date in conventional, steady state, combustors has been hindered by clogging of fuel lines and injectors, poor atomization, incomplete combustion and tar deposits on the combustor walls. To increase the use of heavy fuel oils, the understanding of their combustion characteristics in different environments must be improved, and practical combustors capable of burning these fuels efficiently and in an environmentally acceptable manner need to be developed.

To completely burn a liquid fuel, the fuel must be evaporated, mixed with available oxygen and reacted during its residence time in the combustor. To attain these goals, the liquid fuel must be atomized into a spray of small droplets and burned in a high temperature environment where the rates of mass, momentum and heat transfers are high. Considerations of operating conditions in steady and pulse combustors suggest that unsteady combustors may be better suited for the combustion of liquid fuels because it is well known[3] that pulsations increase the rates of the above mentioned transport processes that control the combustion process. Furthermore, there exists evidence in the literature[4] that pulsations of sufficient magnitude can produce secondary atomization of liquid droplets into a collection of smaller droplets, which evaporate and, thus, burn faster. These conjectures are supported by studies[5,6] showing that pulse combustors are characterized by a highly intense combustion process. The current investigation was undertaken to explore the advantages of burning liquid fuels in general and heavy fuel oils in particular in a pulse combustor. A Rijke type pulse combustor was chosen for this study because of recent demonstrations[7] that it can burn various coals with high combustion efficiencies and low pollutant emissions. This study started with the development of a Rijke pulse combustor that can burn a variety of liquid fuels[8]. This development was followed by an investigation of the combustion of light fuel oils in the developed combustor[8] and by the study of the combustion of heavy fuel oils whose results are reported in this paper.

The design of the Rijke pulse combustor is based upon the characteristics of the acoustic Rijke tube, which was discovered by Rijke in 1859. While the acoustic Rijke tube consists of an open vertical tube with a heat source (e.g., an electrically heated metal gauze) located at a distance of L/4 from the tube entrance, where L is the tube length, the Rijke pulse combustor consists of a similar tube with a combustion process stabilized at the L/4 location. In both cases, heat addition to the gas at the L/4 location excites the fundamental acoustic mode of the tube, which is characterized by pressure nodes at both ends and a pressure antinode at the tube mid-point. While the acoustic Rijke tube must be vertical in order to establish a mean flow in the tube (due to buoyancy), the Rijke pulse combustor can be set at any desired orientation and a fan is generally used to establish a mean flow of desired magnitude, which depends upon the needed air/fuel ratio.

The main problem that needs to be addressed when the combustion of a new fuel in a Rijke pulse combustor is contemplated is the development of means for stabilizing the combustion process in the vicinity of L/4 location. This has been accomplished with solid fuels, such as coal and wood, by burning these fuels on a metal grid located at a distance of L/4 from the combustor entrance[9,10]. Gaseous fuels were burned successfully in a Rijke pulse combustor by simply injecting the fuel into the air stream in the vicinity of the L/4 location because the combustion of gaseous fuels consists of mixing and chemical reactions only. In contrast to gaseous and solid fuels, it is considerably more difficult to constrain the combustion of a liquid fuel to a small region near the L/4 location because the combustion process is considerably more complex involving atomization, mixing and chemical reaction of a moving (i.e., injected) liquid stream. This goal was accomplished under the present program by developing a unique air/fuel injection system that injects the reactants tangentially into the combustor tube in order to generate a strong swirling flow inside the combustor that stabilizes the combustion process in the vicinity of the injection point. While the mechanism that controls the operation of the developed Rijke pulse combustor is not clearly understood and it is currently under study, it is believed that the combustion process is stabilized by a recirculation zone generated by the swirling flow near the injection location[11,12].

Initially, the combustion of kerosene and light fuel oils in the developed Rijke pulse combustor was investigated. This was followed by an investigation of the combustion of heavy fuel oils. The goals of this investigation were to determine the ranges of fuel input rates and air/fuel ratios for which pulse combustion operation was possible. Furthermore, the dependence of the combustor performance, in terms of the pulsations amplitude, combustion efficiency and pollutant emissions, upon the combustor operating conditions was investigated. Finally, the effect of insulating the combustor wall with a refractory liner was investigated to determine the effect of the liner upon the magnitude of the excited pulsations and the temperature distribution inside the combustor. The results of these studies are described in this paper. The experimental setup used in this study is described in the next section. This is followed by discussion of the combustor performance and the paper closes with a brief Summary and Conclusions section.

Experimental Setup

Schematics of the Rijke pulse combustor utilized in this study and its tangential air/fuel injection system are shown in Figures 1 and 2, respectively. Two identical Rijke pulse combustors, one with a refractory lining and one without, were developed. Each combustor consists of a 9 feet long, 5.76 inches inside diameter, vertical, combustor tube. Combustion occurs in a region located approximately at a distance of L/4 from the combustor inlet that is located on top of the combustor tube. A decoupling chamber is attached to each end of the combustor tube to assure that acoustic pressure nodes exist at these planes. The combustion air is divided into three streams. The axial stream enters the combustor through the upper decoupler while the swirling and atomizing streams enter the combustor with the fuel through the air/fuel injection system that is located at the L/4 location. Fuel is supplied to a plenum around a central tube that supplies the atomizing air stream. The atomizing air stream entrains the liquid fuel through a series of orifices in wall of the central tube as it moves through it. The swirling air stream is supplied into the injection duct through two separate lines and it is used to generate a swirling flow inside the combustor tube that is necessary for stabilizing the combustion process. The swirling and atomizing air streams mix with the fuel inside the injection duct and the resulting mixture is injected tangentially into the combustor. The combustion products leave the combustor through the lower decoupling chamber.

The fuel feed system consists of a fuel tank, a pump with an adjustable relief valve and a regulating valve. The relief and the regulating valves are used to control the pressure and flow rate of the fuel which passes through a rotameter prior to entering the combustor. The fuel tank contains a heating element that preheats the fuel oil to a temperature that permits pumping of the fuel into the

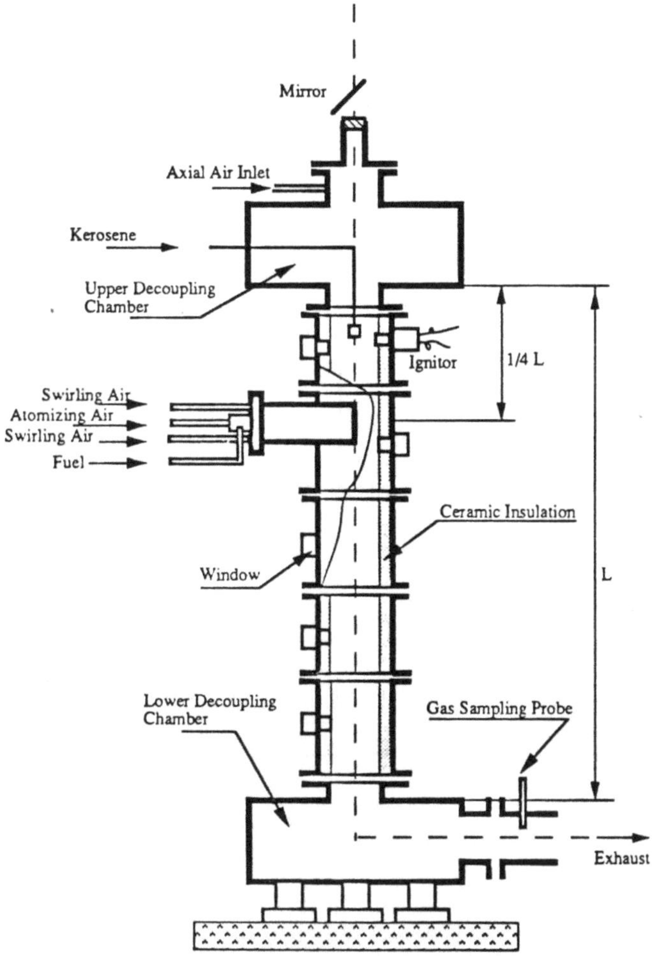

Figure 1. A schematic of the liquid fuel burning, Rijke type, pulse combustor with the tangential air/fuel injection system.

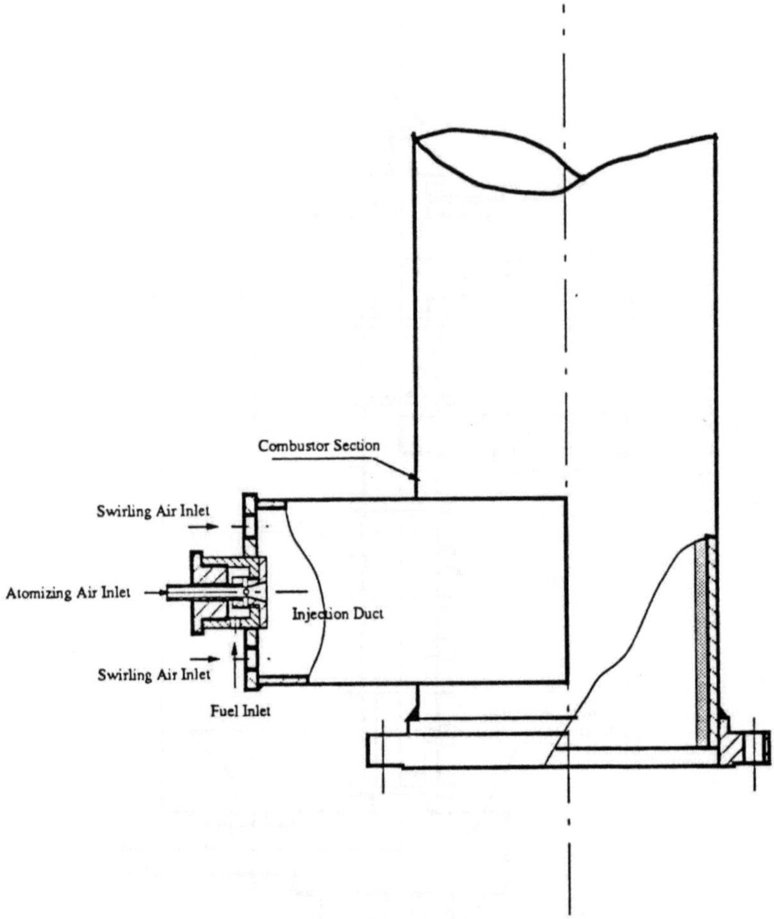

Figure 2. A schematic of the developed tangential air/fuel injection system.

injection system. To assure a continuous fuel flow, the fuel line is heated by electrical tapes.

Ignition is attained by use of a methane-oxygen pilot flame. Methane and oxygen are fed into a small stainless steel ignitor chamber where they mix and are ignited by a spark plug. The resulting flame is stabilized outside a small orifice at one end of the ignitor, which is attached to the combustor wall. To start a test, a methane-oxygen pilot flame is stabilized at the ignitor orifice inside the combustor. Next, air and a kerosene spray are injected into the combustor where they mix and are ignited by the pilot flame. The kerosene flame is then used to preheat the combustor until its walls attain a predetermined temperature. At this time the investigated heavy fuel oil-air mixture is injected into the combustor where it is ignited by the kerosene flame.

The combustor performance under different operating conditions was determined by measuring the amplitude of the excited pulsations and the composition of the exhaust flow. The amplitudes of the excited pulsations were measured by a piezoelectric pressure transducer which was installed at the combustor mid-point where the amplitude of the acoustic pressure of the combustor fundamental mode is at or near its maximum value. The pressure transducer was attached to a "semi-infinite" tube, which provides a flat frequency response and protects the transducer from the hot combustor wall. A sampling probe was installed inside the exhaust decoupler and used to continuously pump a small fraction of the exhaust gases for chemical analysis. The removed samples were transported into a chemical analysis system where the concentrations of CO, O_2, CO_2, NO_x, SO_2 and hydrocarbons in the combustion products were determined. The measured concentrations were then used to determine the combustion efficiency and pollutant emissions under the investigated test conditions. Finally, thermocouples were installed at several locations along the combustor centerline to determine the temperature distribution inside the combustor.

The data measured in a given test were digitized and stored in a Hewlett Packard computer for further reduction and analysis after the test.

Results and Discussions

Operating Limits

Potential applications of a given combustor often depend upon the range of air/fuel ratios over which it can be operated stably. For example, the air/fuel ratio affects the temperature of the exhaust products, which determines the process temperature and exhaust flow losses. The air/fuel ratio may also affect the emissions of soot, CO and NO_X. Furthermore, certain processes may be only carried out in hot reducing atmospheres that may be generated by a burner operating under fuel rich conditions. Consequently, this investigation initially focused on the determination of the operational limits of the developed Rijke pulse combustor. Two types of operational limits were observed in this study; that is, steady and pulsating operation limits. The former describes the range of air/fuel ratios at which the combustor can operate in a steady mode and the latter the range of air/fuel ratios at which the combustor can operate in a pulsating mode.

The operation limits of the combustor for a given fuel input rate were determined by observing the characteristics of the combustor operation at various air/fuel ratios. Typical results obtained in tests with fuel oil No. 5 supplied at a rate of 3.68 g/sec are presented in Figure 3, where the dependence of the amplitude of pulsations upon the air/fuel ratio is described. It shows that the developed pulse combustor can be operated in a pulsating mode over a wide range of air/fuel ratios that include both fuel rich and lean operating conditions. The excited pulsations amplitudes vary between 140 and 160 dB with maximum amplitudes excited near stoichiometric operating conditions. As the air/fuel ratio α is decreased below stoichiometric conditions, the amplitude slowly decreases until $\alpha \approx 0.75$. As α is further decreased, the amplitude of pulsations decrease sharply and it drops to zero at $\alpha \approx 0.61$, which is the lower limit of pulse combustion operation. As α is further decreased, the combustor operates in a steady mode until combustion stops completely at $\alpha \approx 0.5$. A different behavior is observed when the combustor is operated at values of α higher than stoichiometric. As α increases above stoichiometric operating conditions the amplitude of pulsations decreases slowly until combustion stops suddenly at $\alpha \approx 1.35$, which is the upper limit of the steady and pulsating operating ranges.

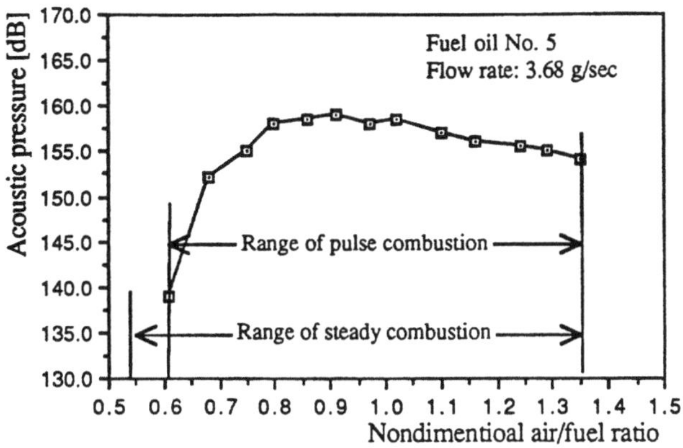

Figure 3. The operating limits of the pulse combustor measured in tests with fuel oil No. 5 supplied at a rate of 3.68 g/sec.

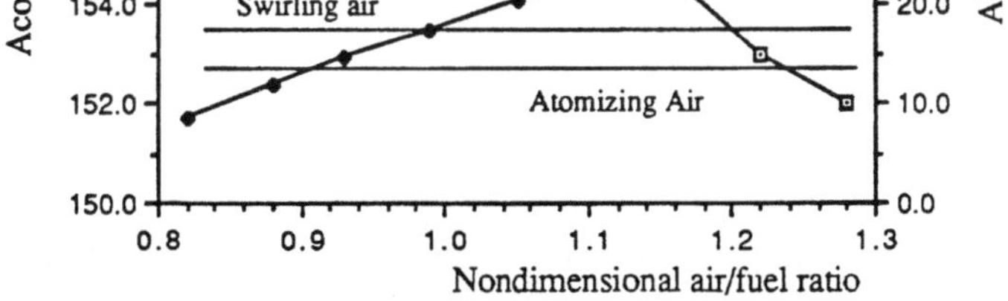

Figure 4. Dependence of the amplitude of the acoustic pressure upon the nondimensional air/fuel ratio measured in tests conducted with a 3.68 g/sec fuel oil No. 5 input rate. Note that the atomizing and swirling combustion air flow rates were kept fixed while the axial combustion air flow rate was increased.

Figure 3 describes ranges of air/fuel ratios for which the combustor can be operated in a steady and pulsating mode. It shows that these two ranges overlap and that for .61 ≤ 1.35 the combustor can be operated in either a pulsating or a steady mode. As is discussed in the next section, the amplitude of the excited pulsations at a given value of α depends upon the relative flow rates of the three combustion air streams and steady operation can be obtained for an appropriate combination of these flow rates.

The final important observation provided by Figure 3 is that the developed pulse combustor can be operated in steady and pulsating modes over a fairly wide range of air/fuel ratios including fuel rich and lean operating conditions.

Dependence of the Pulsations Amplitude upon the Combustion Air Streams Flow Rates

It is highly desirable to have a capability for controlling the pulse combustor amplitudes. For example, it is generally known[13] that the heat transfer rate increases as the amplitude of pulsations increases. Consequently, when the combustor is used, for example, as an immersion heater, operating at a maximum amplitude will decrease the amount of required heat transfer area. Furthermore, the increased heat losses from the combustion zone will decrease its temperature and, thus, thermal NO_x formation. Also, operating at higher amplitudes will increase mixing and, thus, reaction rates within the combustor and may result in lower CO and soot emissions. Finally, it is quite possible that NO_x emissions may be minimized at certain amplitudes of pulsations. Consequently, the dependence of the excited pulsations amplitude upon the combustor operating conditions was investigated and the results are reported in this section. Specifically, the dependence of the combustor performance upon the axial, swirling and atomizing air flow rates was investigated. This study was conducted because consideration of the nature of these flows suggested that the axial air flow rate can affect the location of the combustion zone; the swirling air flow rate can affect the strength of the recirculation in the combustion region; and the atomizing air flow rate can affect the atomization of the liquid fuel. Since the combustor driving mechanism was expected to depend upon the location and characteristics of the combustion process, it was also expected that the amplitude of the excited pulsations and combustor performance would depend upon the relative magnitudes of these three combustion air streams.

Initially, the effect of the axial air flow rate upon the pulsations amplitude was investigated. In these tests, the air/fuel ratio was increased by increasing the axial flow rate and keeping the swirling and atomizing air flow rates constant. Typical results, obtained with a 3.68 g/sec flow rate of fuel oil No. 5, are presented in Figure 4. These results show that as the axial combustion air flow rate was increased the amplitude of the pulsations decreased monotonically. It is conjectured that the amplitude of pulsations decreased because the increase in the axial air flow rate increased the mean axial velocity in the combustor, which forced the combustion process to be stabilized downstream of the L/4 location. Since maximum driving in a Rijke pulse combustor occurs when the combustion heat release occurs in the vicinity of the L/4 location, shifting the combustion zone away from this location would result in a less effective driving process and, consequently, lower amplitudes, as observed in Figure 4.

In the next series of tests, the effect of increasing the atomizing air flow rate upon the pulsations amplitude was investigated. In these tests, the air/fuel ratio was increased by increasing the atomizing flow rate and keeping the axial and swirling air flow rates constant. These tests were also conducted with a 3.68 g/sec flow rate of fuel oil No. 5 and the results are presented in Figure 5. These results show that the amplitude of the pulsations increased slowly as the nondimensional air/fuel ratio α increased until it reached a maximum at $\alpha \approx 1.0$. Thereafter, the amplitude decreased rapidly as α increased. For these tests, stable pulse combustion operation could only be attained for $.92 \leq \alpha \leq 1.09$. For values of α beyond this range, the pulsations became unstable and were characterized by constantly changing frequencies and amplitudes. This range was much narrower than the $0.8 \leq \alpha \leq 1.3$ range for which stable pulse combustion operation was attained while changing the axial air flow rate, see Figure 4.

Consideration of the physics of the problem suggests that the atomizing air flow rate affects the behavior of the pulse combustor in two ways. First, it affects the characteristics of the generated droplet spray and, second, it changes the overall air/fuel ratio of the combustor. It is believed that for a certain fuel input rate there exists an optimum range of atomizing air flow rates, where good atomization of the liquid fuel and good mixing of the fuel and air occur. If the atomizing air flow rate is too low, adequate atomization of the fuel is not obtained, which adversely affects the quality of the combustion process. On the other hand, when the atomizing air flow rate is too high, it produces a large angle conical spray. When this occurs, a large fraction of the injected fuel strikes the walls of the injection duct, resulting in fuel accumulation on the walls and incomplete combustion. For the data presented in Figure 5 the flow rate of the atomizing air flow was nearly doubled while the overall nondimensional fuel/air ratio α only changed from 0.92 to 1.09. Because the atomization of the liquid fuel relies entirely upon the atomizing air flow rate, it is likely that for the investigated, fixed, fuel input rate the significant changes in the atomizing air flow rate produced large changes in the characteristics of the generated fuel spray, which may explain the observed behavior.

The dependence of the excited amplitude of pulsations upon the swirling air flow rate is described in Figure 6. These data were also obtained with a fixed, 3.68 g/sec, flow rate of fuel oil No. 5 and the nondimensional air/fuel ratio α was increased by increasing the swirling air flow rate and keeping the axial and atomizing air flow rates constant. Figure 6 indicates that when α was less than 1.0, the amplitude of pulsations increased as the swirling air flow rate increased. The pulsations amplitude reached a maximum at $\alpha \approx 1.0$, and it decreased thereafter as α became larger than 1.0. For this series of tests, stable pulse combustion operation was attained for $0.78 < \alpha < 1.11$.

Variation of the swirling air flow rate changes the amount of tangentially injected combustion air relative to the axial air. Tangential injection of air and fuel results in the formation of a swirling flow inside the combustor. It is expected that a strong swirl produces a recirculation region near the injection location, which plays an important role in stabilizing the combustion process. The formation of this recirculation zone depends, however, upon the ratio of the angular and axial momenta of the flow. The data presented in Figure 6 indicate that for a given fuel flow rate and fixed axial and atomizing air flow rates, there exists an optimum swirling air flow rate which maximizes the amplitude of pulsations.

In summary, the results presented in this section indicate that the pulsations amplitudes may be maximized and controlled by controlling the relative flow rates of the three combustion air streams.

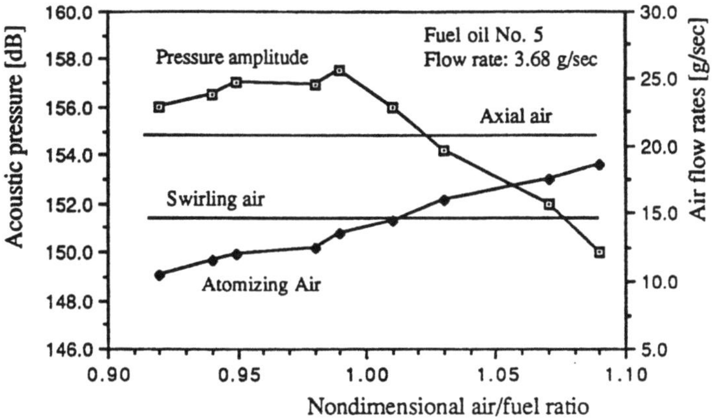

Figure 5. Dependence of the amplitude of the acoustic pressure upon the nondimensional air/fuel ratio measured in tests conducted with a 3.68 g/sec fuel oil No. 5 input rate. Note that the axial and swirling combustion air flow rates were kept fixed while the atomizing combustion air flow rate was increased.

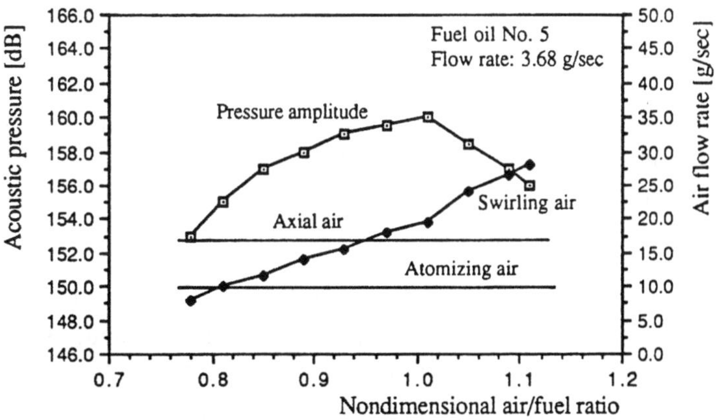

Figure 6. Dependence of the amplitude of the acoustic pressure upon the nondimensional air/fuel ratio measured in tests conducted with a 3.68 g/sec fuel oil No. 5 input rate. Note that the atomizing and axial combustion air flow rates were kept fixed while the swirling combustion air flow rate was increased.

Combustion Efficiency and Pollutant Emissions

The combustion efficiency and pollutant emissions are critical performance parameters that currently determine the acceptability of newly developed combustors. To reduce fuel cost, the combustor must be operated at air/fuel ratios that optimize its thermal efficiency. This information can be obtained from data describing the dependence of the combustion efficiency upon the air/fuel ratio. On the other hand, in order to meet regulated emissions limits, it is necessary to determine the dependence of the combustor CO, soot and NO_X emissions upon the combustor operating conditions. Such information can then be used to determine the combustor operating conditions which maximize fuel savings and minimize emissions. In addition, such data can be used to develop an understanding of the processes that control the combustor operation. To determine these optimum operating conditions, the dependence of the combustion efficiency and pollutant emissions upon the air/fuel ratio was investigated and the results are reported in this section. This information was obtained by determining the dependence of the exhaust flow CO and NO_X concentrations upon the air/fuel ratio. The former was used to determine CO emissions and the combustion efficiency. The CO concentration could be used to determine the combustion efficiency, using a carbon balance, because no soot was present in the exhaust and the measured CO concentration determined the extent of incomplete combustion.

The dependence of the combustion efficiency, obtained with 2.7-3.0 g/sec input rates of fuel oil No. 6, upon the nondimensional air/fuel ratio is described in Figure 7. The results presented in Figure 7 indicate that the combustion efficiency increased as α increased, reaching values higher than 99% for α around 1.10. Since it is generally very difficult to burn fuel oil No. 6, it is very satisfactory to note that it could be burned in the developed combustor with nearly 100% combustion efficiency with only 10% excess air.

The dependence of the exhaust flow NO_X concentration, obtained with 2.7-3.0 g/sec input rates of fuel oil No. 6, upon the nondimensional air/fuel ratio is described in Figure 8. These data show that the NO_X concentration increased as the air/fuel ratio α increased reaching a maximum at $\alpha \approx 1.05$. The concentration of NO_X remained at its maximum value as α was further increased beyond $\alpha \approx 1.05$.

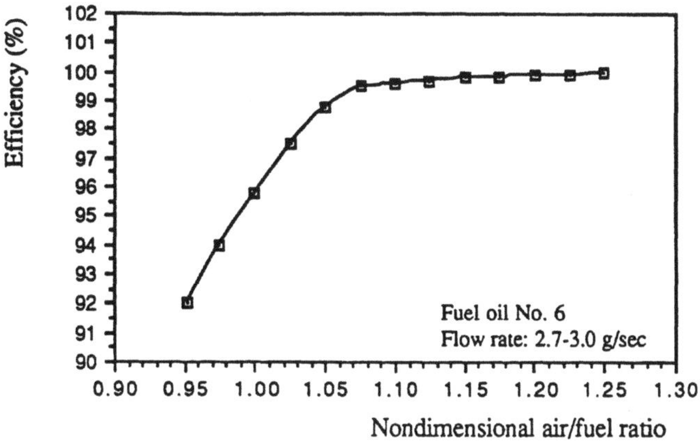

Figure 7. Dependence of the combustion efficiency upon the air/fuel ratio measured in tests conducted with 2.7 - 3.0 g/sec fuel oil No. 6 input rate.

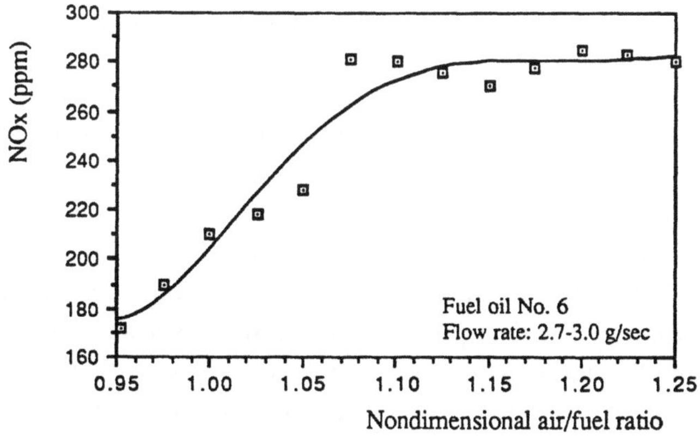

Figure 8. Dependence of NO_x concentrations in the exhaust flow upon the air/fuel ratio measured in tests conducted with 2.7 - 3.0 g/sec fuel oil No. 6 input rate.

Effects of a Ceramic Liner upon the Combustor Performance

The effect of a ceramic liner upon the combustor performance was investigated because such liners may be needed in some applications. For example, a ceramic liner may be needed to protect the combustor walls when the latter is constructed from a lower grade steel (e.g., carbon steel) and/or when the combustor wall is not adequately cooled. On the other hand, when the combustor is used as an immersion heater, an insulating ceramic liner may not be necessary because it would impede the heat transfer through the combustor wall. In this study the effect of a ceramic liner upon the combustor performance was investigated by comparing the performance of two identical Rijke pulse combustors with and without a ceramic liner.

Of specific interest was to determine the effect of the insulating ceramic liner upon the amplitude of pulsations, which, as discussed earlier, can significantly affect the pulse combustor performance and its applications. At the onset of this study it was anticipated that the addition of a ceramic liner would reduce heat losses and introduce acoustic attenuation. The latter was expected to damp the pulsations and, thus, reduce their amplitudes. On the other hand, it was expected that reducing the heat losses would increase the temperatures inside the combustion zone, which would increase the rates of the reaction and heat addition processes, resulting in increased driving and pulsations amplitudes. The tests conducted under this study aimed to determine the net effect of these two offsetting processes by comparing the temperature distributions and pulsations amplitudes in similar insulated and uninsulated pulse combustors under identical operating conditions.

To meet the objectives of this study, tests were conducted with fuel oil No. 2 in the insulated pulse combustor that is shown in Figure 1. The results obtained in these tests were compared with data obtained in previous tests[8], which were conducted under the same operating conditions with a similar uninsulated pulse combustor. The results obtained in these tests are compared in Figures 9 and 10. An examination of the data presented in Figure 9 shows that the amplitudes of pulsations in the uninsulated combustor were higher than the corresponding amplitudes in the insulated combustor over the whole range of tested air/fuel ratios α. Furthermore, a comparison of the measured axial temperature

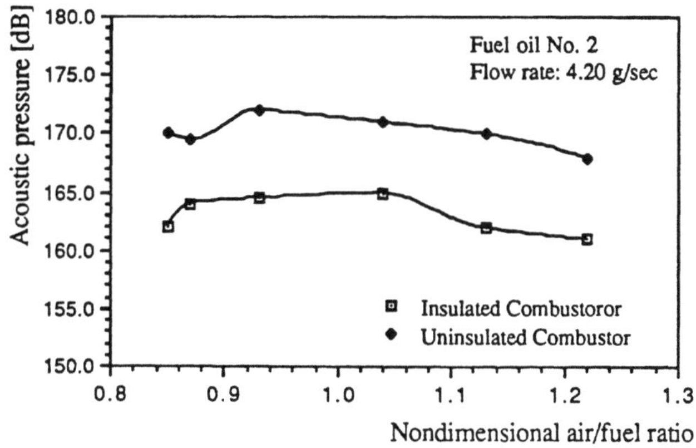

Figure 9. A comparison of the dependence of the amplitudes of the acoustic pressures in the insulated and uninsulated combustors upon the air/fuel ratio measured in tests conducted with a 4.2 g/sec fuel oil No. 2 input rate.

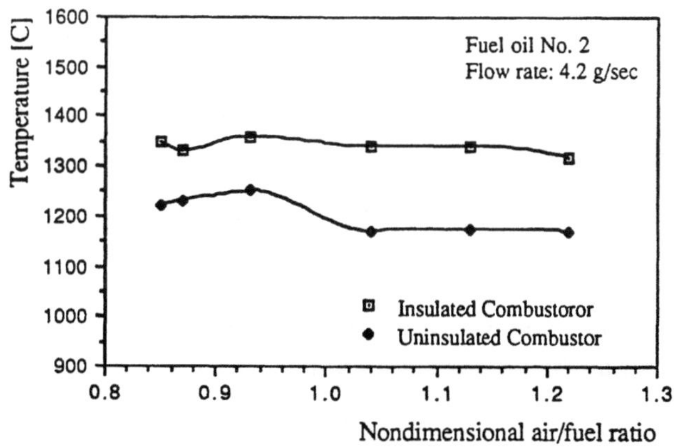

Figure 10. A comparison of the dependence of the maximum temperatures inside the insulated and uninsulated combustors upon the air/fuel ratio measured in tests conducted with a 4.2 g/sec fuel oil No. 2 input rate.

distributions in both combustors, see Figure 10, shows that, as expected, the temperatures were higher in the insulated combustor over the whole range of investigated air/fuel ratios.

Trends similar to those exhibited in Figures 9 and 10 were also obtained in tests with fuel oil No. 5. Both sets of data indicate that the attenuation provided by the ceramic liner dominates any additional driving which might have been introduced by the higher reaction rates in the insulated combustor. These results indicate that unless there is a need to minimize heat losses from the combustor and/or protect the combustor walls, no apparent advantage is gained by the addition of a ceramic liner.

Summary and Conclusions

This paper describes the results of an investigation of the combustion of heavy fuel oils in a Rijke type combustor. It was conducted in a Rijke pulse combustor that utilized a unique, recently developed, tangential fuel/air injection system. Test results show that the developed Rijke type pulse combustor can burn heavy fuels over wide ranges of fuel input rates and air/fuel ratios, including fuel rich and lean operating conditions. In addition, the combustor can be operated at high combustion efficiencies with very low excess air values. Furthermore, exhaust flow concentrations reveal that NO_X emissions are low over the whole range of investigated air/fuel ratios and that CO and particulates emissions are low for fuel lean operation. Comparisons of the performance of the insulated combustors with that of the uninsulated one, under identical operating conditions, reveal that the addition of a ceramic liner increases the maximum temperature inside the pulse combustor and it decreases the amplitude of pulsations.

In conclusion, the results of this study indicate that:

1. A Rijke combustor which incorporates the newly developed tangential air/fuel injection system is capable of burning heavy fuel oils over a wide range of air/fuel ratios and fuel input rates.

2. This pulse combustor attains high combustion efficiencies with relatively little excess air, which decreases exhaust flow losses and increases the thermal efficiency.

3. The exhaust flow NOx concentrations are low over the whole range of tested air/fuel ratios.

4. The net effect of adding a refractory liner is to decrease the combustor heat losses and the amplitudes of the excited pulsations, which may be detrimental in many applications where large amplitudes are needed. Consequently, unless there is a need to protect the combustor walls or decrease heat losses, the addition of a ceramic liner is not recommended.

References

1. Statistical Abstract of the United States, 1988, 108th Edition, Bureau of the Census, U. S. Department of Commerce.

2. Walter Goldstein, "From One Crisis to the Next," Energy Policy, Vol. 17, No. 1, Feb., 1989.

3. Putnam, A. A., Belles, F. E., and Kentfield, J. A. C., "Pulse Combustion," Progress in Energy and Combustion Science, Vol. 12, No. 1.

4. Kumagai, S. and Isoda, H. (1955). In Fifth Symposium (Int) on Combustion. The Combustion Institute, Pittsburgh, Pa.

5. Audibert, E. (1924). Revue de l'Industrie Minerale, No. 73.

6. Katsnel'son, B. D., Marone, I. Ya. and Tarakanovskii, A. A., " An Experimental Study of Pulsating Combustion," Teploenergetika, 1969, 16 (1) 3-6.

7 Wang, M. R., Daniel, B. R., and Zinn, B. T., "Combustion of Unpulverized Coal in Rijke Type Pulsating Combustor," Proceedings of the 1st Annual Pittsburgh Coal Conference, Pittsburgh, PA., September 1984.

8. Zinn, B. T., Daniel, B. R., Shani, S., Bai, T. and Zhang, Z., "Investigation of the Characteristics of Liquid Fuel Burning, Rijke Type, Pulse Combustors," Final Report for DOE Contract No. DE-AS04-85A131881, Feb., 1989.

9. Wang, M. R., Daniel, B. R. and Zinn, B. T., "Combustion of Unpulverized Coal in a Rijke Type Pulsating Combustor," Proceedings of the 1st Annual Pittsburgh Coal Conference, Pittsburgh, PA, September, 1984.

10. Zinn, B. T., Daniel, B. R., and Sheshadri, T. S., "Application of Pulsating Combustion in the Burning of Solid Fuels," Proceedings of the Symposium on Pulse Combustion Technology for Heating Applications, November, 1979.

11. Ruey-Hung Chen and James F. Driscoll, " The Role of Recirculation in Improving Fuel-Air Mixing within Swirling Flames," Twenty-Second Symposium (International) on Combustion, The Combustion Institute, 1988.

12. Gupta, A. K., Lilley, D. G., and Syred, N., Swirl Flows, Abacus Press, 1984.

13. Fedorov, B. N., "Experimental Study of the Effect of Acoustic Oscillations on Heat Transfer in a Gas Flow," Juzherno-Fisicherkii Zhurnal, Vol. 32, No. 1

Matching of Fuel Injection Equipment to Diesel Engines

R Ziarati BSc,PhD,FIProdE, MIMechE, ECng
Faculty of Engineering and Computer Technology
Birmingham Polytechnic Birmingham
United Kingdom

A Veshagh BSc,PhD,MASME, MSAE
Advanced Technology Centre
University of Warwick
Coventry
United Kingdom

SYNOPSIS

This paper describes the development and validation of two computer models developed for the matching of fuel injection equipment to diesel engines. Both models included mathematical representation of fuel atomisation and ignition delay together with a simple simulation model of engine cycle. The effect of fuel quality was taken into account on a qualitative basis while empirical relationships were employed for the prediction of ignition delay under various operating conditions.

Both models were tested against experimental results including tests using varying quality fuels. All the calculations were carried out on a step-by-step basis to predict engine performance parameters over the whole cycle.

The success of the models were primarily based on the use of a number of empirical coefficients or correlation factors in the formulae used in the models. However in order to enhance the simulation capabilities of the models, their independence on such empirical coefficients had to be removed. The two models were therefore merged into a new model combining the "jet-mixing" calculation with those of "droplet burning" with a potential for predicting combustion emissions.

INTRODUCTION

In an earlier project (1) two Diesel simulation programs were developed one based on a "jet mixing" model and the other on a "droplet burning" model for simulating the penetration and combustion of fuel spray. Both models were employed separately to study the effect of various fuel injection equipment (f.i.e.) parameters or fuel spray and also to match the f.i.e. to DI Diesel engine of different types (2). However, no attempts were made to combine the two models into one comprehensive simulation package such that the combined effect of jet mixing and droplet burning could be taken into consideration.

This paper reports on the recent developments of the combined simulation model.

R. S. L. Lee · J. H. Whitelaw · T. S. Wung (Eds.)
Aerothermodynamics in Combustors
IUTAM Symposium Taipei, Taiwan, 1991

THE COMBINED SIMULATION MODEL

The new simulation program is primarily on "jet mixing model" together with an engine cycle model based on modified air standard cycle, Fig 1. the combustion formulation assumes that air entrainment is controlled by an elemental and multizone jet and, after impingement, by an elemental wall jet, Fig 2. The fuel-air mixing within the jet (micro mixing) is controlled by turbulent diffusion. The jet interaction is determined by the spray envelope geometry and when interaction is detected, the entrained air is assumed to be shared by the sprays; this is achieved by using a single "global" model, Fig 3.

The air entrained by the gas jet at any instant quantifies the larger scale mixing (macro mixing) of injected fuel and air in the chamber. Micro mixing is represented by a simple representation of turbulent diffusion.

The details of full mathematical model is given in (2). The following are further refinements on the model.

The initial jet tip penetration is given by:

$$S = 0.39 \sqrt{\frac{2\Delta P}{\rho_l}}\ t$$

After the jet breakup, the penetration is determined:

$$S = 2.95 \left(\frac{\Delta P}{\rho_a}\right)^{1/4} \sqrt{D_o t}$$

The mass entrained therefore is calculated from :

$$m_a = \frac{\pi}{3} \tan^2 \left(\frac{\phi}{2}\right) \rho_a \left(S_i^3 - S_{i-1}^3\right)$$

where tan ϕ

$$b = S \tan\phi$$

Assuming that the jet velocity after impingement of the wall is constant across a given cross section, the Glaurt (3) expression, was employed to determine the jet velocity.

The mass entainment in wall jet is determined from:

$$m_a = \rho_a \int_{t_{ow}}^{t_{ow}+\Delta t} Q dt$$

The air and fuel volumetric flows are given by simple diffusion equations stated as:

$$M_a = D\ u_t(m_a - M_a)$$

$$M_f = D\ u_t(m_f - M_f)$$

The engine cycle is divided into:

1i) <u>Close Cycle Calculations</u>
a. Compression period (1-2)
b. Combustion period (2-3)
c. Expansion period (3-4)

ii) <u>Open Cycle Calculations</u>
a. Blowdown period (4-5)
b. Exhaust period (5-6)
c. Overlap period (6-7)
d. Suction period (7-8)
e. Pre-compression (8-9)

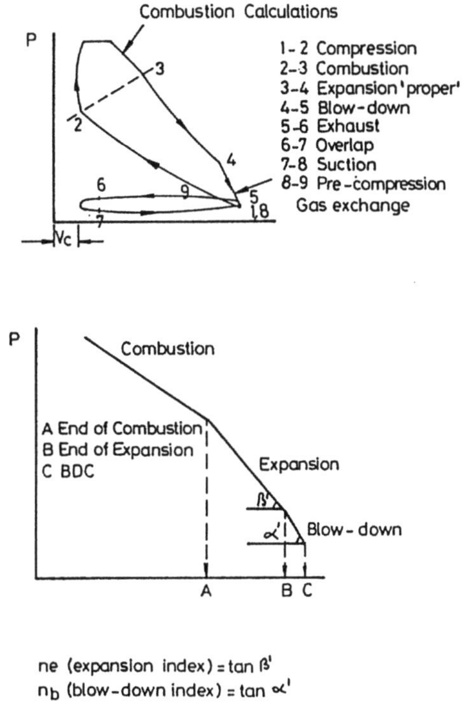

Fig 1. Details of Engine Cycle

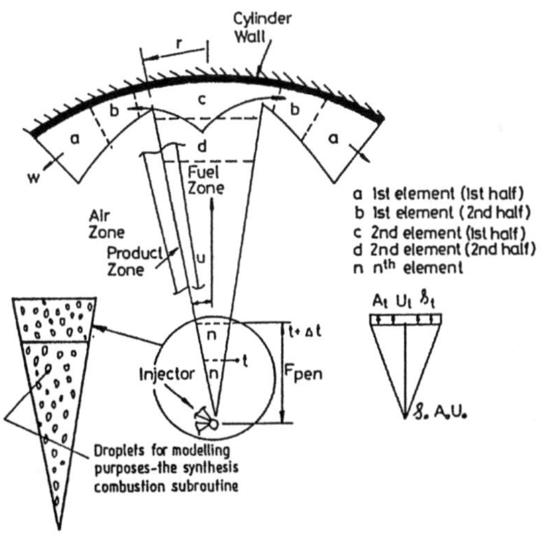

Fig 2. Simplified Presentation of Elemental Free and Wall Jets with No Swirl

The model utilises two types of combustion calculations (A) analysis, (B) synthesis, Fig 4. Analysis is required to interpret a given cylinder pressure diagram of cylinder pressure as a function of crank angle measured on a running engine. The heat release diagram obtained from analysis, is checked against the heat release diagram predicted by synthesis. The procedure assumes a single zone approximation i.e. conditions of pressure, temperature and gas composition are taken as uniform throughout the cylinder.

Synthesis is the reverse procedure. This involves the calculation of cylinder pressure and hence engine performance parameters from combustion equations that leads to the calculation of heat release rate.

A computer program called "Heat Release" is used for the analysis. In this program, the heat release and work output are calculated from the measured values of cylinder pressure together with data on engine geometry and air/fuel ratio.

The basis of the calculation in this program is the application of the first law of thermodynamics to the cylinder contents.

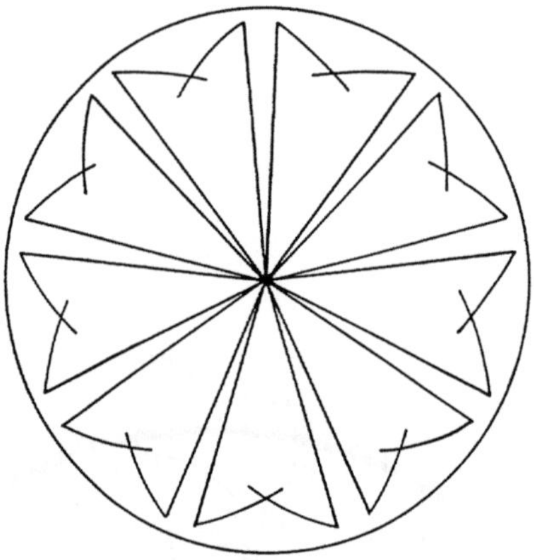

Fig 3. Jet Interaction Model

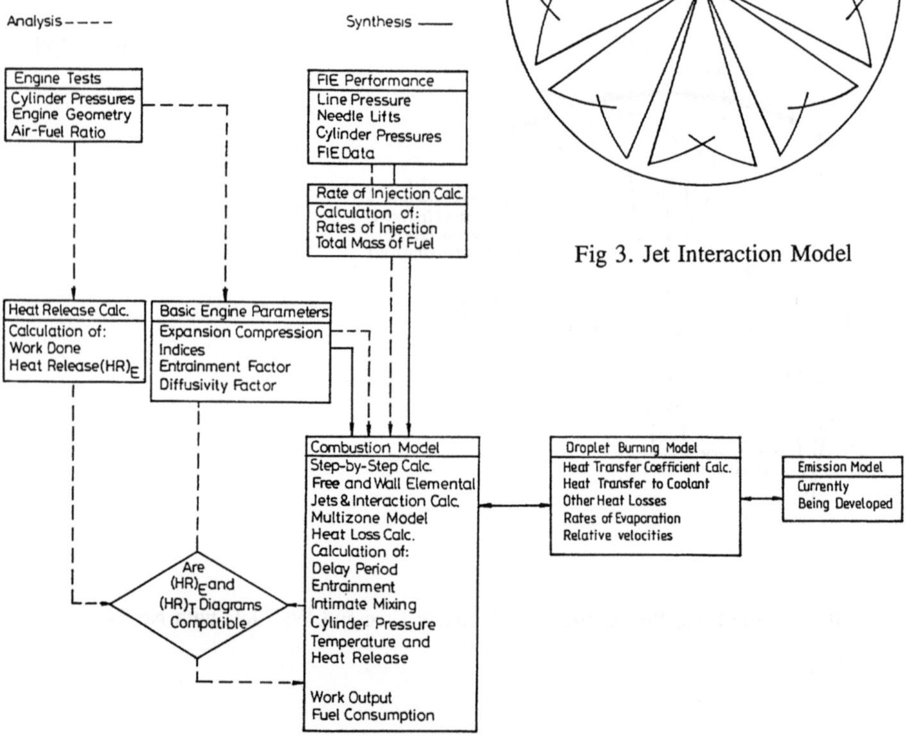

Fig 4. Engine Combustion Model

The synthesis is the main program called "Diesel Engine Simulation". Here for a given set of initial conditions and fuel injection rate diagram, the program calculates pressure and temperature values, heat release, power and fuel consumption for one engine cylinder. The model only considers the thermodynamics and fluid mechanics of the combustion process and does not include chemical kinetics.

It should be noted that the "Diesel Engine Simulation" program (for quantitative considerations) requires the total mass of fuel injected and the rate of fuel injection as input data. These two parameters are obtained from the line pressure and the needle valve lift diagrams. Here, numerical arrays representing the line pressures and the needle valve lifts are fed into a program called "Fuel Injections" and the required output are obtained for any type of nozzle configuration.

The synthesis calculations are supplemented by a "droplet burning" subroutine (3). This subroutine is almost entirely confined to the "closed" period in the cycle and is only used when detailed analysis is required viz when different quality fuels are being tested and/or for emission calculations.

It considers the injection of the fuel span of each element of fuel spray and the droplets within the elements are monitored at given time intervals (usually one degree crankangle). The evaporation of the droplets, the mixing of the vapour with air within each element of the jet and their combustion is examined at each step. The heat released by combustion is calculated by a set of differential equations considering heat transfer to engine block and coolant, leading to the prediction of cylinder pressure variations during the engine cycle.

The subroutine performs the following calculations:

i) heat transfer coefficient calculation using Woschi formula (4)
ii) heat transfer to coolant (5) other heat losses (6)
iii) relative velocity of fuel droplet.

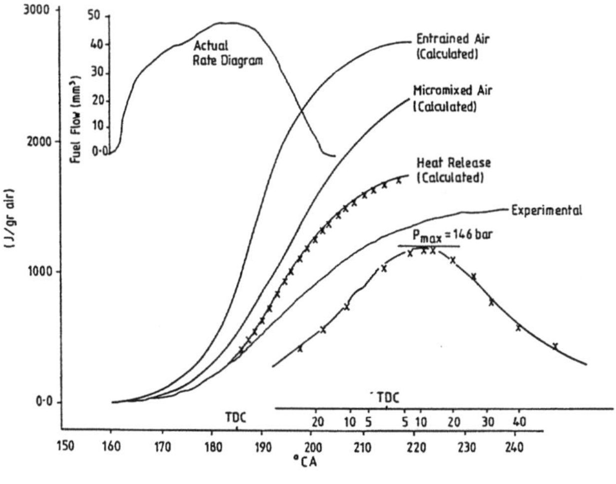

Fig 5. Comparison Between Experimental and Predicted Results for Engine D at Full Load.

The two pressure variations obtained from the synthesis and the above program can be compared and therefore both "rate" and "magnitude" tolerances may be reduced using an iterative techniques. Successful matching can then lead to the calculation of engine performance parameters and emissions.

If at least one set of experimental data on the injection characteristics viz line pressure and needle lift diagrams together with an actual engine pressure oscillogram are available, then an "analysis" evaluation can be carried out (between experimental and predicted data). Good correlation can lead to quantitative prediction of performance parameters and emissions.

Fig 5. shows the experimental and predicted results for a 210 mm diameter bore engine running on diesel oil.

Fig 6. depicts samples engine test results from a 203 mm diameter bore engine running on varying quality fuel at different speeds and loads. The corresponding predicted results are superimposed accordingly.

The droplet burning subroutine eliminated the need for estimating a number of empirical coefficients viz the entrainment factor and diffusivity constants (2).

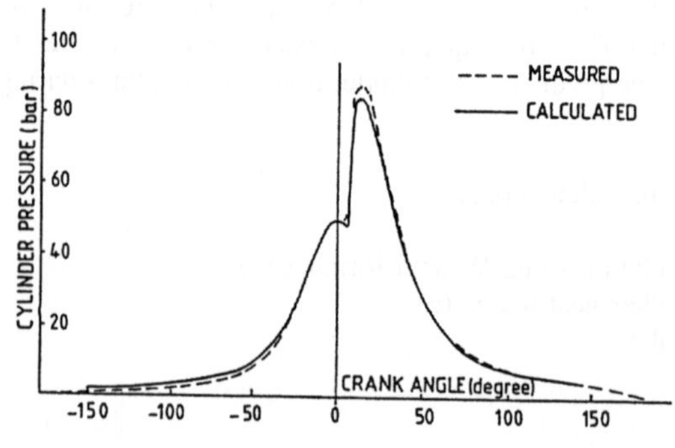

Fig 6a Cylinder Pressure Engine Bore 203 mm Heavy Fuel 60% Load

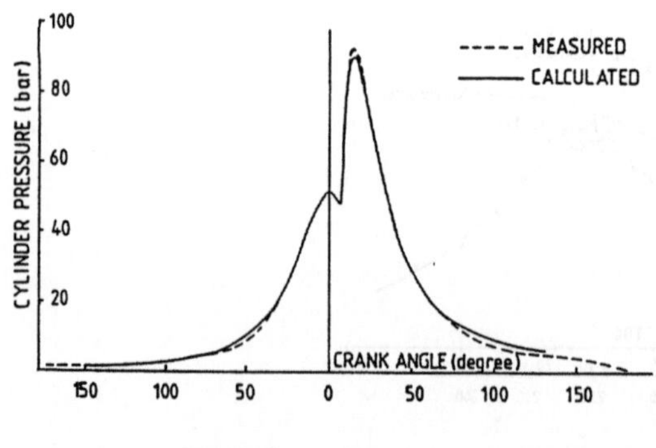

Fig 6b Cylinder Pressure Engine Bore 203 mm Heavy Fuel 60% Load

EMISSION MODELLING PRELIMINARY INVESTIGATION

There are a number of difficulties in predicting exhaust emissions such as:

- complex in-cylinder turbulent flow pattern;

- heterogeneous nature of combustion, where fuel spray penetration and evaporation, air-fuel unburned-burned gas mixing, and combustion can occur simultaneously;

- complicated mechanism of pollutant formation, which is of small-quantity, and of non-equilibrium chemical reacting rate type.

The combustion process of hydrocarbon fuel consists of a series of intermediate reaction steps with different reaction rates, including a number of species (8), (9). Certain stages of combustion such as ignition, wall quenching, fuel oxidation and misfire are known to be primarily responsible for the emission formation. The mechanism of emission formation is based on non-equilibrium reaction rates. Throughout the main part of combustion process the reaction rates are of an equilibrium nature and primarily responsible for the heat release. The combustion model based on the reaction kinetics is still in the developing stage, and requires a substantial amount of computation.

An ideal model for pollutant formation requires emission to be based on other sub-models within the combustion model: penetration and evaporation rates, fuel-air volume flows, unburned-burned gas mixing diffusion, ignition delay and combustion heat release. The resolution and validities of the models would be enhanced by adopting either:

(a) multi-zone model based on a filling-and-emptying method or

(b) multi-dimensional model based on a Computational Fluid Dynamics (CFD) method.

The diesel combustion prediction based on the CFD methods are still limited to the calculations for heat release only, though the CFD code like KIVA provides an option for the combustion model to involve reaction kinetics (10). In the practical methods the equilibrium chemical thermodynamics is used for heat release calculations and an empirical reaction kinetic equation for emission formation. Almost all predictive models for pollutant formation are based on a multi-zone model (11), (12), (13), (14).

Diffusion models to describe the fuel-air mixing process developed recently (2,15) have the disadvantage that these cannot simulate the mixing process accurately to facilitate exhaust emission prediction.

KINETICS OF NO FORMATION

The main source of NO is the oxidation of molecular nitrogen. The widely accepted mechanism for NO formation is originally suggested by Zeldovich, and revised by Bowman (9) as summarised below:

$$0 + N_2 = NO + N$$

$$N + 0_2 = NO + O$$

$$N + OH = NO + H$$

These chemical reactions have large activation energies which results in a strong temperature dependence of NO formation. The calculations should be based on the reaction kinetics. The nitrogen atom concentration (N) is much less that the concentrations of other N species. The formation rate of NO can thus be approximated by the following expression:

$$\frac{d[NO]}{dt} = \frac{2 R_1 \{1 - ([NO]/[NO]_e)^2\}}{1 + ([NO]/[NO]_e) R_1 / (R_2 + R_3)}$$

where

$$R_1 = k_1^+ [0]_e [N_2]_e = K_1^- [NO]_e [N]_e$$

$$R_2 = k_2^+ [N]_e [O_2]_e = K_2^- [NO]_e [O]_e$$

$$R_3 = k_3^+ [N]_e [OH]_e = K_3^- [NO]_2 [H]_2$$

$[]_e$ denotes equilibrium condition.

When $[NO]/[NO]_e << 1$, it can be further simplified:

At low temperature, the reactions of NO formation are virtually "frozen".

HYDROCARBON EMISSION MECHANISM

The prediction of Hydrocarbons relies heavily on the modelling of lean fuel-air mixture auto-ignition and flame propagation combustion as well as rich combustion. Aspects relating to interaction by flame and cylinder will need to be considered. The hydrocarbon oxidation during expansion process can lead to further complications (17).

A reliable model for HC emission requires the following features:

- a high-resolution history of in-cylinder velocity field, fuel-air volume flow, and unburned-burned gas mixing composition.

- a comprehensive combustion sub-model based on the reaction kinetics which can deal with incomplete combustion, wall quenching, misfire and HC burn-out.

CARBON MONOXIDE FORMATION

Carbon monoxide is one of the compounds formed during the intermediate stages of combustion. As combustion proceeds to completion, oxidation of CO to CO2 occurs through recombination reactions between CO and different oxidants.

Scarcity of oxidants and/or temperatures could result in increased CO emission.

The lean-burn nature of diesel combustion leads to low CO emissions under normal operating conditions. Due to this complex nature of CO prediction and the requirement for substantial computing, it was decided to consider CO emission calculations at a later stage.

The soot formation is a non-equilibrium process. Khan et al, (11) suggested the following equations to simulate the soot formation:

$$\frac{dm_s}{dt} = \frac{dm_{sf}}{dt} - \frac{dm_{sc}}{dt}$$

$$\frac{dm_{sf}}{dt} = K_f m_{fg} P^{0.5} \exp\left(-\frac{E_{sf}}{RT}\right)$$

$$\frac{dm_{sc}}{dt} = K_c m_s \frac{P_{02}}{P} p^{1.8} \exp\left(-\frac{E_{sc}}{RT}\right)$$

where m_{sf} is the mass of the soot formed;

m_{sc} is the mass of the soot oxidised;

m_{fg} is the mass of the fuel vapour;

m_s is the mass of the soot;

E_{sf} and E_{sc} are the activation energies.

The K_f and K_c are the constants which are determined from the experiments.

The modelling of Soot and other particulates are presenting a number of challenges yet to be resolved.

CONCLUSION

The combined combustion model has proved to be a unique tool for combustion analysis and diesel engine design. However, this inclusion of emission calculation in the combustion model so far has not produced promising results.

NOTATION

b	fuel jet breadth
D	diffusion coefficient
D_o	nozzle orifice diameter
M_a	air mass fully micromixed
m_a	jet entrained air mass
M_f	fuel mass fully micromixed
m_f	fuel mass injected
Q	volumetric flow rate
S	jet tip penetration
t	time
u_t	jet tip velocity
ΔP	injection pressure difference
ρ_a	density of air liquid interface
ρ_l	density of liquid
ϕ	spray angle

REFERENCES

1. Ziarati R

"An investigation into Fuel Injection Equipment and Combustion Performance of Medium Sized Diesel Engines"
IMechE - Seminar on Practical Limits of Efficiency of Engines, Proc. IMechE London 1986.

2. Ziarati R, Veshagh A

"Mathematical Modelling and Computer Simulation of Medium Size Diesel Engine Running on Varying Quality Fuels"
COMODIA 90 - International Symposium on Diagnostics and Modelling of Combustion in Internal Combustion Engines, Kyoto, Japan, Sept 1990

3. Glaurt M

"The Wall Jet"
JFM. 1956

4. Ziarati R, Veshagh A, Hawskley G

"Interaction and Combustion Modelling"
EC Non-Nuclear Research
Programme Report, Jan 1988

5. Woschni G

"A Universally Applicable Equation for the Instantaneous Heat Transfer Coefficient in the Internal Combustion Engine"
SAE Paper No 670931, 1967

6. Alcock J F

"Thermal Loading of Diesel Engines" IMechE., Trans. Vol 77, 1965

7. Annand W J D

"Heat Transfer in the Cylinder of Reciprocating Internal Combustion Engines"
Proc. IMechE., Vol. 177, No 56, 1963

8. Hirayasu H, Kadota T

"Fuel Droplet Size Distribution in Diesel Combustion Chamber"
SAE Paper No. 740715, 1974

9. Laidler K J

"Chemical Kinetics". 2nd ed., McGraw-Hill, New York, 1965

10. Westbrook C K and Pitz, W J

"Chemical Kenetics Modelling of Combustion of Practical Hydrocarbon Fuels"
SAE paper 890990, 1990

11. Amsden A A, Ramshaw J D, O'Rouke P J and Dukowics J K

"KIVA" A Computer Program for Two and Three Dimensional Fluid Flows with Chemical Reactions and Fuel Sprays"
LA-10245-MS, Los Almos Nat. Lab. 1985

12. Bastress E K, Chng K M and Dix D M

"Models of Combustion and Nitrogen Oxide Formation in Direct and Indirect Injection Compression-Ignition Engines"
SAE Paper 719053, 1971

13. Khan I M and Greeves G

" A method of Calculating Emissions of Soot and Nitric Oxide in Diesel Engines".
SAE Paper 730169, 1973

14. Shahed S M, Chui W S and Yumlu V S

"A Preliminary Model for the formation of Nitric Oxide in Direct Injection Diesel Engines and Its Application in Parametric Studies"
SAE Paper 730083, 1973

15. Hiroyasu H, Kodata T and Arai M

"Use of a Spray Combustion Modelling to Predict Diesel Engine Efficiency and Pollutant Emissions Part 1: Combustion Modelling"
Bulletin of the JESM, Vol 26, n.214, 1983

16. Nishida K and Hiroyasy H

"Simplified Three-Dimensional Modelling of Mixture Formation and Combustion in a D.I. Diesel Engine"
SAE Paper 890268, 1989

17. Bowman C T

"Kinetics of Pollutant Formation and Destruction in Combustion"
Prof Energy Combust. Sci. Vol 1. 1977

Part V

Gas Turbine Combustors

The Turbulence and Mixing Characteristics of the Complex Flow in a Simulated Augmentor

Chao, Y.C, Leu, J.H., Yang, T.H.

Institute of Aeronautics and Astronautics
National Cheng Kung University, Tainan, Taiwan 70101, R.O.C.

Lin, C.K.

Chung-Sun Institute of Science and Technology, Taichung, Taiwan, R.O.C.

Summary

The flow and the fuel-air mixing of the aircraft augmentor equipped with the pilot burner and the V-gutter are studied experimentally in a simulated augmentor. The laser Doppler velocimetry system and an aspirating concentration probe are used to measure the turbulent velocity and the mixture fraction in the augmentor flow. Due to the effect of the interacton of the swirl component and the axial velocity of the pilot burner flow, high turbulent fluctuation and intensive mixing are found in the lower portion of the pilot burner flow. This feature is further confirmed to provide a convenient path for the propagation of the pilot flame to the V-gutter to initiate the main combustion there. Also, turbulent mixing is evidently shown to be the dominant mixing mechanism in the augmentor flow.

Introduction

In modern aircrafts, afterburners or augmentation devices are added to the gas turbine engines to increase the thrust during take-off or emergent maneuver. The oxygen content in the high speed gas stream leaving turbine blades is burnt with additional fuel in this secondary combustion device to increase the gas exit velocity and thrust. Steady burning in this high speed gas stream is very difficult unless sources of continuous ignition or proper flame stabilization devices are present in the combustion chamber. Flame stabilization is usually achieved in the wake of a bluff body, e.g. V-gutter, placed in the gas stream. In addition to the flame stabilization problem, ignition is usually harder to achieve. The primary problems of afterburner ignition are the start-up ignition and the high altitude relight and pilot burners are used extensively in commentional aircrafts. A pilot burner is one of the solutions to these problems and pilot burner are used extensively in conventional aircrafts. However, in modern aircrafts in order to reduce the total weight, the pilot burner is usually of "V" shape, similar to the V-gutter with high angle swirl vans inside. The V-shaped pilot burner can be used to serve

R. S. L.Lee · J. H.Whitelaw · T. S.Wung (Eds.)
Aerothermodynamics in Combustors
IUTAM Symposium Taipei, Taiwan, 1991

partly as a flame stabilization device. This special design of the V-shaped pilot burner produces a complex flow field with strong interaction of high swirl and axial velocities in the wake of the pilot burner. Furthermore, the flowfield will be further complicated by using both the V-gutter and the pilot burner. The turbulence and mixing characteristics are of vital importance to the ignition and the stabilization problems. Thus, the main purpose of this research are to measure the turbulent velocity and mixing characteristics of the complex flow field in a simulated augmentor with the V-shape pilot burner and the V-gutter.

Studies of the bluff-body stabilization and the afterburner characteristics can be found as early as in the 40's. Longwell [1] and Williams [2] studied the bluff-body turbulent flame stabilization and propagation. Recently, Sullerey et al. [3] characterized the turbulent near wake of bluff bodies of different kinds. Stwalley and Lefebvre [4] examined irregular shaped flame holders. Zukoski [5] offered an instructive review of the afterburner. Taylor and Whitelaw [6] studied the velocity characteristics in the near wake of the bluff body. Others [7~9] studied the interaction of the swirling component with the axial coaxial jet. However, the V-shaped pilot burner turbulence characteistics which are generated by the strong interaction of the high swirl and the axial components in the wake, are still open issues and deserve careful studies.

Experiments

The experimental set-up consists of an air blower driven by a DC motor, a settling chamber and a simulated augmentor test section. Air in the settling chamber is first passed through a slowly divergent duct with an angle of 7°, and a honeycomb unit and 3 fine mesh screens are used to further reduce the fluctuations. Figure 1 shows schematically the arrangement of the test section of the simulated augmentor. The front portion of the center body is used to contract the flow leaving the mesh to further reduce the turbulent fluctuation and the rear portion is to simulate the diffuser of the augmentor. The pilot burner is a 1/3 scale of the actual size. The pilot burner is of V-shape with eight swirl vanes inside. The angle of the swirl vane relative to the axial direction is zero at the pilot burner inlet and increases gradually to 75 degree at the outlet. About 8% of the augmentor air passes through the swirl vane to generate the high tangential component of the pilot burner flowfield. As shown in Fig.1, two spray bars, the main and the secondary spray bars, are used to inject helium into the augmentor air stream to simulate the complicated turbulent mixing in the augmentor flowfield. The main spray bar forms 30 degree inclination to the vertical direction and with eight 1.6 mm holes and 5 mm hole separation on one side of the bar to inject the helium laterally into the augmentor air stream. The main spray bar is located in the front portion of

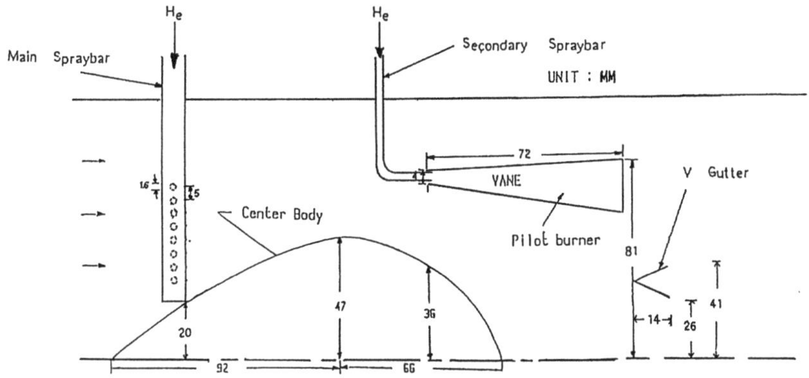

Fig. 1: Schematic of the simulated augmentor test section

the center body to reduce the local injection effects when the mixture reaches the pilot burner and the V-gutter where the main interest of the present study focuses. The secondary spray bar is for the pilot burner to simulate the turbulent mixing of the pilot burner flow. To study the mixing characteristics, the secondary spray bar is inserted between the swirl vanes at the pilot burner inlet. Both the laser Doppler velocimetry (TSI 9100-7) system and a concentration probe (TSI 1441-18) are used to measure the velocity components, the turbulence and the mixture fraction of the augmentor flow, as shown in Fig. 2. The laser Doppler anemometry system uses a 3W argon ion laser as the light source and a Bragg cell after the beam splitter to shift the frequency of one beam by 40M Hz, thereby permitting the determination of both the amplitude and the direction of the velocity components. The light beams are also expanded to improve the signal quality and intersect at the focal point to form a 88μm $\times$ 1.62 mm ellipsoidal probe volume. The scattered light from the probe volume is collected by the lens through a pin hole and is converted into electrical signals by the photomultiplier and processed by the counter. The data from the counter are further processed in a PC computer to perform the statistical analysis weighted by their residence time and 4096 velocity samples are used for each turbulent statistics. The seeding in the flow required to produce the scattering light intensity is generated from paraffin oil with nominal droplet diameter of 1μm. In order to eliminate the noise and difficulties encountered in the LDV measurements due to the curvature of the augmentor wall, sixteen circular windows of 30 mm diameter are attached to both sides of the wall. They are sealed with thin flat optical glasses to allow accurate intersection of the laser beams and to

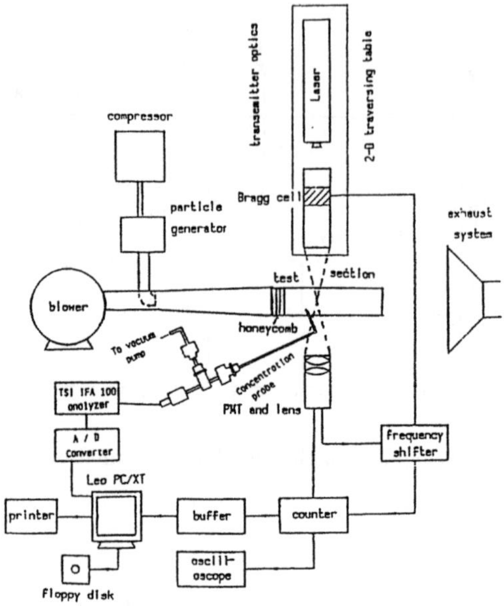

Fig. 2: Experimental setup and instrumentations

reduce the noise in the scattered light intensity. In the present experiment, helium is used to study the turbulent mixing of the complicated augmentor flow. The binary mixing of air and helium is measured by an aspirating concentration probe made up of a hot-film sensor in a tube with a sonic throat. This type of probe has been used with success by Brown and Rebollow [10] and Ahmed and So [11]. Both mean and fluctuating binary mixture concentration can be measured with high resolution. The two-wire probe used by Way and Libby [12] is not suitable for measurements in such a complicated flowfield where the flow direction is usually unknown in advance. However, the aspirating probe, by sucking in a small amount of air using a vacuum pump operated at the chock condition, does not suffer this difficulty. Since the flow at the throat of the probe is at the chock condition, the voltage output of the hot film can be argued to be depended on the binary mixture fraction of the flow. The probe is calibrated in the potential cone of prepared jet flows with different mixture fractions. The probe signals are processed by an anemometer (TSI IFA100) and consequently digitized by a 12-bit analog-digital converter for statistical analysis in a PC computer, see Fig. 2. If the statis pressuure in the flowfield is different from the calibration condition or the pressure in the flowfield fluctuates significantly, the probe output must be calibrated for the pressure dependence. The pressure calibration curve, accouding to Chao and Yang [3], can be written as

$$E^2/(T_w - T_e) = A + B(P - 1) + C(\rho^{\frac{1}{4}} P^{\frac{1}{4}})$$

where A, B, C are constants, P is the static pressure, E^2 is probe output, T_w and T_e are the temperatures of the wire and the flow and ρ is the density of the binary mixture. Figure 3 shows the calibration curve for the probe used and the probe outputs for different pressure collapses into a single curve so that measurements in the situations where pressure is fluctuating due to flow can be achieved by using the aspirating probe together with a pressure sensing device using this general calibration curve.

Results and discussions

To provide a comprehensive view of the features of the turbulence and mixing of the augmentor flow, the results of experiments will be presented by first showing the LDV results of the velocity and the turbulent characteristics of the pilot burner flow without the V-gutter and then with the V-gutter for comparison to facilatate a global view of the augmentor flow. The concentration results will also be presented in the same sequence and the concentration results will also be correlated to the LDV turbulence results to distinguish the effect of turbulence on the mixing in the augmentor flow. The axial velocity distribution for the pilot burner flow in the augmentor is shown in Fig. 4. The dashed curves in the vicinity of the pilot burner indicate the zero axial velocity

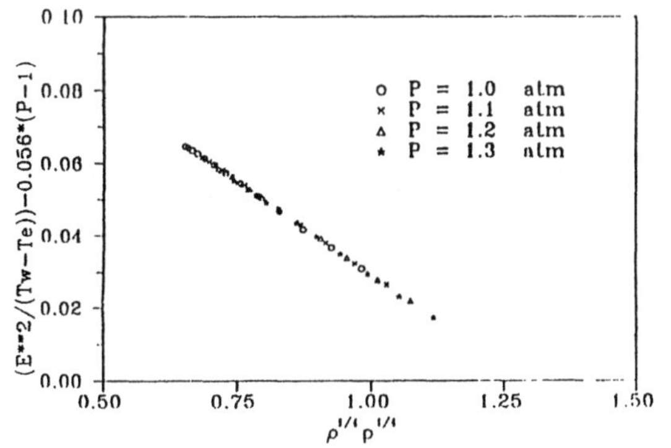

Fig. 3: The calibration curve for the concentration probe including pressure dependence

contour. The wakes due to the pilot burner and due to the central diffuser characterize the near field axial velocity distribution and the axial velocity profile becomes more uniform around Z/D of 0.4. If one carefully examines the velocity distribution in the vicinity of the pilot burner, one can easily find that the axial velocity distribution near the lower zero velocity region around Y/D of 0.6 is steeper than that near the upper zero velocity region around Y/D of 0.8. This phenomenon can further be identified in the turbulent fluctuation profiles in Fig. 5 where local peak fluctuations are found near the lower portion of the pilot burner corresponding to the lower zero-velocity contour in Fig. 4 while that in the upper portion is not so pronounced. This phenomenon is believed to be due to the effect of the interaction of the pilot burner swirl component and the strong axial component due to the acceleration of the V-gutter of the pilot burner which occurs in the upper portion of the pilot burner wake and this was also seen in a preliminary flow visualization study in the water tunnel by Chao and Chen [14]. The effect of the swirl component in the shean layer of the pilot burner wake is to reduce turbulent fluctuation of the axial velocity. In other words, stronger turbulent mixing can be expected in the lower portion of the pilot burner wake. The concentration measurement results also verify these findings and we will discuss this later in the concentration results. In the presence of a V-gutter in Fig. 6, the pilot burner recirculation zone becomes smaller and the wake due to the central diffuser disappears. The axial velocity distribution recovers earlier around Z/D of 0.35 in Fig. 6. On the other hand in Fig. 7, the peak turbulent fluctuations downstream of the pilot burner becomes less pronounced. The local peak fluctuations in the wake of the V-gutter is "symmetric". Globally, the turbulent fluctuation in Fig. 7 is stronger than that in Fig. 5. It reveals that stronge

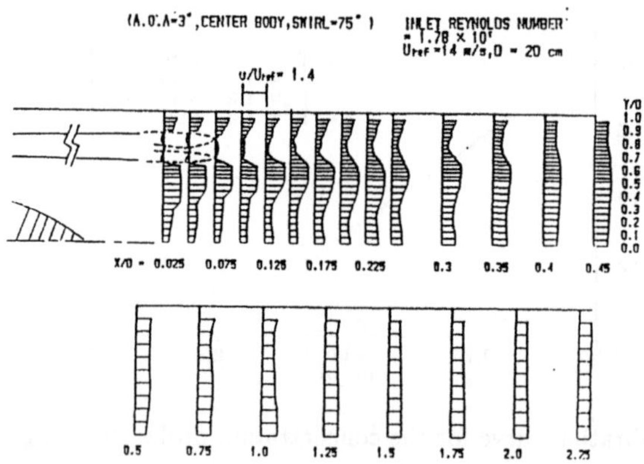

Fig. 4: The axial velocity distribution for the pilot burner flow

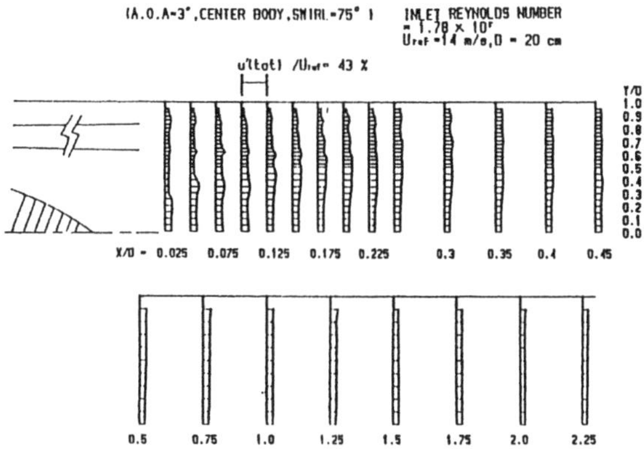

Fig. 5: The turbulent fluctuation distribution for the pilot burner flow

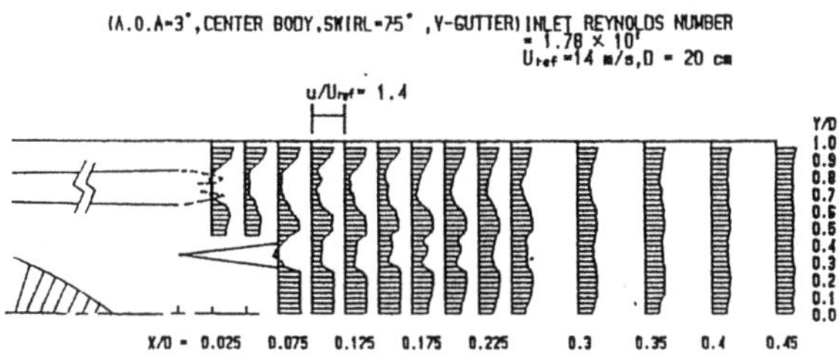

Fig. 6: The axial velocity distribution for the simulated augmentor flow

turbulent mixing in the lower portion of the pilot burner wake and strong mixing in the wake of the V-gutter provides a convenient path for the pilot flame to propagate to the V-gutter and the strong swirl component helps to stabilize the pilot flame in the wake of the pilot burner.

For the concentration measurements, the typical results of the cross-sectional concentration contours at $Z/D = 0.1$ from the pilot burner exit of the case of mixing of the secondary spray with the augmentor air stream without V-gutter is shown in Fig. 8. The swirl component of the pilot burner flow deflects the helium stream along the pilot burner annulus as the stream mixes with the air. The contours clearly show that in the upper portion the mixing of helium with air is limited and the contour line is limited to beneath the pilot burner outer wall while strong mixing is found in the lower

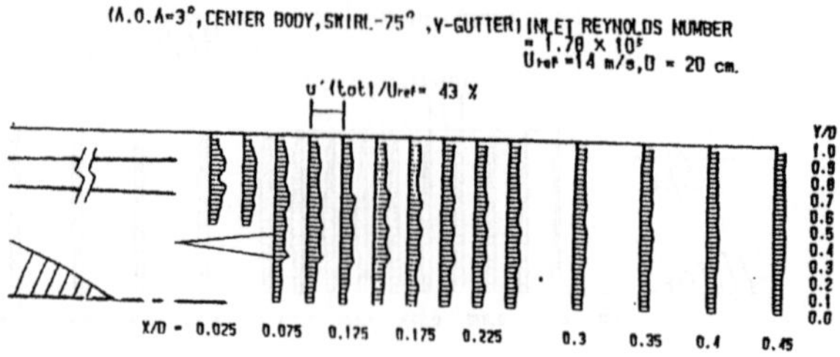

Fig. 7: The turbulent fluctuation distribution for the simulated augmentor flow

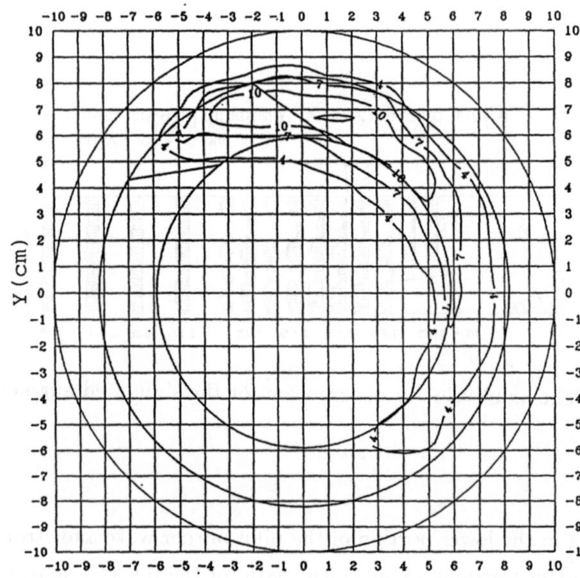

Fig. 8: The mean cross-sectional concentration contour at $Z/D = 0.1$ of the pilot burner flow with the secondary spray only

portion. This phenomenon is in good agreement with the velocity measurement results in Fig. 5 that high turbulent fluctuation in the lower portion results in better mixing.

In other words, the mixing is mainly due to the turbulence in the lower portion of the pilot burner flow. Figure 9 shows the typical contour at Z/D = 0.1 of the case of mixing of helium from the main spray bar with air without V-gutter. It is evident from the contour plot that turbulence in the lower portion of the pilot burner flow can entrain the main spray helium into the pilot burner flow and convects along the annulus. Thus it provides a convenient path for the pilot flame to propagate from the pilot burner to the main combustion area. It is very important for the ignition problem of the augmentor. In the case with the presence of the V-gutter as shown in Fig. 10, the V-gutter separates the main helium stream into two concentration contour centers and stronger turbulent mixing occurs between the lower portion of the pilot burner and the V-gutter. Figure 11 shows the case of complicated turbulent mixing of the augmentor flow with the presence of the V-gutter and the main and the secondary sprays. It is a superposition of the concentration results of Figs. 8, 9 and 10, but the effect is more pronounced with the presence of the V-gutter. If one carefully examines the contours, one can clearly find a strip of concentration distribution suitable for combustion near the X = 0 plane which provides a convenient path for the propagation of the pilot flame to the main combustion of the V-gutter. Another feature of this kind of augmentor turbulent flow is that

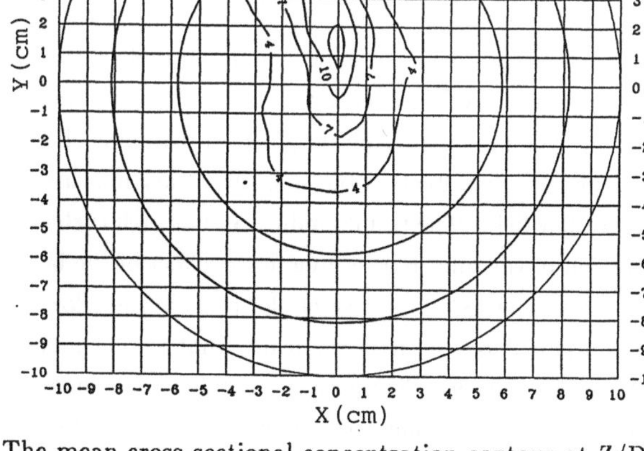

Fig. 9: The mean cross-sectional concentration contour at Z/D = 0.1 of the pilot burner flow with the main spray only

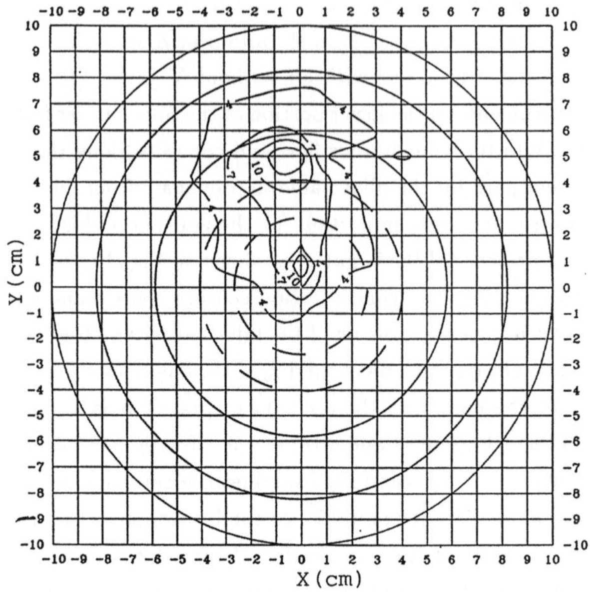

Fig. 10: The mean cross-sectional concentration contour at Z/D – 0.1 of the simulated augmentor flow with the main spray only

the reduced turbulent fluctuation and mixing in the upper portion of the pilot burner result in prohibited combustion between the augmentor wall and the pilot burner outer wall which will help to reduce the danger of hot spots near the augmentor wall.

Conclusions

The complicated augmentor turbulent flow and mixing characteristics with the presence of the pilot burner and the V-gutter are investigated experimentally using the laser Doppler velocimetry system and an aspirating concentration probe to measure the veelocity aomponents, the turbulent fluctuations and the binary mixture fraction distributions in the augmentor flowfield. The results inducate that high turbulent fluctuation occurs in the lower portion of the pilot burner due to the effect of the interaction of the swirl component and the strong axial velocity of the pilot burner flow. Further from the concentration measurement results it is confirmed that strong mixing is found in the lower potion of the pilot burner. This feature provides a convenient path for the pilot flame to propagate to the main combustion region of the V-gutter. The good correlation of the results of the turbulence and concentration measurements also indicates that turbulent mixing is the dominant mechanism in the augmentor flow.

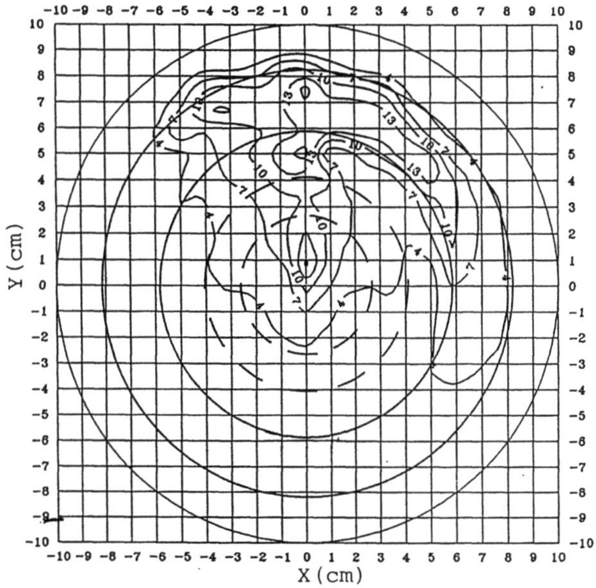

Fig. 11: The mean cross-sectional concentration contour at Z/D = 0.1 of the simulated augmentor flow with both the main and the secondary sprays

References

1. Longwell, J.P., Flame stabilization by bluff bodies and turbulent flames in ducts, 4th International Symposium on Combustion (1952) 90-97.

2. William, G.C., H.C. Hottel and A.C. Scurlock, Flame stabilization and propagation in high velocity gas stream, 3rd Symposium on Combustion and Flame and Explosion Phenomena (1949) 21-40.

3. Sullerey, R.K., A.K. Gupta and C.S. Moorthy, Similarity in the turbulent near wake of bluff bodies, AIAA Journal, 13 (1975) 1425-1429.

4. Stwalley, R.M. and A.H. Lefebvre, Flame stabilization using large flame-holders of irregular shape, AIAA-87-0469, 1987.

5. Zukoski, E.E., After burners, Chapter 21 in the Aerothermodynamics of Aircraft Gas Turbine Engines, ed. by G.C. Oates., AFAPL TR 78-52, 1978.

6. Taylor, A.M.K. and Whitelaw, J.H., Velocity characteristics in the turbulent near wakes of confined axisymmetric bluff bodies, J. Fluid Mech., 139 (1984) 391-416.

7. Ribeiro, M.M. and J.H. Whitelaw, Coaxial jets with and without swirl, J. Fluid Mech., 96 (1980) 769-795.

8. So, R.M.C., S.A. Ahmed and H.C. Mongia, Jet characteristics in confined swirling flow, Exp. Fluids, 3 (1985) 221-230.

9. Brondum, D.C. and J.C. Bennett, Numerical and experimental investigation of nonswirling and swirling confined jets, AIAA-86-0040, 1986.

10. Brown, G.L., and Rebollo, M.R., A small fast-response probe to measure composition of a Binary Gas Mixture, AIAA J., 10 (1972) 649-652.

11. Ahmed, S.A. and So, R.M.C., Concentration distributions in a model Combustor, Exp. Fluids, 4 (1986) 107-113.

12. Way, J. and Libby, P.A. Application of hot wire anemometry and digital techniques to measurements in a turbulent helium jet, AIAA J., 9 (1971) 1567-1573.

13. Chao, Y.C. and Yang, Y.S., The calibration and measurement of the concentration using an aspirating probe, Proceedings of the 14th National Conference on Theoretical and Applied Mechanics, Chungli, Taiwan (1990) 1569-1578.

14. Chao, Y.C. and Chen, T.C., Studies of the flow field characteristics of a pilot burner in the after burner, Proceedings of the 5th National Conference on Mechanical Engineering, CSME, Taipei (1988) 253-264.

Scalar Characteristics in a Liquid-Fuelled Combustor with Curved Exit Nozzle

S. K. CHOW AND J. H. WHITELAW

Department of Mechanical Engineering
Imperial College of Science, Technology and Medicine
Exhibition Road
London
England

1. INTRODUCTION

This investigation was carried out to improve understanding of the processes which determine the performance of a model of a reverse-flow annular combustor of the type used in small gas turbines. The model combustor incoporated a T-vaporiser for fuel supply, Sotheran (1984), and air supplied through dilution holes and film-cooling slots.

Chow, Senda and Whitelaw (1989) measured the scalar characteristics of a similar model combustor, but without the curved exit nozzle, representing the primary and dilution zone only with preheated air, at 515K, and kerosene as fuel and the same experimental conditions were employed in this study. The extent to which the nozzle influenced the combustion characteristics within the combustor were examined by comparing results measured in configurations with and without the exit nozzle.

The following section provides a brief description of the geometry of the combustor and instrumentation with consideration of experimental uncertainties. Results are presented and discussed in Section 3 and summary conclusions are stated in the final section.

R. S. L. Lee · J. H. Whitelaw · T. S. Wung (Eds.)
Aerothermodynamics in Combustors
IUTAM Symposium Taipei, Taiwan, 1991

2. FLOW CONFIGURATION AND INSTRUMENTATION

Air at near atmospheric pressure and 515K was supplied to the combustor with fuel delivered through an airblast atomiser comprising two coaxial metal tubes with inner diameters of 0.5mm and 4mm. The air flow through the vaporiser was the minimum required to vaporise the spray and led to an air-fuel-ratio of 2 for the mixture emerged from the vaporiser. Overall air-fuel ratios were 36 and 60 corresponding to the take-off and ground-idle conditions respectively for the Gem-60 engine with the mass flow rates scaled according to the parameter $mT^{1/2}/AP$, where m is the mass flow rate, T is the temperature, P is the pressure at the inlet to the combustor and A is the cross-sectional area of the combustor.

The geometry of the model combustor is shown in Figure 1. It consists of nine film-cooling slots, a set of five primary holes located on the lower wall with two in line with the vaporiser exits and two sets of dilution holes staggered by a half pitch with respect to the primary holes. The first set of four dilution holes is located on the upper wall and the second set of two on the lower wall.

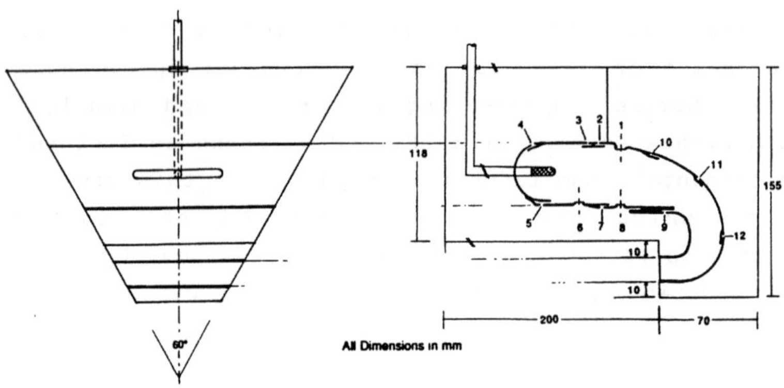

Figure 1 Combustor Geometry

Temperature measurements were obtained with thermocouples made from 80 μm platinum / 13% rhodium and platinum wires. The temperature signal was amplified, digitised and processed in a microcomputer. The uncertainties in the temperature measurements due to conduction and radiation were found to be less than 0.1% and around 5% respectively. The results are close to unweighted mean values as suggested by Toral and Whitelaw (1982) and the comparatively low gas velocities imply that the thermocouple outputs are related to the local static temperature.

Gas samples were extracted through a stainless-steel water-cooled probe with outside and inside diameters of 8.0 and 1.0mm respectively. The sample was sucked through the probe by a pump and passed through to the infrared gas analysers (for CO and CO_2), paramagnetic analyser (for O_2) and flame ionisation detector (for UHC). Uncertainties in the measurement of the various gas concentrations were less than ± 0.5% of the full scale of each analyser and the uncertainty of each gas concentration, associated with sample handling, should be less than 10% of the maximum concentration as suggested by Heitor and Whitelaw (1986) and the measured concentrations of major species are expected to be close to density-weighted averages.

3. RESULTS AND DISCUSSION

The distributions of measured temperature, and the corresponding enthalpy balance temperature, are shown in Figures 2 and 3 for air-fuel ratios of 60 and 36 respectively. In both cases, high temperature regions were found close to the inner wall of the curved exit nozzle with maxima of 1350K and 1650K for the two air-fuel ratios. The enthalpy balance temperatures were generally about 50 to 100K lower than the measured ones which is in accordance with the finding of Chow, Senda and Whitelaw (1989). The calculated pattern factors, based on 78% of the area of the exit plane, are 0.22 and 0.18 for air-fuel ratios of 60 and 36 respectively.

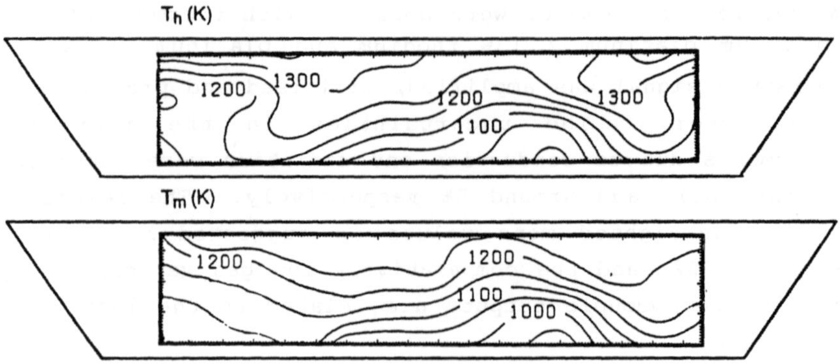

Figure 2 Measured and Enthalpy Balance Temperature in the Exit Plane for Air-Fuel Ratio of 60

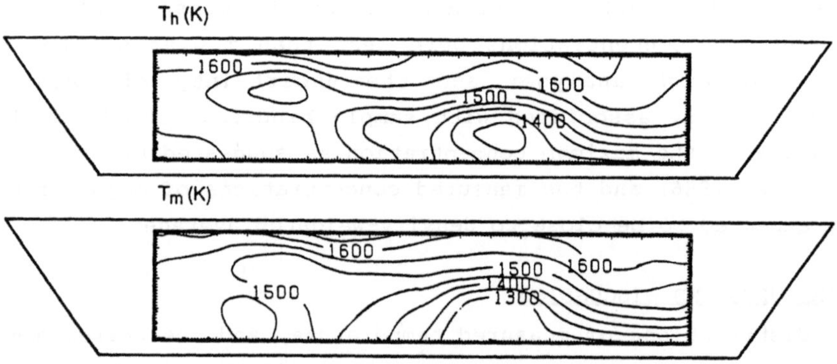

Figure 3 Measured and Enthalpy Balance Temperature in the Exit Plane for Air-Fuel Ratio of 36

Figure 4 shows the distributions of major species concentrations, on a wet basis, and calculated equivalence ratio for the air-fuel ratio of 60. The maximum equivalence ratios, above 0.3, were found close to the inner wall of the curved bend and the equivalence ratios at the outer bend were generally about 0.2. No distinct hot zone was observed across the width of the exit plane and maximum levels of UHC and CO were 0.006

(mole fraction of $C_{12}H_{26}$) and 0.15% respectively. This corresponded to emission indices of CO and UHC of 50 and 5 respectively which led to a combustion efficiency of 98.3%.

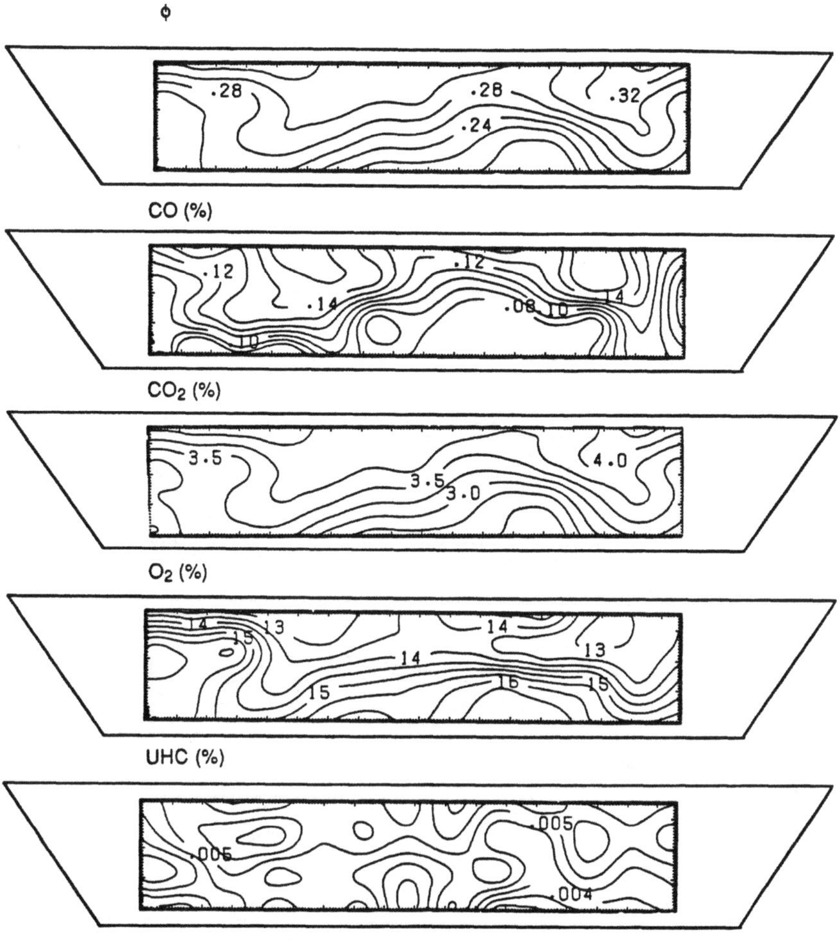

Figure 4 Distributions of Major Species Concentrations in the Exit Plane for Air-Fuel Ratio of 60

The species-concentration distributions for the air-fuel ratio of 36 are shown in Figure 5. The fuel rich regions, with maximum equivalence ratios of 0.48, remained close to the inner wall as in the case for the higher air-fuel ratio. Maximum levels of UHC and CO increased to 0.009 and 0.18% respectively. Compared to the higher air-fuel ratio condition, the emission indices of CO and UHC increased to 68 and 6 respectively and a slight decrease in combustion efficiency to 97.8% resulted. The performance of the combustor with the curved exit nozzle in terms of combustion efficiency, emission indices of CO and UHC and pattern factor is also shown in Table 1.

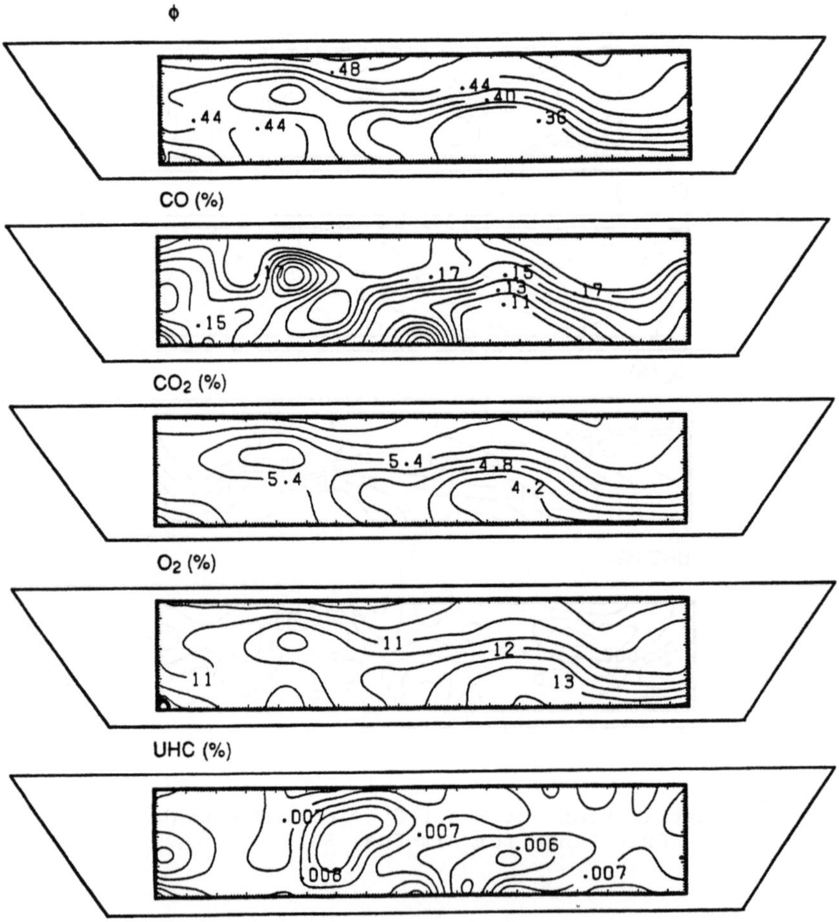

Figure 5 Distributions of Major Species Concentrations in the Exit Plane for Air-Fuel Ratio of 36

The distributions of species concentrations at the exit of the curved nozzle of Figures 4 and 5 for the two air-fuel ratios are compared with those at the exit of the combustor without the nozzle to quantify the effect of the nozzle to the combustion processes. For ground-idle condition, the maximum value of equivalence ratio decreased from 0.42, at the entry to the nozzle, to 0.3, at the exit, due to the additional air from the cooling slots inside the curved nozzle. The two distinct hot zones at the entry disappeared at the exit due to the intense mixing within the nozzle caused by the severe contraction in cross-sectional area (the area ratio between the entry and the exit is 4.25). The high concentrations of fuel at the outer half of the combustor were pushed to regions near the inner wall due, in part, to the secondary flow inside the bend. Moreover, the fluid near the outer wall was diluted by the air coming out from the three film-cooling slots at the outer wall of the nozzle.

The average concentrations of UHC and CO dropped from 0.009 to 0.006% and from 0.18 to 0.12% respectively within the bend in the ground-idle condition. Taking into account the dilution by the extra air, which was about 40% of that coming into the combustor in further upstream regions, from the cooling slots within the bend, the absolute amount of CO and UHC actually decreased by very little inside the bend. Therefore, it can be concluded that the fuel and CO burn-up process almost ceased at the entry of the bend. This is because the temperature in regions of high concentrations of CO and UHC was below 1400K which is the kinetic limit on fuel and CO burn-up rate as suggested by Levebvre (1983).

For take-off condition, the average concentrations of UHC and CO dropped from 0.04 to 0.007% and 0.8 to 0.15% respectively. Contrary to the higher air-fuel ratio condition, the combustion processes continued beyond the entry of the curved bend and the fuel breakup and CO to CO_2 conversion took place as a result of the higher temperature, about 1800K, at that plane since the kinetic limitation of the fuel and CO burn-up were overcome.

4. CONCLUSION

The main conclusions can be summarised by the following paragraphs:

1. For take-off condition, air-fuel ratio of 36, combustion processes continued beyond the entry to the curved nozzle led to lower emission indices of CO and UHC of 68 and 6 and a better combustion efficiency of 97.8%, as compared to the entry of the nozzle, at the exit of the nozzle.

2. An increase in air-fuel ratio from 36 to 60 resulted in an increase in pattern factor from 0.18 to 0.22 and in combustion efficiency from 97.8% to 98.3% at the exit of the nozzle.

ACKNOWLEDGEMENT

The authors would like to acknowledge the financial support from the Procurement Executive of the Ministry of Defence. They, and the paper, have benefitted from many useful discussions with colleagues at Rolls Royce, the Royal Aircraft Establishment at Pyestock and Imperial College.

REFERENCE

1. Chow, S.K., Senda, M. and Whitelaw, J.H. (1989), "Combustion Characteristics of a Liquid-Fuelled Model Annular Combustor", presented at the Ninth International Symposium on Air Breathing Engine, Athens.

2. Heitor, M.V. and Whitelaw, J.H. (1986), "Velocity, Temperature and Species Characteristics of the Flow in a Gas Turbine Combustor", Comb. and Flame, 64, 1-32.

3. Lefebvre, A.H. (1983), Gas Turbine Combustion, McGraw Hill text book.

4. Sotheran, A. (1984), "The Rolls Royce Annular Vaporiser Combustor", J. Engng. for Gas Turbines and Power, 106, 88.

5. Toral, H. and Whitelaw, J.H. (1982), "Velocity and Scalar Characteristics of the Isothermal and Combusting Flows in a Combustor Sector Rig", Comb. and Flame, 45, 251.

Table 1 Combustor Performance

Air-fuel ratio	Pattern factor	EICO (g/kg fuel)	EIUHC (g/kg fuel)	Combustion efficiency (%)
36	0.18	68	6	97.8
60	0.22	50	5	98.3

Structure of the Primary Region of a Model Gas Turbine Combustor

Sunao NAKAMURA
Engineering Research Center, NKK Corporation

Katsuya HYODO
Nippon Telegraph and Telephone Corporation

Osamu KAWAGUCHI
Faculty of Science and Technology, Keio University

SUMMARY

Detailed measurements on flow patterns, gas compositions and temperatures in the primary region of a gas turbine combustor were conducted for a wide range of operating conditions using a simplified model combustor. The flow patterns estimated from the data of velocity measurement by LDV verified the relation of the size of recirculation zone with the swirl strength and the primary air/fuel ratio. The actual equivalence ratio of the internal region of the recirculation region changes remarkably with the change of swirl strength at a constant primary air/fuel ratio and it brings about flame stability with swirl. The enthalpy flows in the recirculation zone and the main flow were also estimated and the stabilization mechanism based on the concept of heat supply from the recirculating flow to the main flow has been made clear.

1. INTRODUCTION

Detailed data in the combution region of a gas turbine combustor have been obtained by using a model combustor to interpret the structure of combustion region. In the previous work the effects of secondary air on the exhaust emissions have been verified[1][2]. The authors report here the profile of gas flow, temperature and gas composition in the combustor for various operating conditions and discuss on the structure of the combustion region.

2. MODEL COMBUSTOR

The model combustor used for the experiments is shown in the Fig. 1. It is a can-type straight flow combustor with a liner of 125 mm in diameter and 350 mm in length. A peeping window with a quartz glass is provided at the liner wall to pass laser

R. S. L. Lee · J. H. Whitelaw · T. S. Wung (Eds.)
Aerothermodynamics in Combustors
IUTAM Symposium Taipei, Taiwan, 1991

beams for velocity measurements. The downstream side is open to the atmosphere.

The combustion air is supplied through the annular path after passing through the swirl generator to give a settled swirl strength. Pure propane for industrial use (97% purity) is injected at the injection angle of 90 degrees from the multi-hole injector with 24 holes.

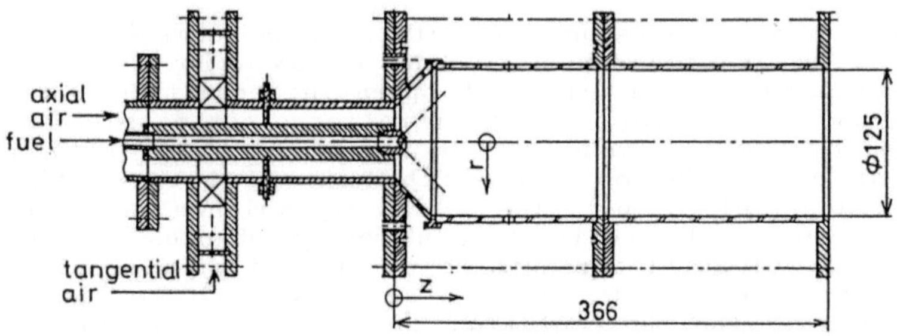

Fig. 1. Model combustor for the experiments

3. MEASUREMENT

The laser doppler anemometry using a two-beam back scattering type optical system has been used to measure the flow in the combustion region. The beams from a 25 mW He-Ne laser pass through the quartz window of the liner. MgO particles less than 2 μm in diameter were introduced directly to the combustion region as the scattering particles with a stainless tube of 2 mm in outer diameter.

The gas temperature was measured by a silica coated Pt/Pt-Rh13% thermocouple of 50 μm in original diameter, which was inserted from the downstream of the combustor. Correction of the error of measured values was not made.

The combustion gas was sampled with a water-cooled stainless steel probe and analyzed by a gas chromatograph.

Experimental parameters adopted were the amount of mass flow of combustion air, Ma (kg/s), the swirl number of combustion air, S, and the air/fuel ratio, Ma/Mf. The standard condition was fixed to Ma = 43×10^{-3} kg/s (reference velocity at inlet condition U = 3.15 m/s), S = 1.0 and Ma/Mf = 15.7.

4. EXPERIMENTAL RESULTS

4.1 Effects of Swirl Strength

Figure 2 shows the stream lines for three different swirl numbers at the standard condition except of swirl number. The stream lines are expressed by using the normalized stream function.

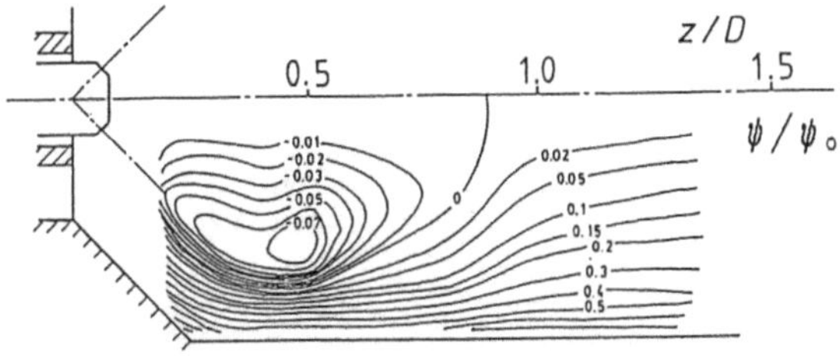

(a) S = 0.7

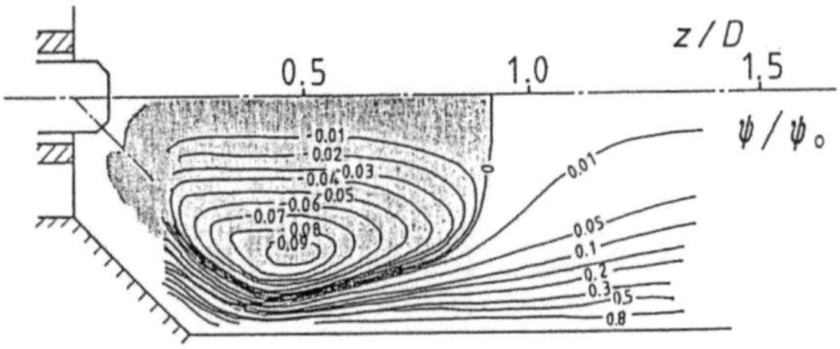

(b) S = 1.0

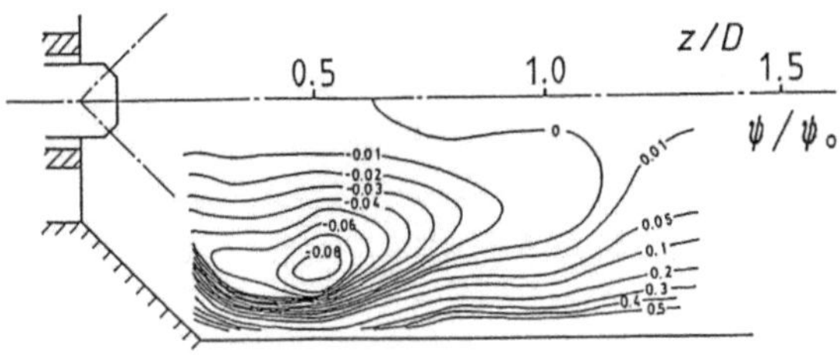

(c) S = 1.5

Fig. 2. Stream lines for various swirl numbers

At the standard condition a recirculation zone (R. Z.) developes in the combustion region and a stable flame is formed. The amount of recirculating gas is 7 to 9 % of the total mass flow. The radial position of the outer boundary of the R. Z. is 0.85 in r/R, and the downstream position of R. Z. is 0.9 in z/D.

On taking notice of the effects of swirl strength, the R. Z. is formed in the downstream region near the fuel injector at the swirl number larger than 0.6, which causes stable combustion in the range of swirl number from 0.7 to 1.5. With increase in the swirl number the vortex center shifts to the outer side and downstream side due to the increase of angular momentum of combustion gas. The size of R. Z. grows as the swirl number increases and at S = 1.5 it becomes 1.5 times compared with that at S = 0.7.

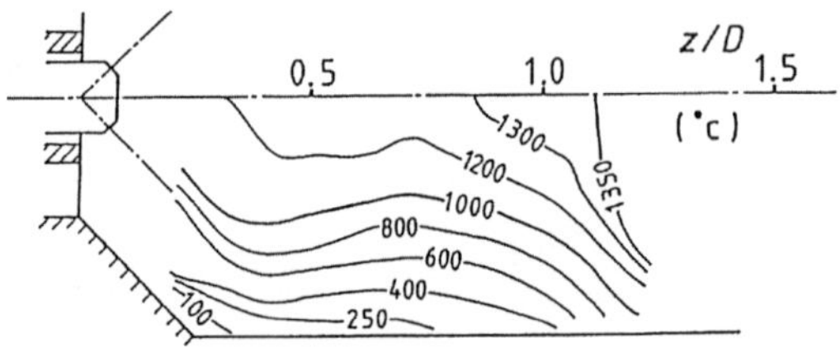

(a) S = 0.7

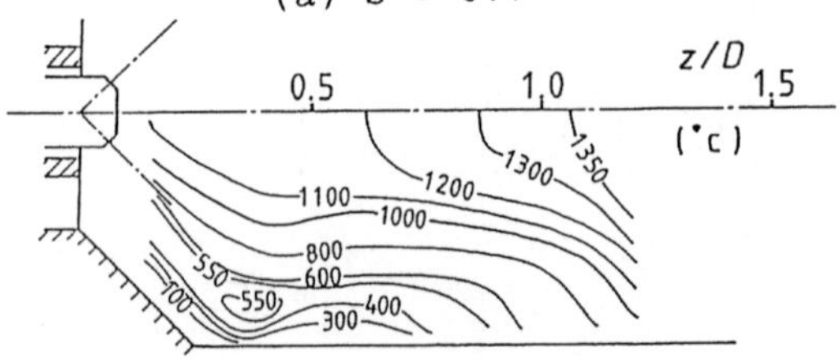

(b) S = 1.0

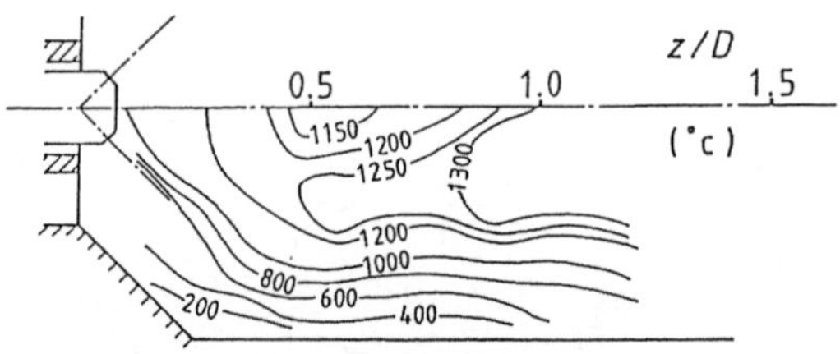

(c) S = 1.5

Fig. 3. Isothermal lines for various swirl numbers

Figure 3 indicates the isothermal lines in the combustion region for three different swirl conditions. It can be infered from the figures that the flame expands outside and shift to the upstream. At this range of swirl strength, however, since a recirculation zone develops sufficiently near the fuel injector, the temperature profiles do not differ so much each other and a high temperature zone exists near the injector.

The profiles of the axial component of flow velocity and its fluctuation, the temperature at the cross section of z/D=0.4, where is the central position of R. Z., are indicated in Fig. 4 and 5, respectively, for three different swirl conditions.

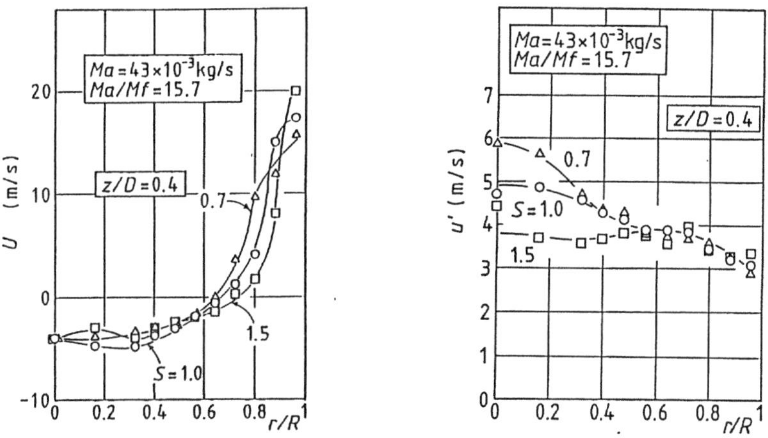

(a) Axial velocity component (b) Velocity fluctuation

Fig. 4 Axial velocity and its fluctuation distributions for various swirl numbers at z/D=0.4

There exist no large differences of velocity according to the swirl conditions but, on increasing the swirl strength, the reverse flow zone expands outside and the velocity gradient in the main stream region increases. The velocity fluctuation becomes less at the reverse flow zone with the swirl increase but in the forward flow zone it tends to increase. The tangential velocity component becomes higher at every radial position with a swirl increase, as a matter of course.

The profiles of gas temperature in the recirculation region

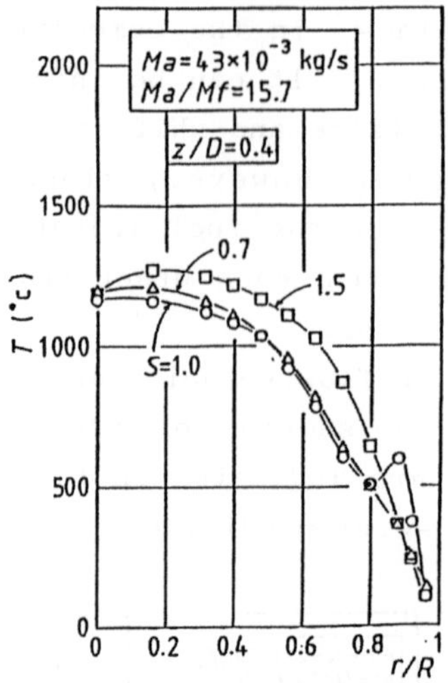

Fig. 5. Temperature distributions for various swirl strengths at z/D=0.4

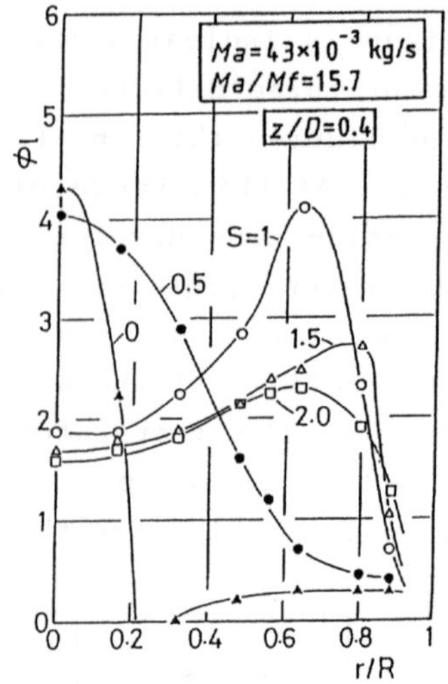

Fig. 6. Local equivalence ratio distributions for various swirl strengths at z/D=0.4

do not present much difference due to the change of swirl strength, but in the high swirl condition (e.g. S=1.5), the temperature rises in the innner region of the recirculation zone, while the main flow has a tendency to reduce its temperature.

The distributions of local equivalence ratio estimated from the data of gas composition are also compared for different swirl strengths, as shown in Fig. 6. The inner region of the recirculation zone presents a rich condition regardless of swirl strength and, as the swirl becomes strong, the region comes near the stoichiometric condition. On the contrary, the main flow around the recirculation zone presents a lean condition for all swirl strengths while it becomes rich with the increase of swirl strength. The results of local equivalence ratio correspond to those of temperature distributions mentioned above.

Since the volume of the R. Z. grows with a swirl increase and the axial velocity in the main stream flowing downstream around the R. Z. increases, the fuel injected to a high swirl air flow is brought to the outer side and most of the fuel flows in the boundary region of the R. Z.. Though it brings about lowering of the local equivalence ratio of the R. Z. and temperature rise, a part of the fuel flows downstream with the main flow without sufficient chemical reactions in the high shear flow. Consequently, an excessive swirl lowers the combustion efficiency.

The temperature rise in the R. Z. with swirl increase is also caused by the increase of residence time in the R. Z. due to the increase of tangential velocity component. The residence time in the R. Z. estimated by the measurements of reduction process of light emission of flame by a sudden cut-off of temporarily supplied sodium carbonate powder to the R. Z. is indicated in Fig. 7 as the function of swirl strength with the representative temperature of the R. Z. and the length of the R. Z. The temperature of the R. Z. rises together with residence time in the R. Z. according to the increase of swirl strength.

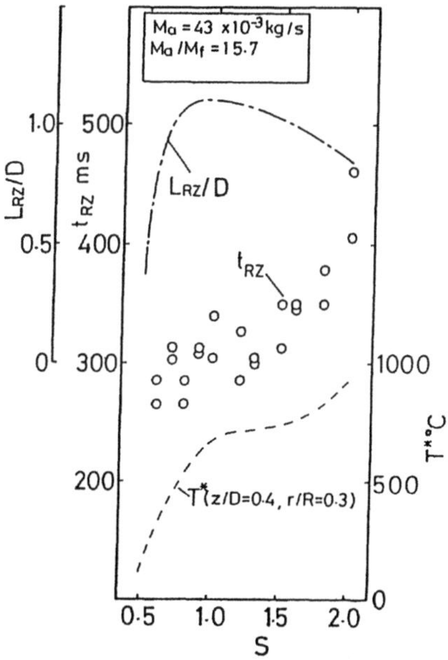

Fig. 7. Residence time, t_{RZ}, representative temperature, T*, and length of the R. Z., L_{RZ} as the function of swirl number

4.2 Effects of Air/Fuel Ratio

Figure 8 indicates the stream lines of the combustion region for three different air/fuel ratio conditions. Except the air/fuel ratio the conditions are fixed to the standard values.

On increasing the air/fuel ratio from 15.7 to 26.2, the length of the R. Z. becomes somewhat short, but at a very lean condition of Ma/Mf=52.3 the flame expands as a cup-like shape and the length of R. Z. increases and nears that of the cold flow.

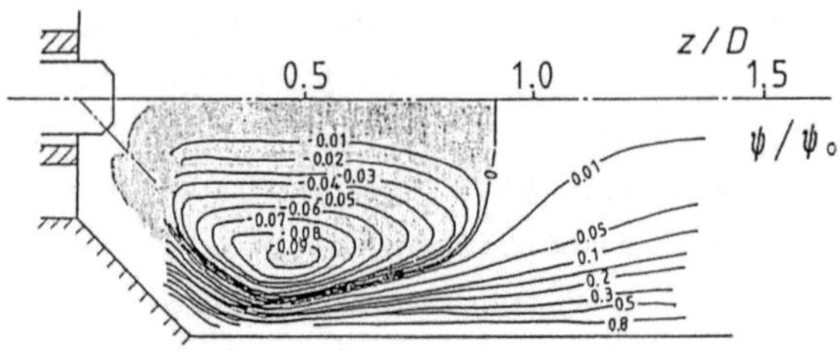

(a) Ma/Mf = 15.7

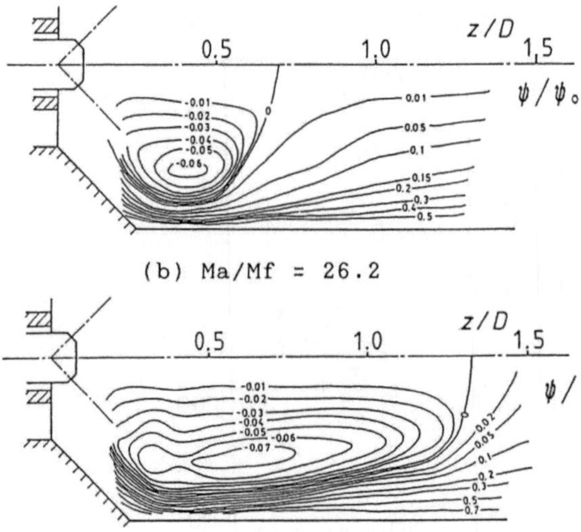

(b) Ma/Mf = 26.2

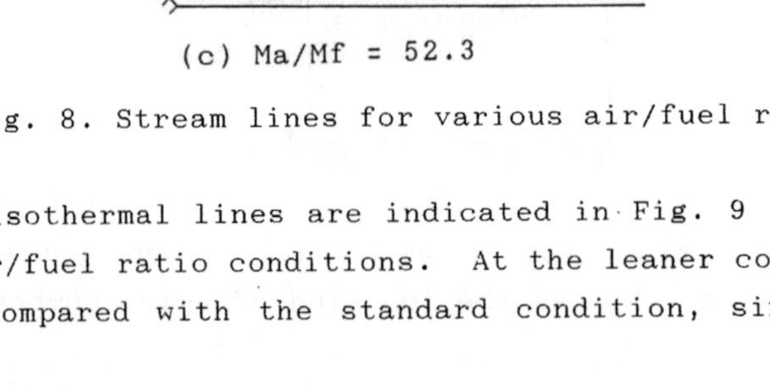

(c) Ma/Mf = 52.3

Fig. 8. Stream lines for various air/fuel ratios

The isothermal lines are indicated in Fig. 9 for three different air/fuel ratio conditions. At the leaner condition, Ma/Mf = 26.2, compared with the standard condition, since the fuel injected diffuses immediately into the air flow and reacts rapidly, the temperature rises in the upstream region. The temperature inside the R. Z. reaches 1500 °C. At the leaner condition, Ma/Mf = 52.3, the temperature becomes high along the flame expanding like a cup-shape but it lowers in the R. Z, whose length becomes longer as the results.

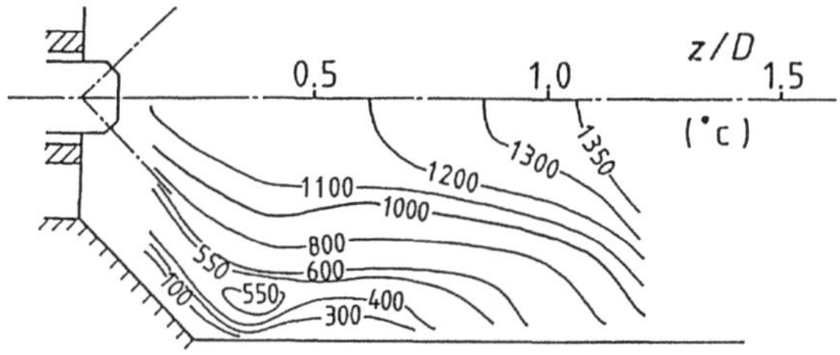

(a) Ma/Mf = 15.7

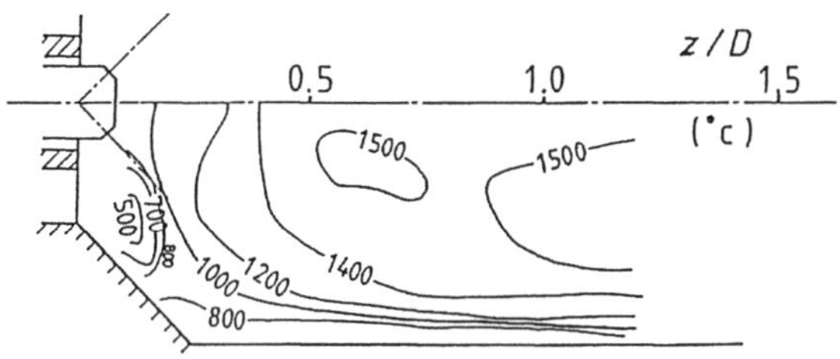

(b) Ma/Mf = 26.2

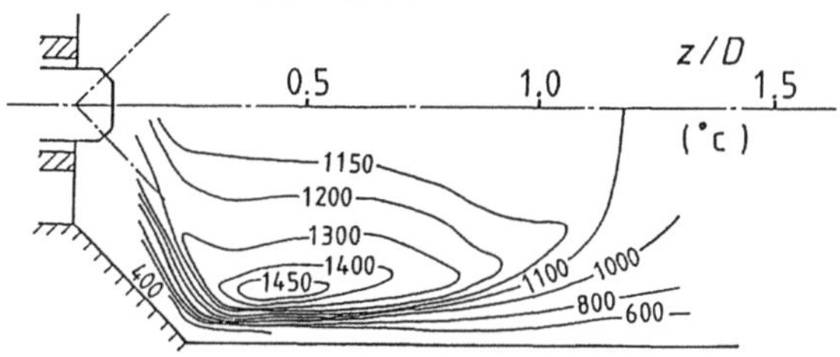

(c) Ma/Mf = 52.3

Fig. 9. Isothermal lines for various air/fuel ratios

The distributions of axial velocity component and its fluctuation at the cross section of z/D = 0.4 are shown in Fig. 10 for three different air/fuel ratio conditions, and those of combustion gas temperature in Fig. 11.

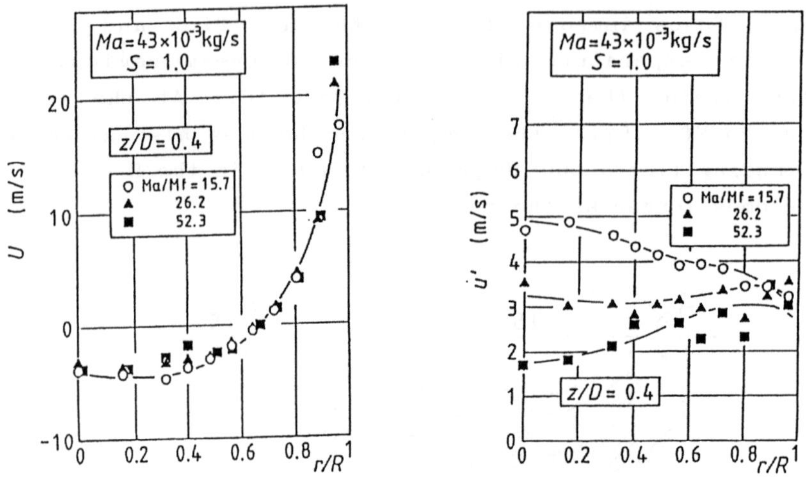

(a) Axial velocity components (b) Axial velocity fluctuations

Fig. 10 Axial velocity and its fluctuation distributions for various air/fuel ratios at z/D=0.4

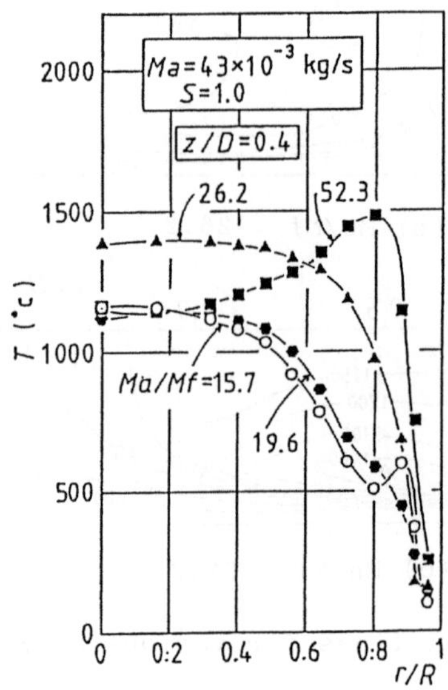

Fig. 11. Temperature distributions for various air/fuel ratios at z/D=0.4

The axial velocity presents about the same value for all air/fuel ratios but the fluctuation increases more at the leaner condition. The tangential velocity component does not change so much with the change of air/fuel ratio but at an extremely lean condition the velocity and its fluctuation decrease because of less heat release.

The temperature is low through the whole radial position at the center of the R. Z. for a lower primary air/fuel ratio than the stoichiometric value (e.g. Ma/Mf=10.5) because of lack of oxygen. With the increase of primary air/fuel ratio, the temperature of the R. Z. rises, but at a very lean condition (e.g. Ma/Mf=52.3) a cup-like flame is formed and only the region near the liner wall presents a high temperature.

On taking notice of local equivalence ratio, the inner region of the R. Z. is richer than the settled primary value for all the air/fuel ratio conditions, and when the primary air/fuel ratio is stoichiometric or rich, the equivalence ratio of the inner region of R. Z. reaches 2 or more and chemical reactions

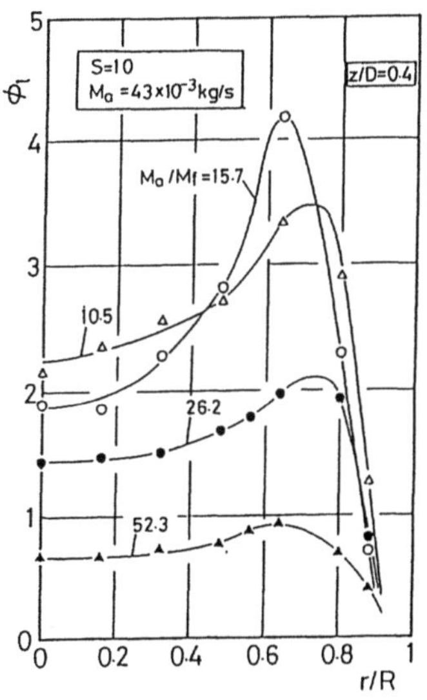

Fig. 12. Local equivalence ratio distributions for various A/F at z/D=0.4

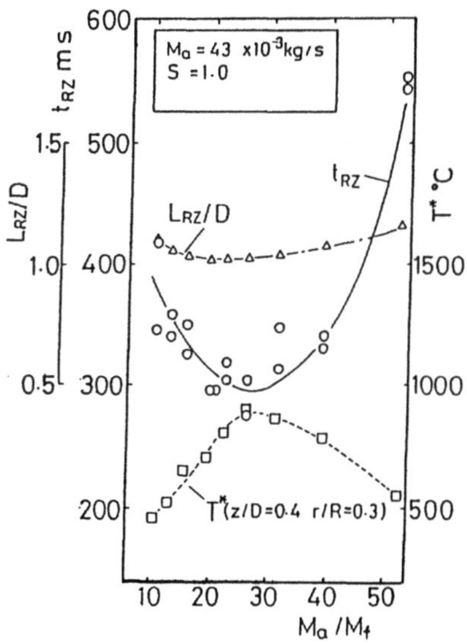

Fig. 13. Residence time, t_{RZ}, temperature, T*, and length of R. Z., L_{RZ}, as the function of A/F

cannot proceed sufficiently. At a leaner condition than the stoichiometric condition the equivalence ratio approaches unity and chemical reactions are promoted. At a extremely lean condition (e.g. Ma/Mf=52.3) the region becomes too lean and chemical reactions are suppressed.

Figure 13 shows the residence time at the R. Z. as the function of air/fuel ratio together with the length and the representative temperature of the R. Z. When the A/F changes, the residence time is inversely affected by the temperature change of R. Z. due to thermal expansion and presents the minimum at the highest temperature condition.

4.3 Flow Lines of Each Gas Component

The mass flow of each gas component can be obtained from the velocity, temperature and gas composition data by using the same calculating process as the stream function. Mass flow lines of C_3H_8, O_2, CO and CO_2 are indicated in Fig. 14 for the standard condition. The shaded area indicates the R. Z. obtained from the stream lines.

Mass exchange by eddy diffusion comes into existence at the shear flow region between the R. Z. and the main stream region. The boundary of R. Z. taking notice of oxygen exists inside the R. Z. boundary of the total gas flow, which is due to the oxygen supply from the main flow to R. Z. Propane expands to some extent by the existence of R. Z. and, after reaching near the liner wall, seems to diffuse into the combustion gas flow. There exists propane of fairly high concentration in the upstream of the main stream, and chemical reactions producing CO and CO_2 seem to occur to a certain extent in the upstream. At the stoichiometric air/fuel ratio condition, fair amount of CO is detected in the gas flowing at the downstream of the main combustion region, which means that the chemical reactions are not completed only around the R. Z.

4.4 Flame Stabilizing Mechanism

The enthalpy flow rate across the central cross section of the R. Z. which passes the vortex center was estimated from the

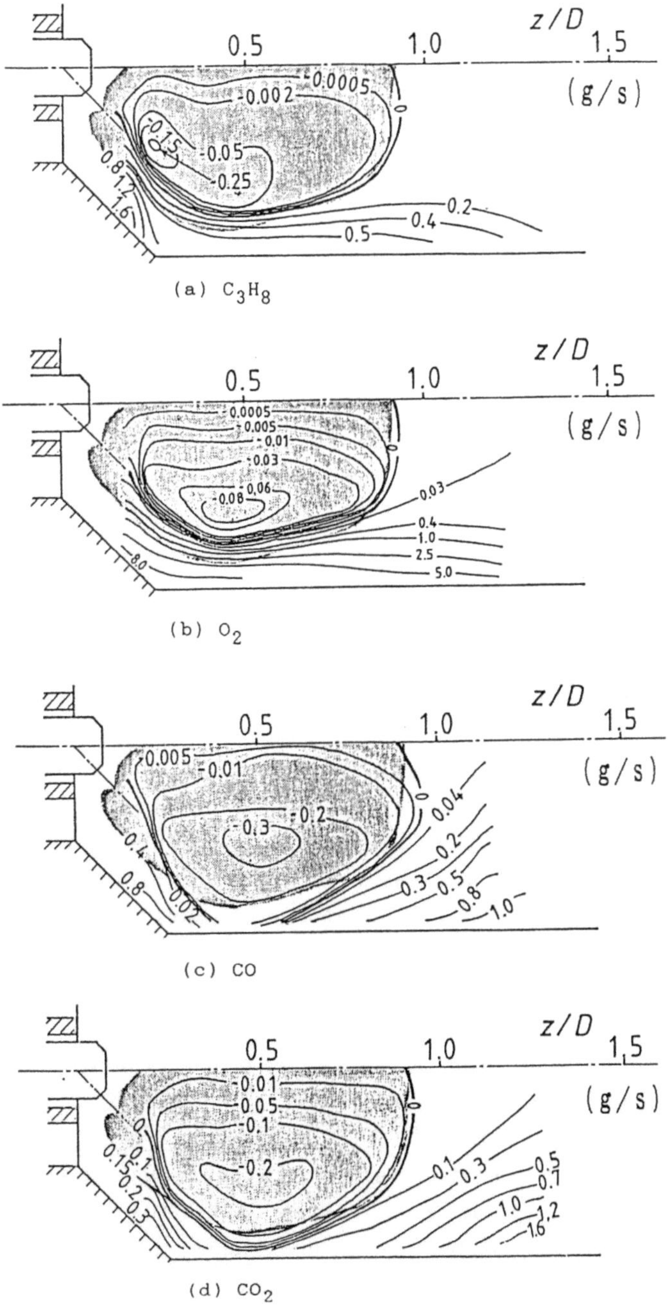

Fig. 14. Equimass flow lines for C_3H_8, O_2, CO and CO_2 at the standard condition

measured values of flow velocity, gas composition and gas temperature. The enthalpy flows in the backward and forward direction following to the gas flow in the R. Z.

Due to calculations based on the experimental results the backward enthalpy flow exceeds the forward one. The difference between the two is equivalent to the heat supplied to the main combustible gas stream to ignite and react it.

Both the forward and backward enthalpy flows increase with the increase of swirl number from 0.7 to 1.5 and also the difference of the two increases from 2.6 to 4.3 % of the heating value of the supplied fuel. It means that the air swirl contributes to stabilize the flame at the combustion region.

When the air/fuel ratio is increased to 52.3, the backward enthalpy flow in the recirculation zone decreases but the forward one changes only a little. As the results, the difference of the two decreases with air/fuel ratio increase but the relative quantity normalized by the supplied heating value is kept almost constant to stabilize the flame.

5. CONCLUSIONS

Detailed measurements of temperature, velocity and gas composition in the combustion region of a model combustor were conducted and following results has been obtained.

(1) The size, shape and velocity distribution of a recirculation zone depend on the air swirl and primary air/fuel ratio.

(2) At a stable combustion condition, a fuel jet with suitable penetration reaches near the liner wall and diffuse in the air flow and reacts to form a high temperature region in the upstream. When the swirl is weak, the inner region of the R. Z. becomes far richer than the fixed primary air/fuel ratio.

(3) In the R. Z. the backward enthalpy flow exceeds the forward enthalpy flow, and the enthalpy difference is consumed to heat and ignite the main flow around the R. Z. to contribute to the flame stabilization.

REFERENCES

(1) Nakamura, S. et al., Proc. 1983 Tokyo Gas Turbine Congress, Vol.1.
(2) Nakamura, S. et al., Journal of GTSJ, Vol. 10-37 (1982) (in Japanese)
(3) Heitor, M. V. et al., Combustion and Flame, Vol. 64-1, p.1-32 (1986)

Development of Gas Turbine Combustors for Low BTU Gas

I. Fukue, S. Mandai and M. Inada
Takasago Machinery Works and Takasago R & D Center
Mitsubishi Heavy Industries, Ltd., Japan

1. INTRODUCTION

Large-capacity combined cycles with high-temperature gas turbines burning petroleum fuel or LNG have already been operated in large numbers in both domestic and foreign countries and have achieved good results. On the other hand, as the power generation technology utilizing coal burning the coal gasification combined plants are also under research. Since coal-gasified gas is lower in BTU and harder to burn than conventional fuel, development of this combustor is one of the important problems to be solved. To develop the combustor burning low BTU gas in our company, we have proceeded the study for blast furnace gas, i.e. by-product gas in the ironworks and have succeeded in practical use of the multi-can type combustor burning low BTU gas in the combined plant supplied to Chiba Works, Kawasaki Steel Corporation.

The following technical problems for this combustor have been overcome, and through the demonstration tests with actual gas the combustor has been made fit for practical use:

(1) Development of multi-can type combustor for large-capacity gas turbine
(2) Wide-range stable combustion technology by means of air by-pass mechanism
(3) Combustor cooling construction of high efficiency

This paper introduces the combustion technology obtained through development of the above combustor burning low BTU gas.

2. DEVELOPMENT OF COMBUSTOR

2.1 Development of Multi-Cannular Type Gas Turbine Combustor Burning Low BTU Gas

Recently in the ironworks most of the externally-dependent energy sources are covered by coal. This coal energy is converted to such various kinds

R. S. L. Lee · J. H. Whitelaw · T. S. Wung (Eds.)
Aerothermodynamics in Combustors
IUTAM Symposium Taipei, Taiwan, 1991

of by-product gas as blast furnace gas, coke oven gas, converter gas, etc. in the processes where coal is used in iron-manufacturing processes. Table 1 shows the typical composition of by-product gas, combustion velocity as an index of the combustion quality (laminar flow maximum combustion velocity) and the combustible range. In the table also the values of LNG, i.e. typical high BTU fuel, are shown for the purpose of comparison.

Table 1 Typical composition and combustion quality of by-product gas

		Blast furnace gas	LDG	Coke oven gas	Methane (for ref.)
Component (Vol. %)	H_2	1.9	1.9	52.8	-
	CO	19.2	61.4	5.8	-
	CH_4	-	-	27.8	100
	C_2H_4	-	-	3.8	-
	CO_2	22.1	15.4	1.9	-
	N_2	52.8	17.3	3.9	-
	H_2O	4.0	4.0	4.0	-
Calorific value (kcal/Nm)		630	1910	4450	8530
Laminar flow combustion velocity	(cm/sec)	≈ 0	56	72	37
Combustible range (Air ratio)	Rich	0.84	0.28	0.46	0.54
	Lean	1.32	2.78	3.70	1.79
	Lean/Rich	1.6	9.9	8.0	3.3

Among the above product gas, since blast furnace gas contains plenty of inert gas in spite of its abundant product volume, its calorific value is as low as 600-800kcal/Nm^3, and moreover it is remarkably poor in combustion quality. Therefore, its application was limited to such combustion equipment of relatively-low combustion load rates as boilers and so on.

On the other hand, in West Europe, U.S.A. and Japan, after World War II it was tried to apply blast furnace gas to gas turbine fuel. Several gas turbines burning blast furnace gas were constructed and put in practical use. However, in many of them, because of the low combustion quality of blast furnace gas large combustors of the single cylinder type with low combustion load rates were arranged.

Fig.1 shows the cross section of the 20,000kW class gas turbine burning blast furnace gas. In addition the cross section of the same class multi-cannular type gas turbine is shown for the purpose of comparison. The conventional gas turbine burning blast furnace gas was so designed that the combustor diameter was increased to lower the combustion load rate and its length was increased to lengthen the staying time of combustion gas. Thus, the size of the combustor became large and nearly equal to that of the turbine main body as shown in Fig.1.

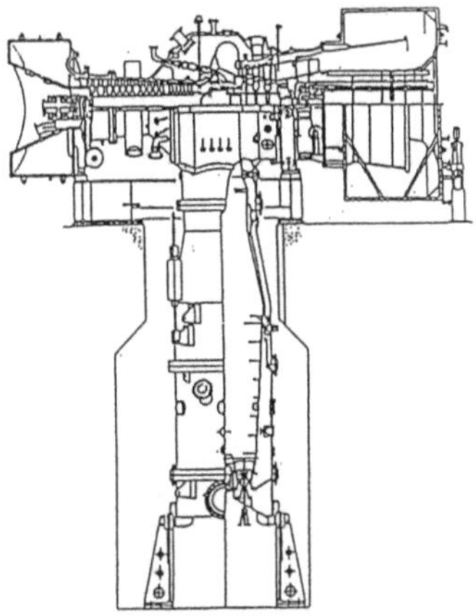

Single cylinder type combustor

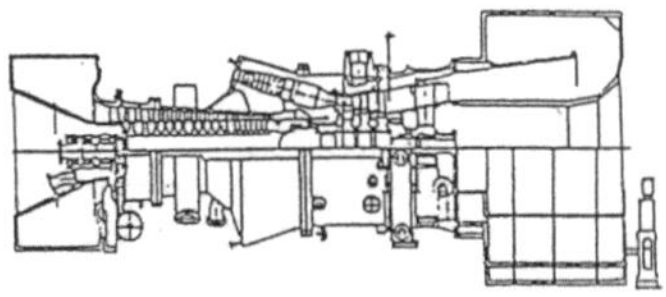

Multi-cannular type combustor

Fig.1 Cross section of gas turbine burning blast furnace gas

For temperature increasing for the purpose of increasing the efficiency of the gas turbine proper or the combined efficiency for combination with the steam turbine, the gas turbine inlet temperature was limited to about 1000°C from the viewpoint of the combustion technology due to the low BTU of blast furnace gas and of the cooling technology of the scroll that guides combustion gas to the turbine.

Further, for capacity increasing of the gas turbine, two or more sets of the large combustors nearly-equal in size to the gas turbine proper are required as mentioned above, when they are designed with the combustion load rates equivalent to the conventional experience values. It is difficult realistically to materialize this from the viewpoint of their arrangement and so on.

In the above background it was indispensable in applying blast furnace gas to large-capacity high-temperature gas turbines to develop such compact multi-cannular type combustors as conventional ones used for high BTU fuel. For this purpose it was required to improve the combustion characteristics by increasing the BTU of blast furnace gas and by increasing the turbine inlet temperature. Then, application of high BTU by-product gas in the ironworks was considered to this BTU increasing purpose. Coke oven gas or converter gas could be used for this purpose, but it was important to limit its use rate as low as possible because of the energy balance in the ironworks.

Fig.2 shows the expanding tendency of the combustible range for increased BTU gas when blast furnace gas is mixed with coke oven gas to increase the BTU.

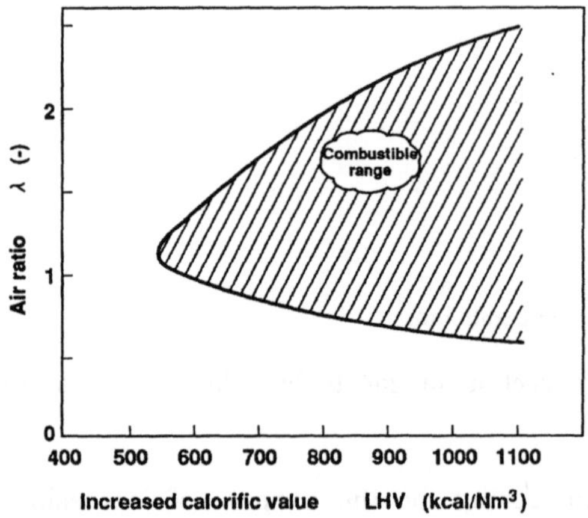

Fig.2 Combustion quality improving effect by increasing BTU of blast furnace gas

As seen in this figure, the BTU should be increased to about 900kcal/Nm^3 at least in order to obtain the combustible range equal to that of conventional high BTU gas. (Refer to methane in Table 1.)

The combustible range more than that of natural gas could be obtained by increasing the BTU, because of the high combustion quality effect of coke oven gas. However, since the BTU increasing rate had to be kept as low as possible from the viewpoint of energy balance in the ironworks as mentioned above, it was necessary to overcome the technical problems in designing the combustor as shown below.

2.2 Development of Combustor that can Keep BTU Increase to Required Minimum and Maintain Stable and High Combustion Efficiency in Wide Gas Turbine Operation Range

Fig.3 shows the change of the combustion efficiency versus turbine inlet temperature when blast furnace gas mixed with coke oven gas to increase the BTU is burnt in the multi-cannular type combustor of the normal size.

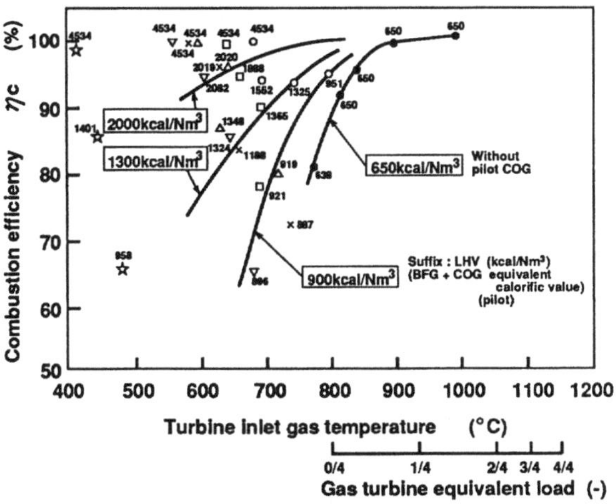

Fig.3 Effect of increased BTU on combustion efficiency

In order to carry out combustion with the stable and high combustion efficiency maintained in all load range of the gas turbine, the BTU had to be increased to about 2,000kcal/Nm³ for the conventional combustor.

Further, we have noticed in Fig.3 that even in combustion of low BTU gas the high combustion efficiency can be attained by maintaining the turbine inlet gas temperature at about 900°C or higher. Thus, we have tried to solve the above technical problems by supplying part of combustor air to the downstream side of the combustion region and also by adopting the air flow regulating variable mechanism that can control the above air flow in accordannce with the gas turbine load (see Fig.4).

Fig.5 shows the combustion efficiency improvement effect of this air flow regulating variable mechanism at the gas turbine no-load condition when the blast furnace gas BTU is increased to 1000kcal/Nm³.

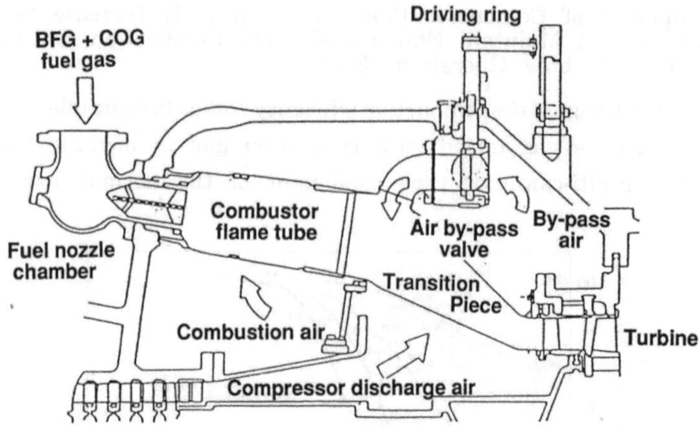

Fig.4 Construction of combustor

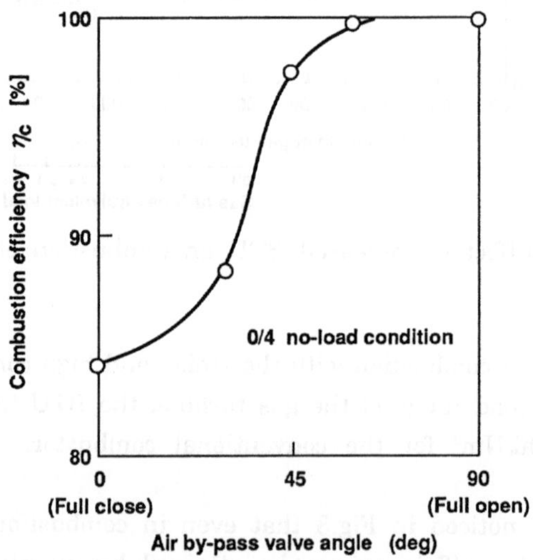

Fig.5 Combustion efficiency improvement effect by means of air flow regulating variable mechanism

The multi-cannular type combustor having this air flow regulating mechanism can keep the BTU increasing rate of blast furnace gas as low as possible and moreover enables the operation that can maintain the stable and high combustion efficiency in the wide operation range from the no-load of the gas turbine to its rated load.

By overcoming the above-mentioned technical problems in developing the multi-cannular type combustor burning low BTU gas, we obtained the prospect to make the large-capacity high-temperature gas turbine burning low BTU gas fit for practical use.

However, in making this combustor fit for practical use, the final verification/confirmation of the following items in the actual device design had to be carried out:

- Shape of fuel nozzle
- Selection of air velocity in cross section of combustor
- Optimization of air distribution by means of air hole arrangement
- Confirmation of direct firing with ignition plug
- Combustion characteristics during firing and speed increasing

For this purpose, before carrying out fundamental designing of the plant, we have carried out the atmospheric pressure combustion test with the full-scale combustor and actual blast furnace gas in the site of Chiba Ironworks, Kawasaki Steel Corporation.

(1) Firing performance -- Fig.6-1

When blast furnace gas is mixed with coke oven gas to increase the BTU to about 1,000kcal/Nm³ or higher, direct firing is possible with an ignition plug.

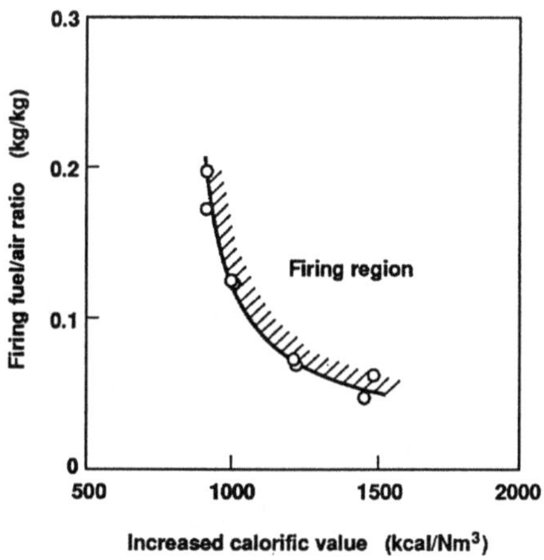

Fig.6-1 Firing performance

(2) Blowing-out performance -- Fig.6-2

It is confirmed that the flame stability at assumed gas turbine load cutoff has no hindrance specially in operating the actual combustor with fuel of 1,000kcal/Nm³ or higher in increased calorific value.

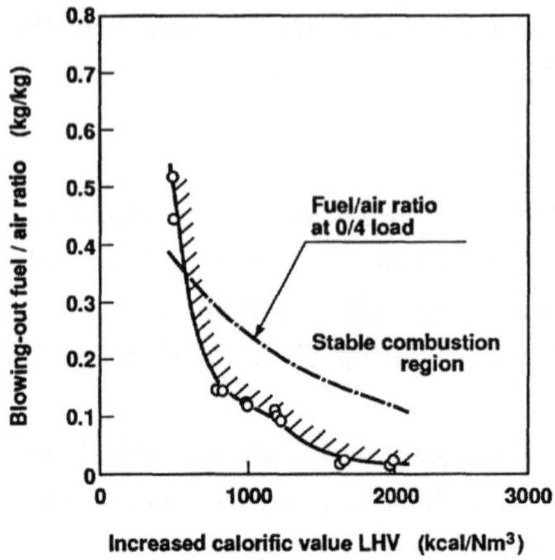

Fig.6-2 Blowing-out performance

(3) Combustion characteristic during speed increase---------------- Fig.6-3

The multi-cannular type combustor with the air flow regulating variable mechanism can increase the speed during the firing and speed increasing operations with fuel of 1,000kcal/Nm³ in calorific value, maintaining the combustion efficiency at 99% or higher.

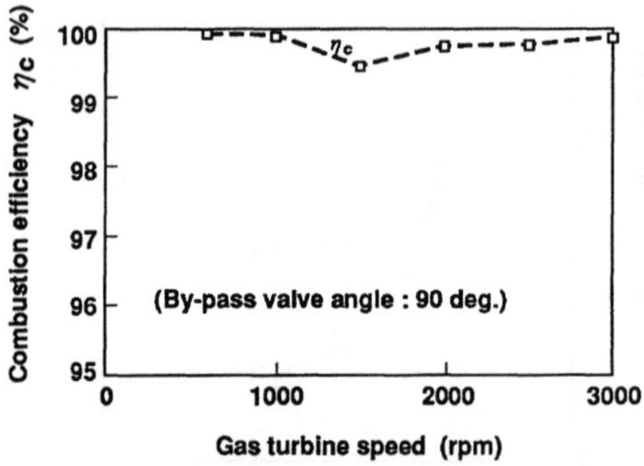

Fig.6-3 Combustion characteristic during gas turbine speed increase

(4) Load characteristic -- Fig.6-4

The combustor can keep stable combustion with the air flow regulating variable mechanism within the range from gas turbine no-load to its rated load, maintaining the combustion efficiency at 99.5% or higher. In addition, low BTU gas having an intrinsically-low flame temperature has the advantage of a low NOx generation rate. Therefore, the combustor can have low NOx combustion of 20ppm in exhaust density (O_2 5% conversion) or lower without such low-NOx countermeasure as water or steam injection into the combustion as shown in the figure.

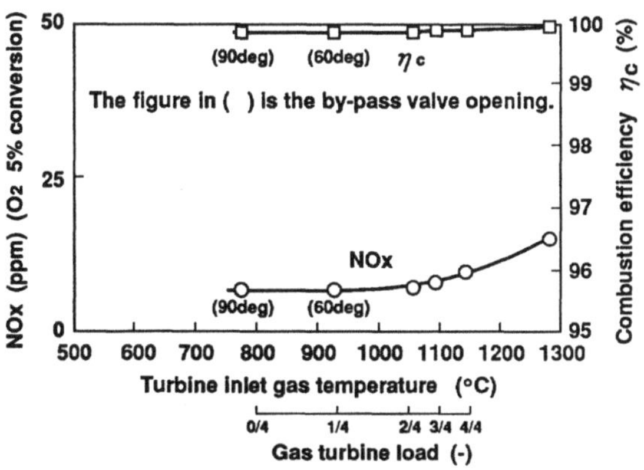

Fig.6-4 Load characteristic: Combustion efficiency and NOx exhaust density

As mentioned above, practical use of the multi-cannular type combustor burning low BTU gas has become possible through overcoming of its technical problems and verification with the full-scale atmospheric pressure combustion tests with actual gas.

POSTSCRIPT

The combustion technology for the combustor burning low BTU gas, which has been developed and made fit to practical use with ironworks by-product gas, can be applied to the coal gasification combined plant. However, in future this should be made fit to practical use further by trying to adopt higher temperature and give lower public nuisances.

(1) Higher temperature → 1300°C class of turbine inlet gas temperature

(2) Lower public nuisances → Decrease of fuel NOx

Measurements of Gas Composition and Temperature Inside a Can Type Model Combustor

W P JONES and A TOBER

Department of Chemical Engineering & Chemical Technology
Imperial College of Science, Technology & Medicine
London, SW7 2BY, England

Summary

Concentrations of major species and temperature have been measured at air–fuel ratios (AFR) of 57 and 91 in a model can–type gas turbine combustion chamber burning kerosene introduced via a pressure jet atomizer. Conditions within the combustor including primary and dilution jet trajectories and mixing rates are inferred from the mean mixture fraction field which is evaluated from the measured species concentrations. A comparison with earlier measurements in the same combustor burning gaseous propane suggests that the flow field and mixing rates are similar for both fuels. The maximum levels of CO are found to be effectively independent of AFR with maximum concentrations of about 6% arising under slightly fuel rich conditions in both cases.

Introduction

The combustor involved in the present experimental study is a model of a can–type combustion chamber typically found in aircraft gas turbines. The combustor is fuelled via a liquid kerosene spray and the flame is stabilised by a recirculation zone driven by a swirler air flow through the head of the combustor and by primary dilution jet flows; the inlet air flow is preheated with an electrical heater. Detailed flow studies in a perspex water flow model of the same combustor have previously been reported in [1], and for the combusting case detailed measurements are available [2], over a range of operating conditions burning gaseous fuel (propane). Further combusting flow measurements in closely similar geometries are described in [3], [4] and

R. S. L. Lee · J. H. Whitelaw · T. S. Wung (Eds.)
Aerothermodynamics in Combustors
IUTAM Symposium Taipei, Taiwan, 1991

[5]. The aim of the present investigation is to obtain measurements of gas composition and temperature so as to allow the effects of a change from a gaseous fuel to a liquid kerosene fuel spray to be quantified. Mixture species concentrations were obtained by probe sampling and temperatures were measured by thermocouple. All experiments were conducted at atmospheric pressure.

Experimental Set Up

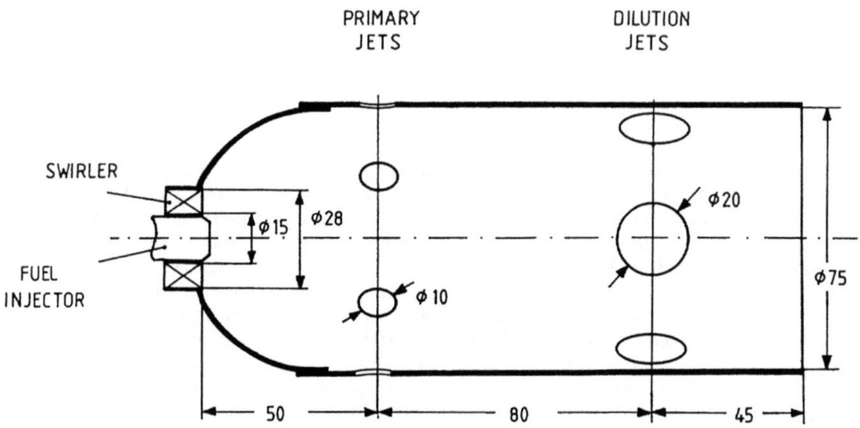

Fig. 1. Combustor Geometry

The combustor together with its dimensions is shown in figure 1. Air is supplied through a plenum chamber surrounding the combustor which itself exhausts freely into the laboratory. The combustor is fabricated from a laminated porous sheet material called Transply which provides cooling of the combustor walls. Air also enters through a vaned "aerodynamic" swirler located concentrically in the hemispherical head of the combustor and through two rows of six dilution holes equally spaced around the circumference of the combustion chamber. These rows of holes are staggered with respect to each other such that the angle between the axial planes in which the holes lie is 30°. The dimensions of the swirler and dilution holes determine the relative air inflow rates and were designed to give a flow split and flow pattern representative of that arising in practical gas turbine combustors. Measurements made with pitot and hot wire probes in the cold flow confirmed the design. Thus 9.9% of the total air flow enters through the swirler, 3.0% through the perforations in the domed head, 13.7% through the first set of holes, here termed primary holes, 15.4% through the barrel cooling and 58% through the second set of holes here called dilution holes. The airflow was preheated with an electrical heater and enters the combustor at a temperature of 400K.

The pressure jet atomizer which produces a hollow cone spray with an included angle of about 70° is located in the centre of and is concentric with the swirler. The fuel used was aviation grade kerosene (JET A–1) and the fuel pressures were 60 bar and 27 bar for the two cases measured. These pressures resulted in a fuel flow rate of 1.33 ml/min and 0.84 ml/min and with a constant air flow rate of 100 g/s resulted in air–fuel ratios (AFR) of 57 and 91 respectively.

Measurement Methods

Temperatures were measured by uncoated fine wire (80 micron) thermocouples (pt/pt13%rh) and the mean values shown were obtained after averaging over a 4 sec period. The measurements were uncompensated for radiation losses or thermal inertia effects. For the higher air–to–fuel ratio case temperature measurements were limited to the dilution zone where probe temperatures were sufficiently below the melting point of platinum to avoid destruction of the probe. For a detailed description see [1]. To avoid condensation of water and fuel vapours in the probe the probe cooling water was kept at a temperature above 150°C in accordance with SAE ARP 1256A. The species measured were CO_2, CO, H_2, O_2 and total unburnt hydrocarbon (UHC). These were sufficient to allow the calculation of the concentrations of the remaining major species, N_2 and H_2O by mass balance. The assumptions made in this calculation were that the gas is composed of the above seven species plus an intermediate hydrocarbon (C_2H_2) only and that the carbon to hydrogen ratio and the oxygen to nitrogen ratio are those of the raw fuel and air respectively at any point in the combustor. The measured and calculated concentrations allowed the profiles of mean mixture fraction, $f\left(=AFR/(AFR+1)\right)$ to be determined and the gas temperatures to be obtained by enthalpy balance.

Results

An overview of conditions inside the combustor can be seen from figures 2 to 4 which show contours of mean mixture fraction and unburnt hydrocarbon and carbon dioxide mole fractions for an AFR of 57. The results for the other AFR case are qualitatively similar. Figures 2(a), 3(a) and 4(a) show contours in an axial plane passing through the primary holes whereas figures 2(b), 3(b) and 4(b) display results in a plane midway between the primary holes and passing through the dilution hole. Measurements were extended into the primary zone and upstream of the primary jets only in the plane bisecting the primary holes as in the other case the jets are more likely to impinge on the probe near entry to the combustor; the probability of the probe distorting the flow is larger in this case.

The contours of mixture fraction, UHC and CO_2 allow some inferences to be drawn of the flow pattern arising in the combustor. High values of mixture fraction and unburnt fuel (the averaged value of f based on the fuel flowrate and swirler and head cooling airflow rates is 0.11) are measured in a conically expanding volume which corresponds roughly to that anticipated from the hollow cone fuel spray provided by the injector. At the measuring position closest to the injector the mixture fraction has a value of about 0.45 but this falls to a value of around 0.13 just upstream of the primary jet entry. While the levels of UHC and CO_2 present indicates that some burnt products are present in this region mixing rates are clearly low; the size of the zone suggests that the mass flowrate cannot be that large in this region. It appears that the air entering through the swirler spreads and closely follows the combustor walls and there is little evidence of any recirculation in the near wall region in the head of the combustor. The contours of f, UHC and CO_2 suggest that the fluid close to the wall is predominantly swirler air and that the velocities are relatively high.

The leaner conditions prevailing in the vicinity of the combustor centreline are separated from the swirler air by the fuel spray. The measurements suggest that this region corresponds to a recirculation zone comprising hot burnt products and into which primary jet air is entrained. The primary

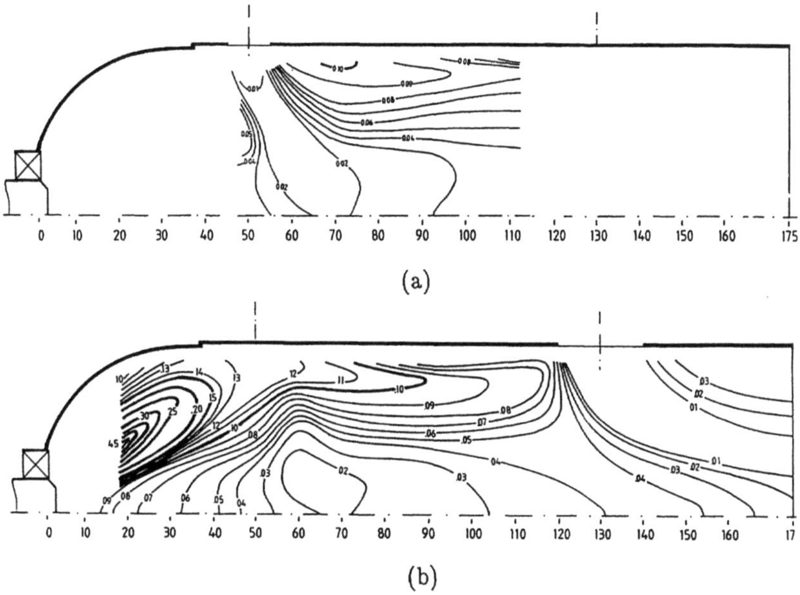

Fig. 2. Mixture Fraction Contours, AFR=57
(a) Primary Jet Plane (b) Dilution Jet Plane

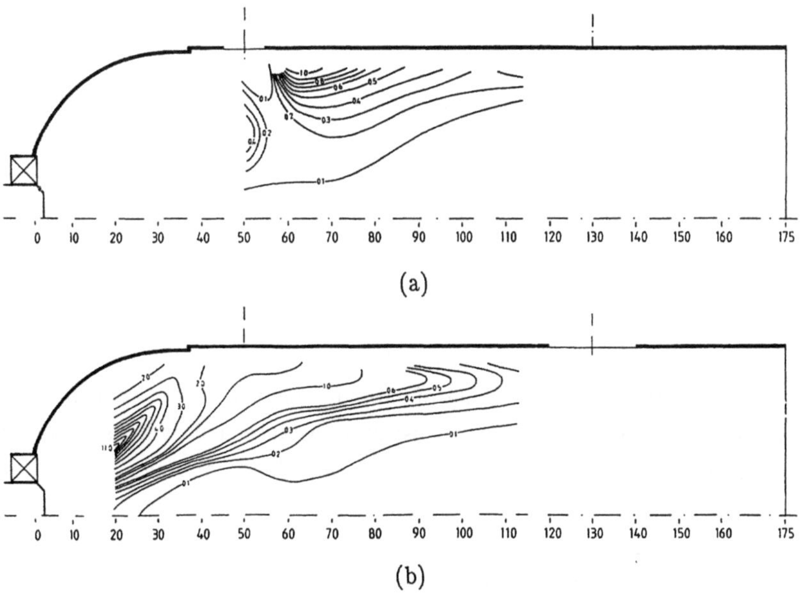

Fig. 3. Unburnt Hydrocarbon Mole Fraction Contours, AFR=57
(a) Primary Jet Plane (b) Dilution Jet Plane

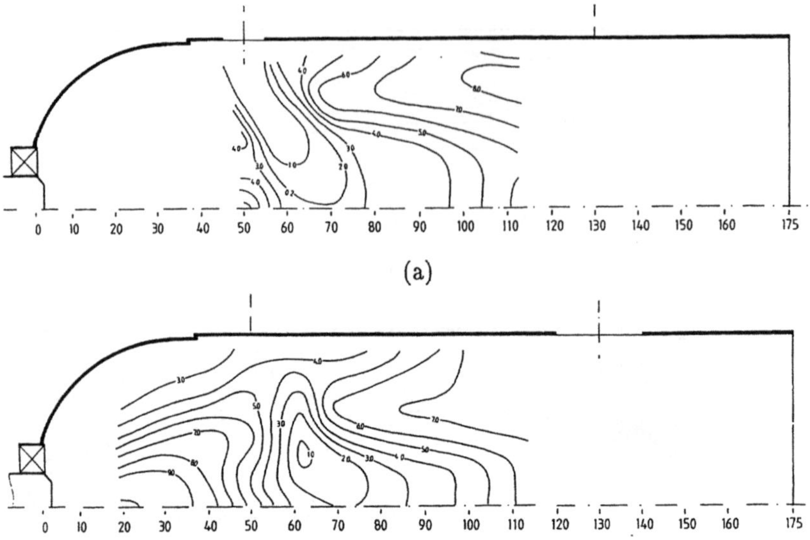

(a)

(b)

Fig. 4. Carbon Dioxide Mole Fraction Contours, AFR=57
(a) Primary Jet Plane (b) Dilution Jet Plane

jets are deflected downstream and impinge at a position corresponding roughly to the minimum value of the mixture fraction on the centreline at one to two jet diameters downstream of the jet entry. Presumably some of the primary zone air is entrained into the recirculation zone and dilutes the fuel–air mixture originating from the fuel jet and swirler air. Downstream of the indicated primary jet impingement point the centre–line values of mixture fraction increase slowly and the contours are consistent with the major portion of the primary jet air flowing downstream. The position of the maximum in the radial profile of mixture fraction moves away from the combustor wall with increasing downstream distance and is consistent with the primary jet air mixing with the co–flowing fluid (and the profiles thus becoming more uniform) and the cooling air flow through the transply walls of the combustor. The mixture fraction contours in the plane of the dilution jets show that they deflect richer mixture fluid towards the centreline. The jets themselves do not impinge, however, and are deflected downstream. Given that the pressure drop across the combustor walls is similar for both primary and dilution jets – the combustor is plenum fed – the difference in trajectory of the two jets can be mainly attributed to the larger diameter of the dilution jets which result in a larger "blockage" to the axial flow. However, both the differences in the approach (axial) velocity and the larger positive radial pressure gradient associated with the swirling flow in the primary zone, which assists jet penetration, also contribute towards the differences in primary and dilution jet penetration.

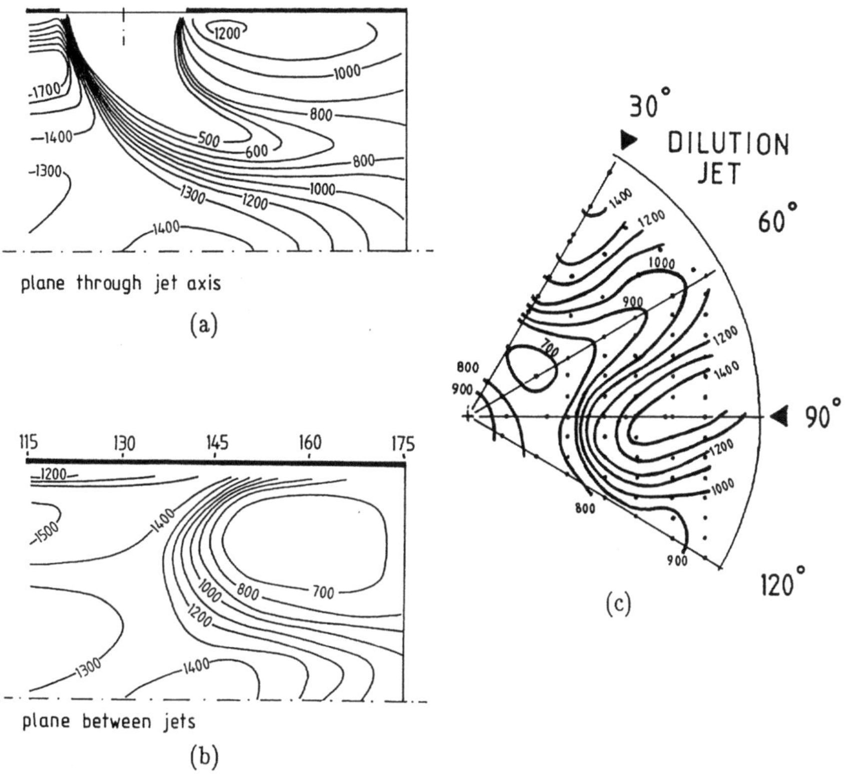

Fig. 5 Temperature (K) Contours
(a) Dilution Jet Plane (b) Primary Jet Plane
(c) Exit Plane

Burning is essentially complete by the dilution jets and the temperature contours shown in figures 5(a), (b) and (c) thus provide an indication of the conditions in this region. Qualitatively the high temperatures are associated with burnt gas with a high fuel–air ratio and the lower temperatures arise from the preheated air entering the combustor; conditions are leaner than stoichiometric everywhere in the dilution jet region. Consistent with the mixture fraction contours the variation in the temperature of the approach flow upstream of the dilution jet between the two planes is not large. Although the pressure drop is somewhat lower than would be typical of a practical can–type combustor it appears sufficient for the jet entry to be roughly perpendicular to the combustor wall though there is indication a contraction of the jet as would be typical for the sharp edged holes of the combustor. The hot gas observed near the combustor walls upstream of the jets can be seen to be both defected towards the centreline and accelerated between the jets. At the same time the highest temperatures fall rapidly as the large temperature differences between the hot gas and cold dilution air lead to high mixing rates. An indication of the acceleration around the jets is provided by the way in which the colder cooling air, which enters with low momentum through the transply walls upstream of the dilution holes, is carried between the jets close to the walls of the combustor. The mixing of hot gas and jet fluid as the jet is bent downstream and evidence of the existence of the counter rotating vortices which are anticipated to arise as the jet is deflected by the cross–flow can be seen from the contours in the plane between the jets; a large area with temperatures below 700K appears about one jet diameter downstream of the rear edge of the jet between the jet trajectory and the combustor wall. The large gradients just upstream of this region suggests a substantial inflow of cool air which displaces the hot burnt gas flowing between the jets. This hot gas can be seen in the dilution jet plane directly behind the jet. The contours do not show clearly whether or not there is a recirculation zone behind the jet but the most probable source of the hot gas is from the flow around the jet; close to the inlet its influence will be similar to a solid cylindrical obstacle placed in the flow. The temperature contours in the exit plane, figure 5(c) shows a "ring" of colder gas with lobes extending towards the combustor wall and hotter regions which lie in line with the jets. The explanation offered here is that the colder "ring" represents the six coalesced dilution jets which do not penetrate to the centreline with the consequence that the hot gas originating from upstream is observed in the centre of the combustor. The lobes are associated with the two sides of the jet which is deformed into a horse shoe shaped cross–section by the counter–rotating vortices which presumably formed behind the jet.

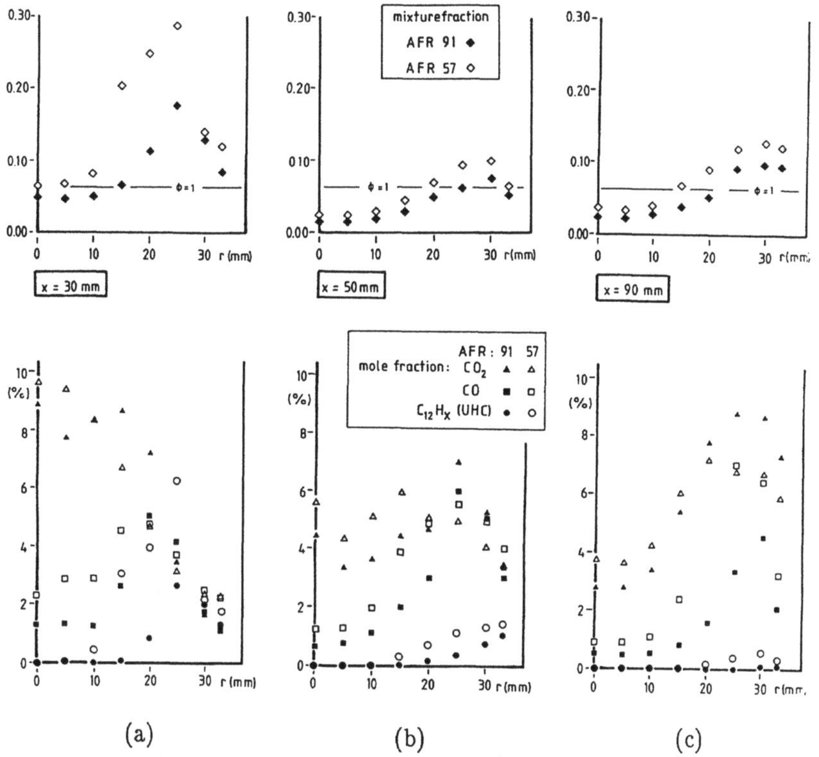

Fig. 6. Gas Composition and Mixture Fraction Profiles at two Air–Fuel Ratios.

The differences between the two AFRs measured are shown in the radial profiles of figure 6a to c. Together with the profiles of mixture fraction, the concentrations of CO_2, CO and UHC are shown and these indicate to what extent the fuel has been consumed at those stations. The profile at x = 30mm (in the primary zone) shows an area of high CO_2, moderate CO and low UHC concentrations around the centreline. At the lower AFR the increase in mixture fraction brings an increase in CO levels and a corresponding reduction in CO_2 concentrations. The concentration of unburnt fuel rises towards the combustor wall but reaches a maximum at about two thirds of the combustor's radius. The reduction from here to the combustor wall together with a fall in both CO and CO_2 shows the coexistence on a mean basis of unconsumed oxygen with fuel and some combustion products. The most obvious difference between the higher and lower AFR cases, where the general pattern is the same, is that for the higher air–fuel ratio, the fuel concentration is more strongly reduced on the inner shoulder of the maximum. In line with the overall lower levels of fuel content a larger part of the fuel is burnt. At stations further downstream the similarity between concentration profiles for the two AFRS is increased. As the area with mixture fraction above stoichiometric values becomes larger in the lower AFR case high concentrations of UHC are found and values for CO exceed those of CO_2. The highest fuel concentrations are now found to occur close to the combustor wall but at the station mid way between the primary and dilution hold (x = 90mm) the effect of the wall cooling air causes a sharp drop as the wall is approached – the closest measurement was 4mm from the wall.

The profiles also show the maximum levels of CO which arise to be about the same for both air–fuel ratios. In both cases the measured maximum concentration is around 6% and occurs under conditions slightly richer than stoichiometric. Consistent with many other measurements in turbulent flames this is substantially less than the maximum concentrations (in excess of 20%) which would be found if the mixture were to be in chemical equilibrium.

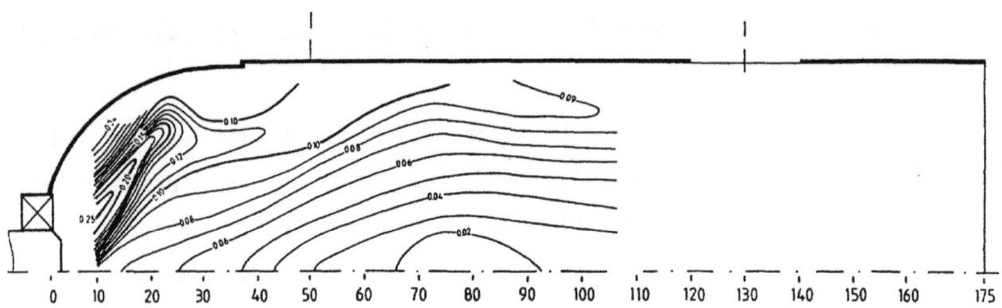

Fig. 7. Mixture Fraction Contours Burning Propane, Tse (1988), AFR=57

Finally, figure 7 shows the mean mixture fraction contour plot taken from Tse (2) for the same combustor fuelled with gaseous propane rather than kerosene. A comparison with the corresponding figure 2(b) indicates some differences close to the injector where the propane mixes more rapidly with the air – in part because of the necessary geometric differences in the two fuel injectors used – and the fuel distribution appears to be more influenced by the swirler flow. However, the differences are not large and conditions within the combustor appear to be remarkably similar, at least qualitatively, for both liquid gaseous fuels. This suggest, perhaps somewhat surprisingly that liquid fuel with its potential to penetrate the air stream does not in fact spread or mix in a significantly different way from the gaseous propane.

Conclusions

Detailed measurements of composition and temperature have been made in a model can–type gas turbine combustor burning kerosene at two air–fuel ratios (57 and 91) and with an inlet air temperature of 400K. The species concentrations (CO, CO_2, H_2, O_2 and unburnt hydrocarbons) are consistent with the overall operating conditions with the richer air–fuel ratio resulting in higher concentrations of carbon dioxide practically everywhere within the combustor. In the primary zone the levels of carbon monoxide and unburnt hydrocarbons are, as expected, correspondingly higher also. It is interesting to note, however, that the maximum levels of carbon monoxide, which as anticipated arise under fuel–rich conditions, are approximately the same at around 6% for both air–fuel ratios; the location of these maxima of course differs in each case. The maximum concentrations of CO are, as expected, also far lower than the values which would be attained if the mixture had reached a state of chemical equilibrium.

The composition measurements also allow some inferences to be drawn on the flow pattern within the combustor. The existence of a central recirculation which is partly fed by the impingement of the primary jets is suggested and there are no indications of a "corner" recirculation zone in the head of the combustor outside the air stream from the swirler. In the downstream region the flow pattern is indicated more clearly by the dilution jets where reaction plays only a minor role. Here too the jet trajectory follows the pattern intended in the design of the combustor and is successful in reducing the temperatures and fuel content of the exhaust gases so that by the exit plane reasonably uniform profiles result. At the exit plane regions of higher temperature ("hot spots") are located at positions in line with (and downstream of) the dilution holes.

Acknowledgements

The authors would like to acknowledge the financial support provided by Rolls–Royce plc and useful discussions with colleagues at Imperial College and staff at Rolls–Royce and RAE/Pyestock.

References

[1] Palma, J.M.L.M. (1988), "Mixing in non reacting gas–turbine combustor flows", Ph.D. Thesis, University of London.

[2] Tse, D.G.N. (1988), "Flow and Combustion Characteristics of model annular and Can Type Combustors", Ph.D. Thesis, University of London.

[3] Jones, W.P. and Toral, H. (1983), "Temperature and Composition Measurements in a Research Gas–Turbine Combustion Chamber", Comb. Sci. and Tech., 31, 249.

[4] Bicen, A.F., Heitor, M.V. and Whitelaw, J.H. (1986) "Velocity and Temperature Measurements in a Can–Type Gas–Turbine Combustor", AGARD–CP–399, paper no. 14.

[5] Jones, W.P. and Tober, A. (1988), "Velocity, Composition and Temperature Fields in an Axisymmetric Model Combustor", Proceedings of the Fourth International Symposium on Applications of Laser Anemometry to Fluid Mechanics, Lisbon, Portugal.

Numerical Model for Predictions of Reverse Flow Combustor Aerothermal Characteristics

P. DI MARTINO, E. NARCISO, G. CINQUE

Alfa Romeo Avio S.p.A.
Research and Development Department
Pomigliano d'Arco (NA) - ITALY

Abstract

A mathematical model of three-dimensional two-phase reacting flows in gas turbine combustion chambers has been developed. The governing time-averaged partial differential equations and the physical modelling for the turbulence, combustion and thermal radiation have been solved by a numerical tecnique based on the finite volume approach. A reverse flow type combustor was used to validate the model. The results are found to be in close agreement with available experimental data. An assessment of the adequacy of the various submodels embodied has been made in view of future improvements.

1. Introduction

Until fairly recently gas turbine combustor designers were almost totally guided by previous experience. New or modified combustors emerged only after lenghty experimental and development programs. However, awareness of the limitations of energy resources as well as pollutant emissions regulations, introduced a new aspect into the design equation which made it difficult to extrapolate from existing design by simple scaling laws. The improvement of performances and the need to satisfy all requirements make necessary large departures from conventional design practice. In such circumstances the contributions which could be made by theoretical procedures for predicting combustor performances are potentially substantial.

Rapid advances in computer technology have provided the means whereby mathematical modelling can be applied to gas turbine combustors problems to facilitate greatly the ardous initial stages of combustor design, thus reducing both time and cost required by experimental methods.

R. S. L. Lee · J. H. Whitelaw · T. S. Wung (Eds.)
Aerothermodynamics in Combustors
IUTAM Symposium Taipei, Taiwan, 1991

In this study a comprehensive mathematical model of three-dimensional fluid mechanics and two-phase turbulent combustion processes is approached in order to give combustor engineers efficient tools of analysis and design.

2. Gas phase model

Steady fully elliptic time-averaged equations describing gas phase, under low Mach number approximation, coupled to the energy and momentum balance equations for the liquid phase, are considered [1]. The k-ε model provides for turbulence closure. The conservation equations solved for the gas phase are those for u,v,w momentum, mass, stagnation enthalpy h, turbulent kinetic energy k and its rate of dissipation ε, mixture fraction f. The general form of these equations is written in cylindrical polar coordinates [1]:

$$\frac{1}{r}\left[\frac{\partial}{\partial x}(r\rho u\phi) + \frac{\partial}{\partial r}(r\rho v\phi) + \frac{\partial}{\partial \Theta}(\rho w\phi)\right] =$$

$$\frac{1}{r}\left[\frac{\partial}{\partial x}\left(r\Gamma_\phi \frac{\partial \phi}{\partial x}\right) + \frac{\partial}{\partial r}\left(r\Gamma_\phi \frac{\partial \phi}{\partial r}\right) + \frac{\partial}{\partial \Theta}\left(\frac{1}{r}\Gamma_\phi \frac{\partial \phi}{\partial \Theta}\right)\right] + S_\phi + S_d \quad (1)$$

where ϕ is any one of the general flow variables, Γ_ϕ is the effective diffusion coefficient for ϕ and S_d and S_ϕ are the sources terms for ϕ from liquid droplets and gas phase, respectively.

The definition of stagnation enthalpy is:

$$h = \int_{T0}^{T} C_p \cdot dT + m_{fu} \cdot H + ½(u^2 + v^2 + w^2) \quad (2)$$

and it is used to calculate the temperature T. In this equation m_{fu} is the unburned fuel mass fraction, H is the heat of combustion and C_p is the specific heat of the mixture at constant pressure.

The chemical reaction is assumed to be infinitely fast and it is restricted to an infinitely small reaction zone, situated at the surface where f attains its stoichiometric value.

Outside this reaction zone the chemical reaction is considered to be completely quenched and the composition is frozen. The further assumptions of negligible heat losses due to radiation and the equality of all diffusive transport coefficients allow all mass fractions of chemical species to be expressed as linear functions of a conserved scalar variable such as the mixture fraction. In this idealized case the conservation equations for chemical species, assuming a single exotermic chemical reaction (Fuel + Oxygen ---> CO_2 + H_2O), are:

$$0<f\leq f_{st} \quad m_{fu}=0; \quad m_{ox}=\left(1-\frac{f}{f_{st}}\right)m_{ox,air} \tag{3}$$

$$1>f>f_{st} \quad m_{fu}=\left(\frac{f-f_{st}}{1-f_{st}}\right); \quad m_{ox}=0 \tag{4}$$

$$f_{st} = \frac{1}{[1 + S(m_{fu}/m_{ox}),air]} \tag{5}$$

where m_{fu} and m_{ox} are the fuel and oxygen mass fraction respectively, S denoting the oxygen/fuel mass stoichiometric ratio and st denoting the stoichiometric value.The mass fractions of H_2O and CO_2 are calculated by stoichiometric relations.

In this simplified model the consequences of turbulent fluctuations on the chemical composition are neglected. It is not correct in priciple but it results in economic of computer storage and time and, in view of the additional uncertainties imposed by the droplet model, it is probably of acceptable precision. A statistical approach to describe the mixture fraction fluctuations will be the subject of a successive study. The density is provided by the equation of state:

$$\rho = P/[RT\Sigma_j(m_j/M_j)] \tag{6}$$

where M_j and m_j are the molecular weight and the mass fraction of species j respectively, and R is the universal gas constant.

The effects of radiative heat transfer are accounted for by reference to the six-flux model, based on the solution of a set of six first-order ordinary differential equations, which provide the source term in the energy balance equation for the gas phase [1]. In computational work second order equations are preferable, so the six differential equations can be transformed to give:

$$\frac{d}{dx}\left[\left(\frac{1}{a+s}\right)\frac{dR_x}{dx}\right] = a\left(R_x - E\right) + \frac{s}{3}\left(2R_x - R_r - R_\Theta\right) \tag{7}$$

$$\frac{1}{r}\frac{d}{dr}\left[\left(\frac{r}{(a+s+1/r)}\right)\frac{dR_r}{dr}\right] = a\left(R_r - E\right) + \frac{s}{3}\left(2R_r - R_x - R_\Theta\right) \tag{8}$$

$$\frac{1}{r}\frac{d}{d\Theta}\left[\left(\frac{1}{a+s}\right)\frac{1}{r}\frac{dR_\Theta}{d\Theta}\right] = a\left(R_\Theta - E\right) + \frac{s}{3}\left(2R_\Theta - R_x - R_r\right) \tag{9}$$

where a is the flux absorption coefficient, s is the scattering coefficient, $E=\sigma T^4$ is the black body emissive power at fluid temperature and σ is the Stefan-Boltzmann constant. Finally R_x, R_r and R_Θ are the composite fluxes [1].

The radiation contribution to the source term of the enthalpy equation is given by:

$$S_{h,rad} = 2a(R_x + R_r + R_\Theta - 3E) \tag{10}$$

The effect of soot are neglected in this analysis. Models including soot in radiation calculation have to be addressed.

Although radiation from flame was neglected in the hypoteses of section 2, if predictions of combustor liner temperatures are to be made, it should be considered. In this study the assumptions were made that the main calculation can still be treated as adiabatic.

Strictly speaking, the introduction of non-adiabatic effects in the enthalpy equation invalidates the single scalar assumption, but several authors do not state how large this effect may be or how it may be justified.

Anyway even if the enthalpy equation contains a heat loss term, it pratically does not affect the gas temperatures, as we have proved performing two calculations, one with radiation terms and the other without. At present the wall heat fluxes calculated when radiation is considered are introduced in a post-processor which evaluates the wall temperatures by considering an energy balance across the liner wall.

3. Droplet model

The liquid-phase equations are based on the Lagrangian formulation of the droplet trajectory, transient heating and vaporization [2]. The entire fuel spray is constructed using a finite number of size ranges obeying a two parameter droplet size distribution, which is assumed to be of Rosin-Rammler type. The equations of motion for each of the droplets, representing the size groups which constitute the spray, are written neglecting all external effects except the drag force [2]. The description of droplet heating and evaporation is added to the computational model using two equations which are solved separately, assuming that during the first phase the droplet is heated to an equilibrium temperature (boiling point), followed by the second phase in which the evaporation process causes a decrease in diameter. These equations may be expressed as

$$\frac{dD}{dt} = -\left(\frac{Cb}{2D}\right)(1+0.23Re^{1/2}) \tag{11}$$

and

$$\frac{dT}{dt} = 6\lambda\,(2+Re^{1/2}\cdot Pr^{1/3})\,\frac{(T - Td)}{\rho_d D^2 C_p L} \tag{12}$$

where Cb is the vaporization rate constant whose value depends on the physical properties of the surrounding medium as well as of the fuel itself; Re is the relative Reynolds number, D is the droplet diameter, ρ_d is the liquid density, λ and C_p are the thermal conductivity and the specific heat at constant pressure of the surrounding gas, L is the latent heat of vaporization, T and Td are the gas and droplet temperatures respectively.

The liquid-phase equations are coupled to the equations describing gas phase through the droplet source terms, which are obtained by calculating what is lost or gained by the spray in terms of mass, momentum and energy as the droplets enter and leave volume elements.

With this liquid phase model it is possible to simulate also gaseous fuels by choosing a very low boiling point and a low value of L in order to obtain a flash evaporation. This is what has been done in the present study, when we simulated the combustion chamber described in the following sections.

4. Numerical solution procedure

The partial differential equations previously described have been transformed into difference equations by integrating over each control volume and solved numerically by using the well established SIMPLE semi-implicit iterative scheme [3]. The diffusion terms were replaced by their central-difference analogs, while upwind differencing was used for the convection terms to ensure numerical stability and convergence of solution procedure at high Reynolds' numbers. The solution of the individual equations sets was obtained iteratively by a form of Gauss-Seidel line-by-line relaxation method. All dependent variables are calculated at mesh grid points centered at an elementary volume except velocity components, which are evaluated at the control volume faces. Wall functions have been used to avoid detailed calculations in the near-wall regions. Simulation of contoured flowpath boundaries was accomplished by the so called "stairstep" boundary treatment into the model.

The gas-spray interaction terms are supplied from the set of simultaneous ordinary differential equations integrated numerically by the fourth order Runge-Kutta method at suitable intervals within the iterative solution procedure.

5. Model validation and results

The previous model, together with appropriate boundary conditions, was used to perform a calculation of the internal reactive flowfield of the GEM-60 reverse flow combustion chamber, for which experimental results are available [4,5]. The geometry of combustor is shown in Fig. 1.

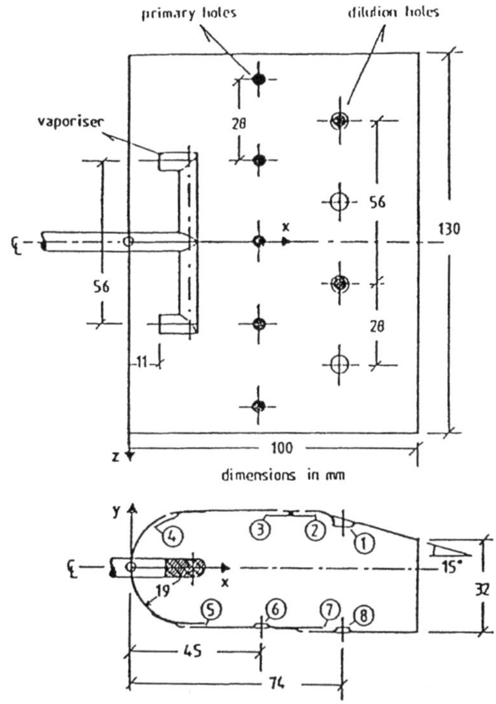

Fig. 1. Geometry of GEM 60 Combustor

The grid used in calculations (Fig. 2) possessed 47x27x27 nodes in the x,r and Θ directions respectively. In the present case the combustor geometry suggested the use of cartesian coordinates (r=1 in the round brackets of equation (1)).

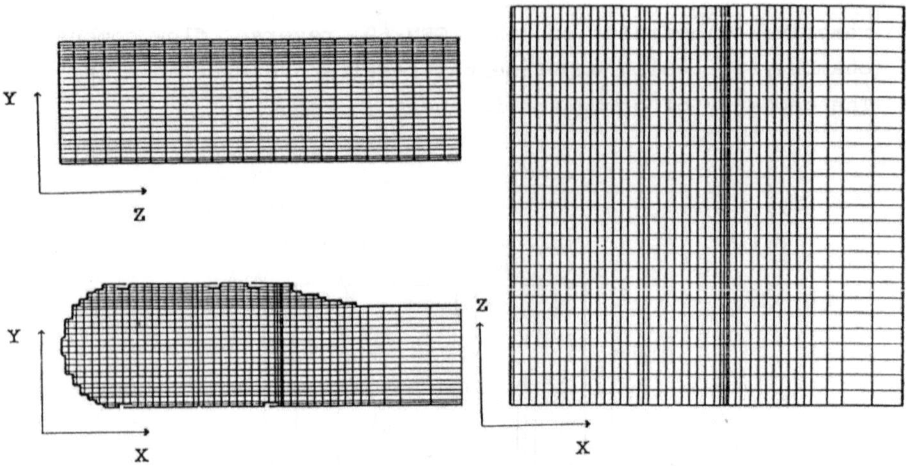

Fig. 2. 47x27x27 Grid Mesh used for Model of GEM 60 Combustor

The convergence history of the residuals is given in Fig. 3.

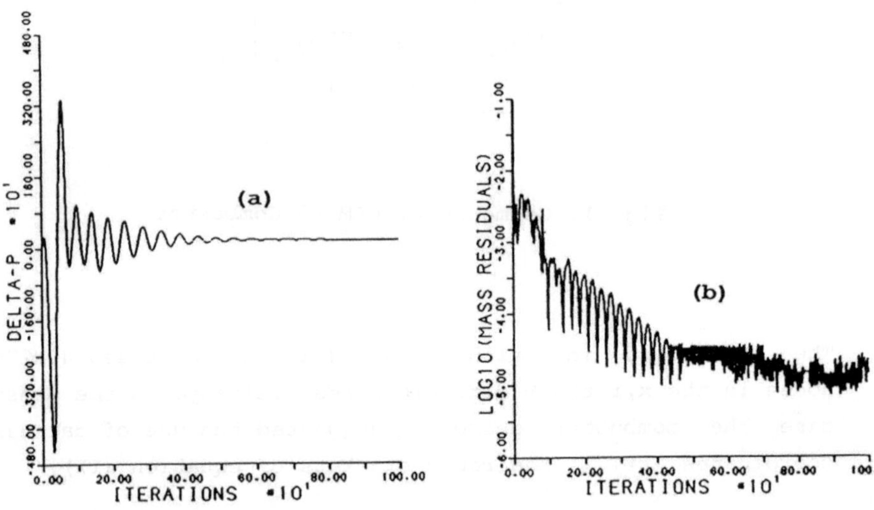

Fig. 3. Convergence histories: (a) pressure difference in the chamber and (b) mass residual

After 1000 iterations the maximum residual of mass is of the order 1.E-05 and pressure difference in the chamber is constant, thus showing that the solution is reasonably converged at this point.

The predicted velocity field in three vertical planes is shown in Fig. 4.

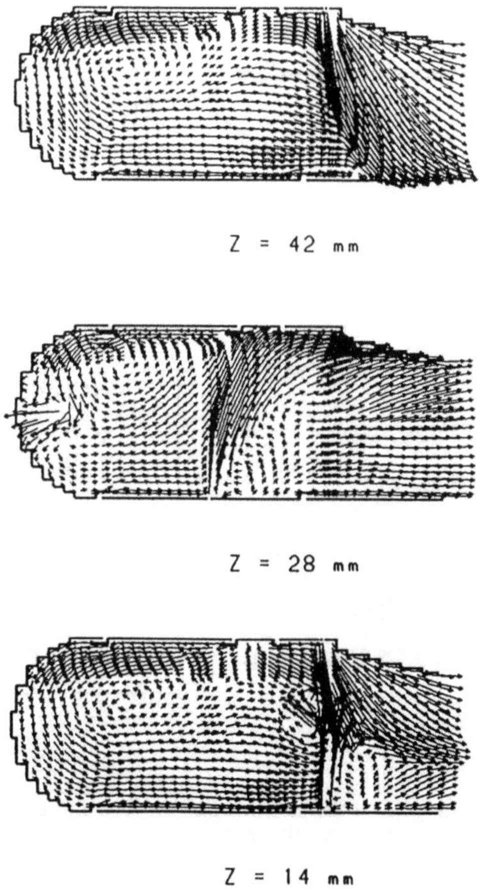

Fig. 4. Calculated velocity field in various vertical planes: Z=14 mm; Z=28 mm; Z=42 mm

A strong recirculation in the primary zone and the penetration of the entering air jets can be observed.
The temperature field in the same planes (Fig. 5) indicates

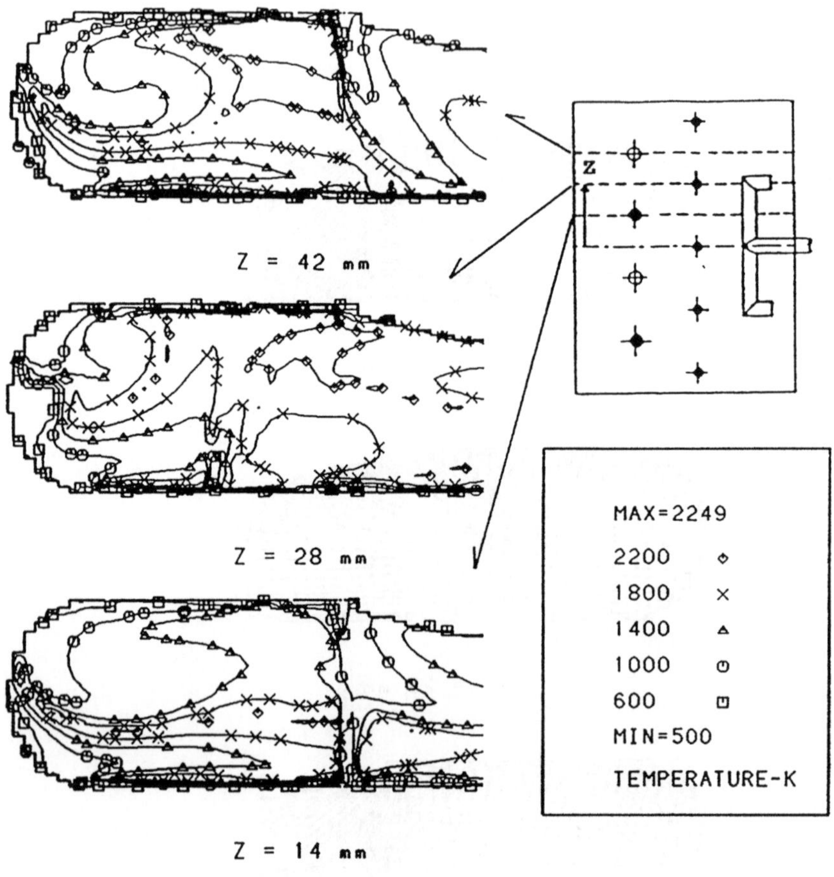

Fig. 5. Calculated gas temperatures in various vertical planes: Z=14 mm; Z=28 mm; Z=42 mm

that the maximum temperatures occur in the upper half of the primary and intermediate zones and are in line with the vaporizer exits, as it can be seen in Fig. 6, where cross-plane distributions of temperatures are reported.

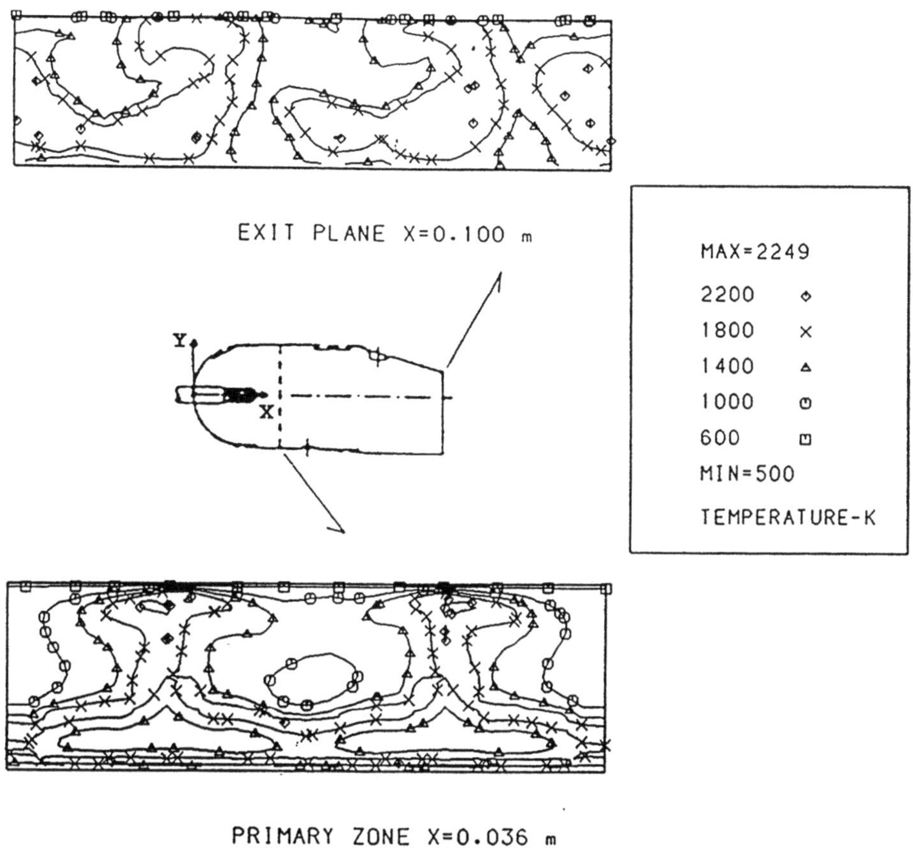

Fig. 6. Calculated cross-plane distributions of temperature: primary zone and exit plane

This general pattern has been maintained at the exit plane with a slight shift in the high temperature regions. The excess fuel which does not burn in the primary zone passes to the diluition zone and combusts with the diluition air, resulting in a temperature increase.
This trend can be observed in Fig 7, where the main chemical species are plotted.
Finally the exit plane temperatures distribution measured at different values of y-coordinate are shown in Fig. 8.
The root mean square (RMS) of relative error between theoretical and experimental results is indicated in each diagram. The trend is fairly well predicted.

6. Conclusions

A two-phase turbulent reactive flow model for a gas turbine combustor has been numerically solved. The computed results reproduce the main features of the reactive flow and show good agreement with experimental data, in view of the many interdependent processes involved. The main critical points of the present study concern the fact that no predictions of carbon oxide and nitric oxides have been performed. Besides the lack of reliable methods to predict the soot concentration and hence the emissivity of luminous flames have prevented a detailed study of radiation. The weak turbulence/chemistry interaction is another point of the model which needs a better treatment. Finally there is the inability to deal with a completely arbitrary three-dimensional geometry without resorting to some irregular boundary treatment.
At present our research group is already working to refine the description of chemical reactions and turbulence by using statistical approach based on probability density functions and full chemical equilibrium assumptions.

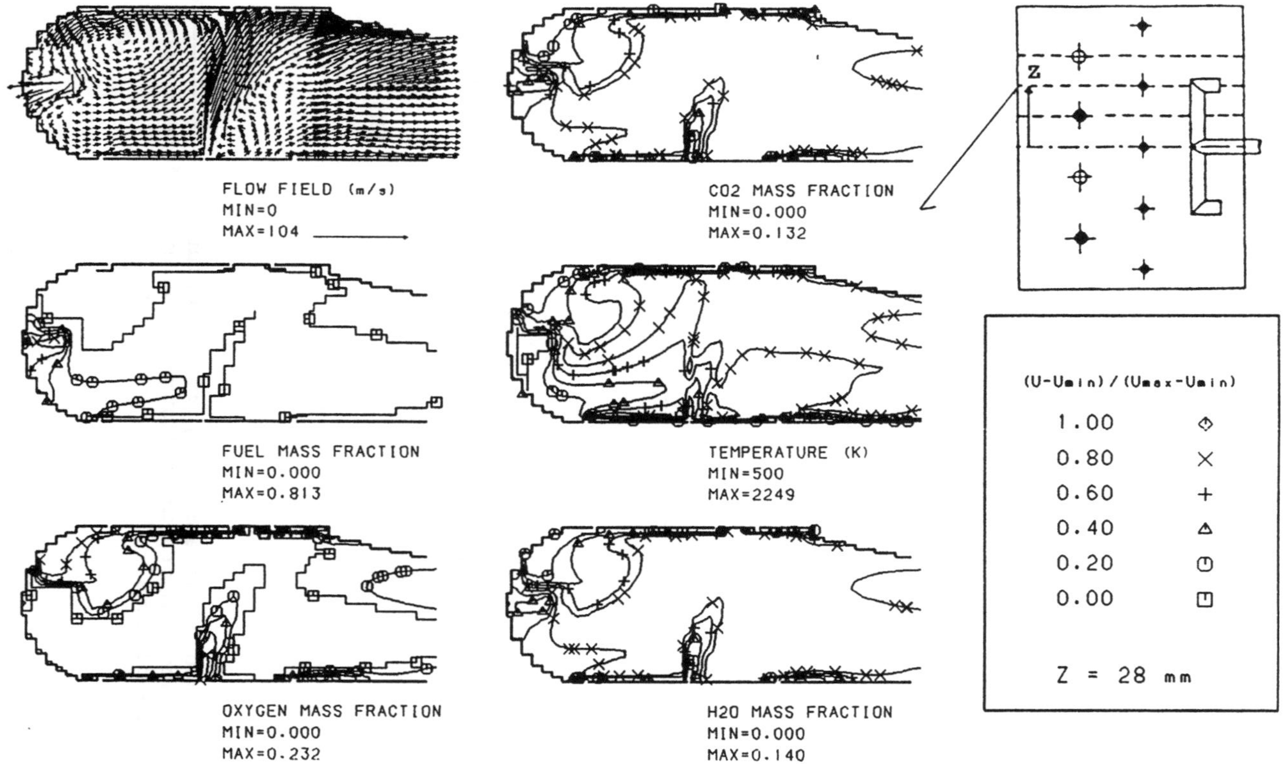

Fig. 7. Calculated velocity field, temperature and chemical species concentrations in a vertical plane: Z=28 mm. Temperature and concentrations are normalized according to the formula shown in the picture

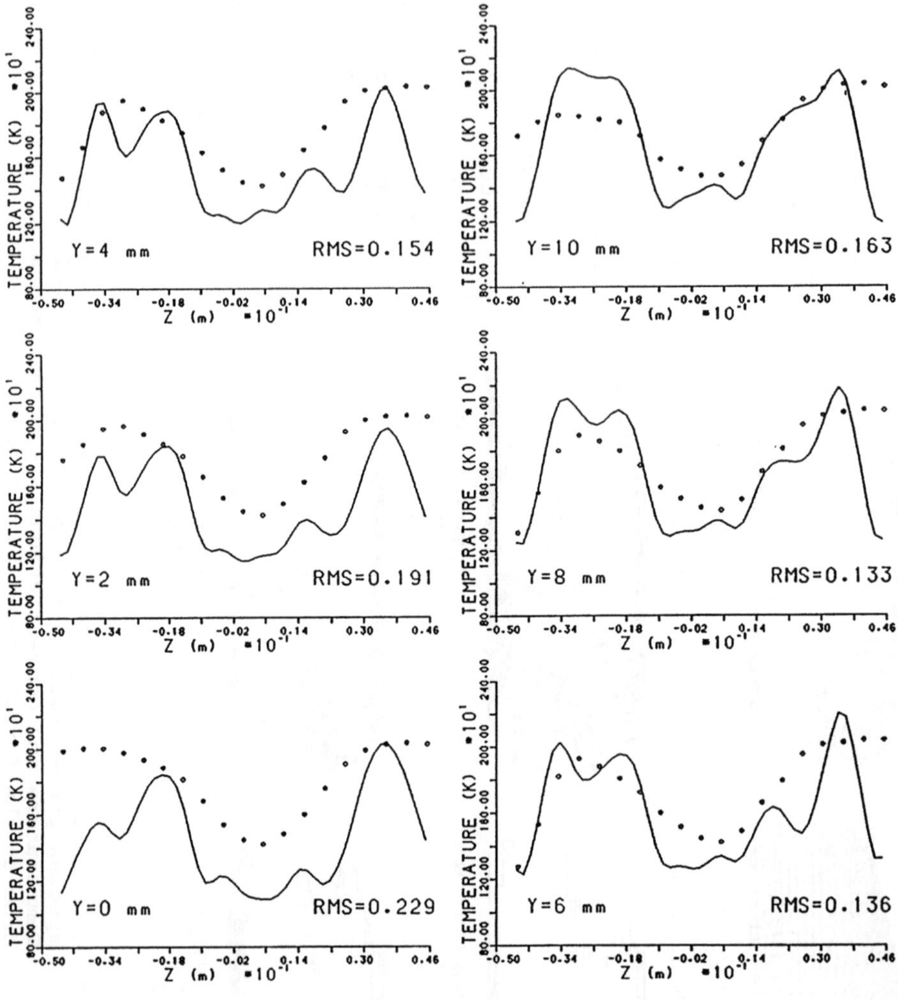

Fig. 8. Comparison of measured (°) vs. calculated (———————) exit-plane temperatures distribution at different values of y-coordinate.

Finally generalized body fitted curvilinear mesh structure for complex geometries will be used to improve the description of irregular boundaries.

We believe that presently mathematical modelling of gas turbine combustors is a practical alternative to empirical methods during the early steps in a combustor design.

References

1. Gupta, A.K.; Lilley, D.G.: "Flowfield Modeling and Diagnostics", Abacus Press, 1985.

2. Kuo, K.K.:"Principles of Combustion", John Wiley and Sons, 1986

3. Patankar, S.V.; Spalding, D.B.: "A Calculation Procedure for Heat, Mass and Momentum Transfer in Three-Dimensional Parabolic Flows", International Journal of Heat and Mass Transfer, 1972, Vol. 15, pp.1787-1806.

4. Bicen, A.F.; Senda, M.; Whitelaw, J.H.:"Scalar Characteristics of Combusting Flow in a Model Annular Combustor", Journal of Engineering for Gas Turbines and Power, Transaction of the ASME, Vol.111, Jan. 1989.

5. Bicen, A.F.; Tse, D.; Whitelaw, J.H.:"Flow Characteristics of a Model Annular Combustor", Propulsion and Energetic Panel 70th Symposium, Crete 1987, AGARD CP-422.

MIX
Papier aus verantwortungsvollen Quellen
Paper from responsible sources
FSC® C105338

FSC
www.fsc.org

If you have any concerns about our products,
you can contact us on
ProductSafety@springernature.com

In case Publisher is established outside the EU,
the EU authorized representative is:
Springer Nature Customer Service Center GmbH
Europaplatz 3, 69115 Heidelberg, Germany

Printed by Libri Plureos GmbH
in Hamburg, Germany